Genetic and Evolutionary Computation

Series Editors:
Wolfgang Banzhaf ⓘ, Michigan State University, East Lansing, MI, USA
Kalyanmoy Deb ⓘ, Michigan State University, East Lansing, MI, USA

More information about this series at http://www.springer.com/series/7373

Wolfgang Banzhaf • Erik Goodman
Leigh Sheneman • Leonardo Trujillo • Bill Worzel
Editors

Genetic Programming
Theory and Practice XVII

 Springer

Editors
Wolfgang Banzhaf
Computer Science and Engineering
John R. Koza Chair, Michigan State
University
East Lansing, MI, USA

Leigh Sheneman
Department of Computer Science
and Engineering
Michigan State University
Okemos, MI, USA

Bill Worzel
Evolution Enterprises
Ann Arbor, MI, USA

Erik Goodman
BEACON Center
Michigan State University
East Lansing, MI, USA

Leonardo Trujillo
Depto Ingenieria en Electronic Electrica
Tecnológico Nacional de México/
IT de Tijuana
Baja California
Tijuana, Mexico

ISSN 1932-0167   ISSN 1932-0175   (electronic)
Genetic and Evolutionary Computation
ISBN 978-3-030-39960-3   ISBN 978-3-030-39958-0   (eBook)
https://doi.org/10.1007/978-3-030-39958-0

This Springer imprint is published by the registered company Springer Nature Switzerland AG.
The registered company address is: Gewerbestrasse 11, 6330 Cham, Switzerland

We dedicate this book to the memory of the co-founder of the Workshop series on Genetic Programming—Theory and Practice, Rick Riolo, who passed away on August 25, 2018.

Foreword

It is a genuine pleasure to write this brief foreword to the collected proceedings of GPTP XVII. It was my privilege to act as opening keynote speaker at the gathering, returning after a 16-year break from playing the same role for GPTP I in 2003. In both cases, I was a fascinated outsider learning about a community that seemed at once oddly similar and yet weirdly different from the computational evolutionary biologists who comprise my own academic tribe (specifically those concerned with the origin and early evolution of life).

On both occasions, I was struck immediately by the potential for the Genetic Programming Theory and Practice (GPTP) community to answer questions that "my people" struggle to frame. How and why did the computational basis of biology evolve to comprise the particular set of rules and pieces which freshmen biologists now strive to memorize, some four billion years later (4 genetic letters, 20 amino acid building blocks of proteins and their interactions)?

But this year, just as in 2003, careful listening soon brought a far deeper conviction that the questions of evolutionary computing are not and should not be limited to those which happen to interest me, or indeed anyone else. There is something too fresh, vibrant, and exploratory about the border formed by introducing evolutionary principles into programming. The diverse works which follow will grow, within the reader, an inescapable sense that it would be to the detriment of human knowledge and technological progress for anyone to presume, at this early stage, any particular purpose or direction for the field. There's simply too much exploration to be done first!

This truth only highlights a more urgent and somber note which must rightfully dominate my remaining words. While it would be nice to write here only a tourist's guide to the series of locations along the border between evolution and computing which populate the following pages, something far more serious dominated the gathering and must be spoken openly. When I, a nosy outsider, asked participants to bring me up to speed on the history of their field "while I was away," one message united all answers: Deep Learning has emerged to pose a deep and perhaps existential threat to our community. The numerous directions in which this particular form of neural network can find answers are undeniable. Equally undeniable is the attractiveness of a simple, reliable, and user-friendly product developed by the financial might and business acumen of Google. But just as, at least within the USA, the emergence of "big box" stores brought reliability, cost savings, and convenience only at the cost of conformity which eroded a far richer consumer ecosystem, so it is very clear from the pages that follow that Deep Learning is flattening something far richer.

Both implicitly and explicitly, the pages which follow demonstrate that Deep Learning is not the answer to every problem. From industry to computing theory, genetic programming and genetic algorithms can help where neural networks and other forms of machine learning struggle. A subtler, deeper message to be found between their lines is one familiar throughout research science. Surprisingly often, it turns out that an answer to the question, as originally posed, is downright unhelpful. We needed, instead, to understand why the question was badly framed. That need not be expressed in the past tense. Any history of science suggests that we progress less by obtaining answers than by forming better questions. Douglas Adams satirized this important truth famously within the Hitchhiker's Guide to the Galaxy when he told the story of an unimaginably advanced civilization which built planet Earth as a supercomputer with which to calculate the answer to life in the universe and everything. Only when this answer arrived in the form of the number 42 did the civilization reflect that perhaps the question had not been well formed.

The truth behind this humor matters when a core limitation of Deep Learning is its lack of transparency. What just happened? How did it reach that answer? Is that really what we needed to know/solve/achieve? In contrast to the black ("big") box of Deep Learning, the diverse "Mom-and-Pop" stores of the GPTP community invite such meta-questions. Through them, we have every reason to believe, a deeper kind of learning proceeds. Let us not wait for Deep Learning to produce all of the answers, only to discover that we now need to dust off, resurrect, or reinvent alternative approaches that it drove extinct along the way. It matters, then, that the community of evolutionary computing spreads this message: through its areas of success and the unexpected insights it uncovers. And if you, the reader, are in any way new to the field represented by GPTP then it matters that you keep reading.

Baltimore, MD, USA   Stephen Freeland
October 2019

Preface

After 16 annual editions of the workshop on Genetic Programming Theory and Practice (GPTP) were held in Ann Arbor, 2019, we saw the workshop venturing out from that location for the first time. This 17th GPTP workshop was held in East Lansing, Michigan, from May 16 to May 19, 2019, at Michigan State University, one of the first land-grant institutions in the USA. It was organized and supported by the NSF-funded BEACON Center for the Study of Evolution in Action, a Science and Technology Center funded by the NSF since 2010.

The collection you hold in hand contains the written final contributions submitted by the workshop's participants. Each contribution was drafted, read, and reviewed by other participants prior to the workshop. Each was then presented at the workshop, and subsequently revised, after the workshop, on the basis of feedback received during the event.

GPTP has long held a special place in the genetic programming community, as an unusually intimate, interdisciplinary, and constructive meeting. It brings together researchers and practitioners who are eager to engage with one another deeply, in thoughtful, unhurried discussions of the major challenges and opportunities in the field. Despite the change in location, the large group of interested individuals at MSU this year resulted in one of the largest groups ever participating in the workshop with approximately 50 regular attendees.

It should be kept in mind that participation at this workshop is by invitation only, and every year the editors make an effort to invite a group of participants that is diverse in several ways, including participants both from academia and industry, junior and senior, local, national, and international. Efforts are also made to include participants in "adjacent" fields such as evolutionary biology.

GPTP is a single-track workshop, with a schedule that provides ample time for presentations and for discussions, both in response to specific presentations and on more general topics. Participants are encouraged to contribute observations from their own, unique perspectives, and to help one another to engage with the presented work. Often, new ideas are developed in these discussions, leading to collaborations after the workshop.

In this year's edition, the regular talks touched on many of the most important issues and research questions in the field, including: opportune application domains for GP-based methods, game playing and co-evolutionary search, symbolic regression and efficient learning strategies, encodings and representations for GP, schema theorems, and new selection mechanisms.

Aside from the presentations of regular contributions, the workshop featured three keynote presentations that were chosen to broaden the group's perspective on the theory and practice of genetic programming. This year, the first keynote speaker was Dr. Stephen Freeland, University of Maryland, on "Alphabets, topologies and optimization." He returned to the workshop after giving a keynote at the first GPTP workshop in 2003, with 15 years of additional research to report on. On the second day, the keynote was presented by Gavin A. Schmidt from the NASA Goddard Institute for Space Studies, on "Some Challenges and Progress in Programming for Climate Science." The third and final keynote was delivered by Indika Rajapakse Associate Professor of Computational Medicine and Bioinformatics, Mathematics and Bioengineering at the University of Michigan in Ann Arbor, on "Cell Reprogramming." As can be gathered from their titles, none of these talks focused explicitly on genetic programming per se. But each presented fascinating developments that connect to the theory and applications of genetic programming in intriguing and possibly influential ways.

While most readers of this volume will not have had the pleasure of attending the workshop itself, our hope is that they will nonetheless be able to appreciate and engage with the ideas that were presented. We also hope that all readers will gain an understanding of the current state of the field, and that those who seek to do so will be able to use the work presented herein to advance their own work, and to make additional contributions to the field in the future.

Acknowledgements

We would like to thank all of the participants for again making GP Theory and Practice a successful workshop 2019. As is always the case, it produced a lot of interesting and high-energy discussions, as well as speculative thoughts and new ideas for further work. The keynote speakers delivered thought-provoking talks from perspectives not usually directly connected to genetic programming.

We would also like to thank our financial supporters for making the existence of GP Theory and Practice possible for the past 16 years. For 2019, as we moved to another location, we needed additional funds raised from different sponsors. We are grateful to the following sponsors:

- John Koza
- Jason H. Moore
- Babak Hodjat at Sentient
- Mark Kotanchek at Evolved Analytics

- Stuart Card
- The BEACON Center for the Study of Evolution in Action, at MSU

A number of people made key contributions to the organization and assisted our participants during their stay in East Lansing. Foremost among them is Constance James, who made the workshop run smoothly with her diligent efforts behind the scenes before, during, and after the workshop. Special thanks go to Michigan State University, particularly the College of Engineering and its Dean, Professor Leo Kempel, for hosting us in the Engineering Conference room, as well as to the Springer Nature Publishing Company, for producing this book. We are particularly grateful for contractual assistance by Melissa Fearon and Ronan Nugent at Springer.

We would also like to express our gratitude to Carl Simon at the Center for the Study of Complex Systems at the University of Michigan for continued support.

East Lansing, MI, USA	Wolfgang Banzhaf
East Lansing, MI, USA	Erik Goodman
Okemos, MI, USA	Leigh Sheneman
Tijuana, Mexico	Leonardo Trujillo
Ann Arbor, MI, USA	Bill Worzel
October 2019	

Contents

Contributors

Michael Affenzeller Heuristic and Evolutionary Algorithms Laboratory (HEAL), University of Applied Sciences Upper Austria, Hagenberg, Austria
Department of Computer Science, Johannes Kepler University, Linz, Austria

Wolfgang Banzhaf Department of Computer Science and Engineering & Beacon Center, Michigan State University, East Lansing, MI, USA

Earl T. Barr CREST, University College London, London, UK

Bogdan Burlacu Heuristic and Evolutionary Algorithms Laboratory (HEAL), University of Applied Sciences Upper Austria, Hagenberg, Austria
Josef Ressel Center for Symbolic Regression, University of Applied Sciences Upper Austria, Hagenberg, Austria

Mauro Castelli NOVA IMS, Universidade Nova de Lisboa, Lisboa, Portugal

Mariana Chan-Ley EvoVisión Laboratory, Ensenada, BC, Mexico

Francisco Chávez University of Extremadura, Badajoz, Spain

Andrei Denissov Sentient Investment Management, San Francisco, CA, USA

Camille Dollé Sentient Investment Management, San Francisco, CA, USA

Emily Dolson Department of Translational Hematology and Oncology Research, Cleveland Clinic, Cleveland, OH, USA

Justin Dyer Sentient Investment Management, San Francisco, CA, USA

Austin J. Ferguson The BEACON Center for the Study of Evolution in Action, Michigan State University, East Lansing, MI, USA

Francisco Fernández de Vega University of Extremadura, Badajoz, Spain

Benjamin Fowler Department of Computer Science, Memorial University of Newfoundland, St. John's, NL, Canada

Ivo Gonçalves INESC Coimbra, DEEC, University of Coimbra, Coimbra, Portugal

Donn Goodhew Sentient Investment Management, San Francisco, CA, USA

Erik Goodman BEACON Center, Michigan State University, East Lansing, MI, USA

Steven Gustafson MAANA Inc., Bellevue, WA, USA

Thomas Helmuth Hamilton College, Clinton, NY, USA

Jose Guadalupe Hernandez The BEACON Center for the Study of Evolution in Action, Michigan State University, East Lansing, MI, USA

Malcolm I. Heywood Faculty of Computer Science, Dalhousie University, Halifax, NS, Canada

Arend Hintze Department of Integrative Biology, Michigan State University, East Lansing, MI, USA
Department of Computer Science and Engineering, BEACON Center for the Study of Evolution in Action, Michigan State University, East Lansing, MI, USA

Babak Hodjat Cognizant Technology Solutions, Dublin, CA, USA

Ting Hu School of Computing, Queen's University, Kingston, ON, Canada
Department of Computer Science, Memorial University, St. John's, NL, Canada

Daniel Junghans The BEACON Center for the Study of Evolution in Action, Michigan State University, East Lansing, MI, USA

Lukas Kammerer Heuristic and Evolutionary Algorithms Laboratory (HEAL), University of Applied Sciences Upper Austria, Hagenberg, Austria
Department of Computer Science, Johannes Kepler University, Linz, Austria
Josef Ressel Center for Symbolic Regression, University of Applied Sciences Upper Austria, Hagenberg, Austria

Stephen Kelly Department of Computer Science and Engineering & Beacon Center, Michigan State University, East Lansing, MI, USA

Douglas Kirkpatrick Department of Computer Science and Engineering, BEACON Center for the Study of Evolution in Action, Michigan State University, East Lansing, MI, USA

Michael Kommenda Heuristic and Evolutionary Algorithms Laboratory (HEAL), University of Applied Sciences Upper Austria, Hagenberg, Austria
Josef Ressel Center for Symbolic Regression, University of Applied Sciences Upper Austria, Hagenberg, Austria

Arthur Kordon Kordon Consulting LLC, Fort Lauderdale, FL, USA

Theresa Kotanchek Evolved Analytics LLC, Midland, MI, USA

Mark Kotanchek Evolved Analytics LLC, Midland, MI, USA

Gabriel Kronberger Heuristic and Evolutionary Algorithms Laboratory (HEAL), University of Applied Sciences Upper Austria, Hagenberg, Austria
Josef Ressel Center for Symbolic Regression, University of Applied Sciences Upper Austria, Hagenberg, Austria

Alexander Lalejini The BEACON Center for the Study of Evolution in Action, Michigan State University, East Lansing, MI, USA

Daniel Lanza University of Extremadura, Badajoz, Spain

Simon Lau Sentient Investment Management, San Francisco, CA, USA

Joel Lehman Uber AI, San Francisco, CA, USA

James McDermott National University of Ireland, Galway, Ireland

Risto Miikkulainen Cognizant Technology Solutions, Dublin, TX, USA
The University of Texas at Austin, Austin, CA, USA

Jason H. Moore Institute for Biomedical Informatics, University of Pennsylvania Philadelphia, PA, USA

Miguel Nicolau University College Dublin, Quinn School of Business, Belfield, Dublin, Ireland

Charles Ofria The BEACON Center for the Study of Evolution in Action, Michigan State University, East Lansing, MI, USA

Gustavo Olague CICESE, Ensenada, BC, Mexico

Edward Pantridge Swoop, Inc., Cambridge, MA, USA

Anil Kumar Saini College of Information and Computer Sciences, University of Massachusetts, Amherst, MA, USA

Marta Seca NOVA IMS, Universidade Nova de Lisboa, Lisboa, Portugal

Hormoz Shahrzad Cognizant Technology Solutions, Dublin, CA, USA

Moshe Sipper Institute for Biomedical Informatics, University of Pennsylvania, Philadelphia, PA, USA
Department of Computer Science, Ben-Gurion University, Beer Sheva, Israel

Andrew N. Sloss Arm Inc., Bellevue, WA, USA

Robert J. Smith Faculty of Computer Science, Dalhousie University, Halifax, NS, Canada

Lee Spector Department of Computer Science, Amherst College, Amherst, MA, USA
School of Cognitive Science, Hampshire College, Amherst, MA, USA

College of Information and Computer Sciences, University of Massachusetts, Amherst, MA, USA

Ryan J. Urbanowicz Institute for Biomedical Informatics, University of Pennsylvania, Philadelphia, PA, USA

David R. White Department of Physics, University of Sheffield, Sheffield, UK

Stephan M. Winkler Heuristic and Evolutionary Algorithms Laboratory (HEAL), University of Applied Sciences Upper Austria, Hagenberg, Austria
Department of Computer Science, Johannes Kepler University, Linz, Austria

Yuan Yuan Department of Computer Science and Engineering & Beacon Center, Michigan State University, East Lansing, MI, USA

Chapter 1
Characterizing the Effects of Random Subsampling on Lexicase Selection

Austin J. Ferguson, Jose Guadalupe Hernandez, Daniel Junghans, Alexander Lalejini, Emily Dolson, and Charles Ofria

1.1 Introduction

Evolutionary computation is often used to solve complex, multi-faceted problems where the quality of a candidate solution is measured according to its performance on a large set of test cases. For these test-based problems, we must somehow meld performances across many test cases to select individuals to serve as parents for the next generation. In many test-based problems, we cannot exhaustively evaluate a candidate solution over the entire space of possible test cases. As a result, it can be challenging to balance the trade-off between using a large enough test set to thoroughly evaluate candidate solutions while keeping the test set small enough to preserve computational resources and rapidly progress through generations.

Lexicase selection is a relatively new parent-selection algorithm developed for genetic programming (GP) and has been demonstrated as an effective tool for solving difficult test-based problems [11, 12, 27]. Many traditional selection strategies for solving test-based problems score potential solutions by aggregating their fitness across all test cases. The lexicase algorithm, however, chooses each parent for the next generation by sequentially applying test cases in a random order, keeping only the best performers on each test case until the population has been winnowed to a single individual. Because the ordering of test cases changes for

A. J. Ferguson (✉) · J. G. Hernandez · D. Junghans · A. Lalejini · C. Ofria
The BEACON Center for the Study of Evolution in Action, Michigan State University, East Lansing, MI, USA
e-mail: fergu358@msu.edu; herna383@msu.edu; junghan2@msu.edu; lalejini@msu.edu; ofria@msu.edu

E. Dolson
Department of Translational Hematology and Oncology Research, Cleveland Clinic, Cleveland, OH, USA
e-mail: dolsonem@msu.edu

© Springer Nature Switzerland AG 2020
W. Banzhaf et al. (eds.), *Genetic Programming Theory and Practice XVII*, Genetic and Evolutionary Computation, https://doi.org/10.1007/978-3-030-39958-0_1

every parent selection event, individuals that perform well on different subsets of test cases are able to co-exist [4, 9].

The drawback of many test-based selection schemes, including lexicase, is that assessing individuals using a large set of test cases can be computationally expensive; this drawback is exacerbated when tests are costly to perform (e.g., robotics simulations). Using a large number of test cases constrains the number of generations we are able to run evolutionary search. Using too few test cases, however, may fail to accurately represent the problem domain and lead to overfitting. To combat this, many techniques dynamically subsample test cases (from a large pool representative of the problem domain) for candidate solution evaluation and selection (see [14, 20] for recent reviews). Indeed, subsampling has been used to reduce computational effort in GP [2, 7] and to improve the generalizability of evolved programs [8, 20].

In this chapter, we characterize the effects of random subsampling on the lexicase parent-selection algorithm. Previous work has shown that lexicase selection performs well when combined with random subsampling. Moore and Stanton applied random subsampling to lexicase selection in the context of an evolutionary robotics problem because evaluating robot controllers on test cases (simulation environments) was too costly to permit exhaustive assessments [23–25]. In [13], we proposed down-sampled and cohort lexicase selection, two variants of standard lexicase that employ random subsampling to reduce the number of per-generation evaluations required by lexicase selection. We demonstrated that both down-sampled and cohort lexicase could yield higher problem-solving success than standard lexicase on a fixed evaluation budget in the context of program synthesis [13].

Here, we explore *why* random subsampling can improve lexicase selection's problem-solving success. Additionally, we characterize the effect of subsampling on diversity and specialist maintenance, both of which have been shown to be important factors behind lexicase selection's efficacy [4, 9, 10, 24]. We show that the improvement in problem-solving success gained from subsampling is due to its facilitation of *deeper* evolutionary searches (i.e., consisting of more generations relative to standard lexicase) given a fixed evaluation budget. Moreover, we show that both down-sampled and cohort lexicase find solutions with less computational effort than standard lexicase. While we predicted that subsampling would degrade diversity, we find no evidence for systematic degradation of phenotypic diversity. However, as the level of subsampling increases, cohort lexicase generates and maintains more phylogenetic diversity than down-sampled lexicase. As expected, we find that random subsampling degrades specialist preservation relative to standard lexicase. Our phenotypic diversity results seem to contradict our specialist preservation findings; this could be because of the particular problems we are using or because of our choice of time to measure phenotypic diversity (at the time a solution was found). Future work will continue investigating how subsampling affects diversity maintenance in an expanded problem domain and with more fine-grained data collection and analysis.

1.2 Lexicase Selection

Spector [27] initially proposed the lexicase parent-selection algorithm for solving modal GP problems where programs may have to output qualitatively different responses to different inputs. To accomplish this, lexicase does not aggregate fitness across test cases like many selection schemes. Instead, for each selection event (where a single parent must be selected), lexicase randomly permutes the test cases in the training set. Each test case is then considered in this permuted order, keeping only those candidate solutions that solve the focal test case (or tie for highest fitness if no candidate solutions solve it). This process continues until either a single candidate solution remains or all test cases have been exhausted. If more than one candidate solution remains, the winner is chosen at random. Each selection event follows this pattern with a different permutation until all parents for the next generation have been selected. Because the order of test cases changes for every parent selection event, individuals that perform well on different subsets of test cases are able to co-exist [4, 9]. This dynamic creates niches in a lexicase population and encourages multiple co-existing solutions that focus on different subsets of test cases. See [12, 27] for a more detailed description of lexicase selection.

Since its conception, lexicase selection has been successfully applied in the field of genetic programming. Such applications include program synthesis [11] and regression [16]. Lexicase selection has also been in other areas such as evolutionary robotics [23], genetic algorithms [22], and learning classifier systems [1].

1.2.1 Applying Subsampling to Lexicase Selection

Several variants of lexicase selection (and lexicase-inspired selection algorithms) exist, such as ϵ-lexicase, truncated lexicase, batch-tournament, batch-lexicase, down-sampled lexicase, and cohort lexicase [1, 13, 21, 28]. Here, we investigate down-sampled and cohort lexicase, both of which leverage random subsampling to reduce the number of per-generation evaluations required for lexicase selection.

1.2.1.1 Down-Sampled Lexicase

Down-sampled lexicase [13] applies the random subsampling technique [8] to lexicase selection. Each generation, down-sampled lexicase selects a random subset of the test-cases in the training set to use for all selection events, guaranteeing that unselected test cases are not evaluated. Here, we use D to represent our 'down-sample factor', where each generation $\frac{1}{D}$ of the training set is used. For example, a D of 5 implies a 20% subsampling rate (i.e., each generation we use one fifth of the training set to evaluate individuals). Down sampling divides the number of evaluations performed each generation by D. Given a fixed budget of evaluations, the computational savings afforded by down sampling allows us to continue our evolutionary search for more generations (or with a larger population size) than standard lexicase selection.

1.2.1.2   Cohort Lexicase

Cohort lexicase selection [13] makes use of the full set of training cases each generation while also ensuring that each prospective solution is evaluated against only a subset of tests. Every generation, cohort lexicase randomly partitions both the population and test-case set into K equally-sized sub-groups (cohorts). Each of the K candidate solution cohorts is paired with a test-case cohort, and each candidate solution is evaluated against only the test cases in its cohort. Thus, the number of evaluations performed each generation (relative to standard lexicase selection) is divided by K. Candidate solutions compete only within their cohort, and within-cohort competition is arbitrated by the test cases in the associated cohort. Because cohorts are shuffled each generation, offspring will be assessed with a different subset of test cases than their parents. Note that the down-sampling factor, D, is identical to the number of cohorts, K, in both the total number of evaluations and the number of training cases a candidate solution sees per generation. Thus, K and D provide equivalent subsampling rates for the two selection schemes, and hereafter, we substitute D for K to simplify comparisons between down-sampled and cohort lexicase.

1.3   Methods

We conducted a series of experiments to characterize the effects of applying random subsampling to lexicase selection. In all evolution experiments, we evolved populations of linear genetic programs to solve four program synthesis problems. Using this setup, we replicated previous results [13], tested the effect of the additional generations afforded by subsampling, and investigated how different types of subsampling affect the computational effort expended to solve problems. Additionally, we analyzed how these subsampling techniques affect both population diversity and specialist maintenance.

1.3.1   Evolutionary System

For each of our evolution experiments, we evolved populations of 1000 linear genetic programs on four program synthesis problems (each described in detail in Sect. 1.3.2). Our linear-GP representation used:

- an instruction set that includes arithmetic, memory management, flow-control, and additional problem-specific instructions
- memory accessed with binary tags [18]
- modules referenced via binary tags [17, 29]

A more detailed description of our GP system (including source code) can be found in the supplemental material [5].

We propagated programs asexually, subjecting offspring to mutations. Single-instruction insertions, deletions, and substitutions were applied, each at a per-instruction rate of 0.005. Modules were duplicated and deleted at a per-module rate of 0.05. We also applied 'slip' mutations [19], which have the possibility of duplicating or deleting sequences of instructions, at a per-program rate of 0.05. Program-tags were mutated at a per-bit rate of 0.001. The run-termination criteria varied per experiment and is included in each experiment description.

1.3.2 Program Synthesis Problems

For all evolution experiments, we evolved programs to solve problems from the general program synthesis benchmark suite [11]. To test our hypotheses, we needed a set of problems known to be challenging but not impossible for GP systems to solve. The general program synthesis benchmark suite comprises introductory-level computer science programming questions, many of which have been solved using lexicase selection [6, 11]. We used the following four program synthesis problems in our experiments: *Smallest*, *Median*, *For Loop Index*, and *Grade*. A description of each problem is given below:

Smallest Programs are given four integer inputs ($-100 \leq input_i \leq 100$) and must output the smallest value. We measured program performance on a pass-fail basis. We limited program length to a maximum of 64 instructions and also limited the maximum number of instruction-execution steps to 64.

Median Programs are given three integer inputs ($-100 \leq input_i \leq 100$) and must output the median value. We measured program performance against test cases on a pass-fail basis. We limited program length to 64 instructions and also limited the maximum number of instruction-execution steps to 64.

For Loop Index Programs receive three integer inputs *start* ($-500 \leq start \leq 500$), *end* ($-500 \leq end \leq 500$), ($start < end$), and *step* ($1 \leq step \leq 10$). Programs must output the following sequence:

$$n_0 = start$$

$$n_i = n_{i-1} + step$$

for each $n_i < end$. We limited program length to a maximum of 128 instructions and also limited the maximum number of instruction-execution steps to 256. Program performance against a test case was measured on a gradient, using the Levenshtein distance between the program's output and the correct output sequence.

Grade Programs receive five integers in the range [0, 100] as input: A, B, C, D, and $score$. A, B, C, and D define the minimum score needed to receive that letter grade. These are specified such that $A > B > C > D$ (i.e., they are monotonically decreasing and unique). The program must read in these thresholds and return the appropriate letter grade for the given $score$, or F if $score < D$. We limited program length to a maximum of 64 instructions and also limited programs' maximum instruction-execution steps to 64. On each test, we evaluated programs on a pass-fail basis.

For these experiments, the *Smallest*, *Median*, and *For Loop Index* problems have an associated training set of 100 test cases, and a separate validation set of 1000 test cases (withheld during fitness evaluations). We used 200 training cases and 2000 validation cases for the *Grade* problem. A program had to solve all test cases in both the training and validation sets to be considered a "perfect" solution. All training and validation sets can be found in the supplemental material [5].

1.3.3   Experimental Design

We conducted five experiments: (1) we replicated a previous experiment [13] to evaluate subsampling's effect on lexicase selection's problem-solving success; (2) we tested whether or not subsampling improves problem-solving success because it facilitates deeper evolutionary searches; (3) we evaluated whether subsampling can reduce the computational effort expended by lexicase selection to solve problems; (4) we tested the effect of random subsampling on lexicase selection, comparing the diversity maintenance of standard, down-sampled, and cohort lexicase; (5) we compared each of standard, down-sampled, and cohort lexicase's capacity to maintain specialist candidate solutions (i.e., programs with low aggregate fitness that solve test cases that the majority of the population fails).

1.3.3.1   Does Subsampling Improve Lexicase Selection's Problem-Solving Success Given a Fixed Computation Budget?

First, we replicated the experiment conducted in Hernandez et al. [13] where both down-sampled and cohort lexicase improved problem-solving success relative to standard lexicase selection. To evaluate whether subsampling improves lexicase's problem-solving success, we evolved programs using down-sampled, cohort, and standard lexicase selection to solve each of the four program synthesis problems (described in Sect. 1.3.2). While the sets of program synthesis problems are not identical, the main difference between the two experiments is that our previous work included a test case that was designed to minimize program size of candidate solutions that solved all normal test cases; this minimizing test case was discarded for all experiments in this work. For a control, we also tested *reduced lexicase*: standard lexicase performed on a statically reduced training set that was randomly sampled at the beginning of the run. Reduced lexicase is similar to down-sampled lexicase, with the exception that test cases remain constant throughout the evolutionary search and are not sampled every generation.

All three of these lexicase variants were tested at five subsampling levels: 100% (identical to standard lexicase), 50, 25, 10 and, 5% ($D = 1, 2, 4, 10$, and 20, respectively). For standard lexicase and each variant, we limited each instance to a maximum computation budget of 30,000,000 evaluations.[1] Thus, standard lexicase ran for 300 generations, and the subsampled variants ran for 300, 600, 1200, 3000, and 6000 generations, respectively. We compared the problem-solving success (i.e., the number of replicates that produced a perfect solution) of each variant to standard lexicase. For each problem, we ran 50 replicates (each with a unique random seed) of each subsampled configuration, and 250 replicates (each with a unique random seed) of standard lexicase (50 replicates for each subsampling level).

1.3.3.2 Does Subsampling Improve Lexicase Selection's Problem-Solving Success Because it Facilitates Deeper Searches?

Both down-sampled and cohort lexicase perform fewer test case evaluations per generation than standard lexicase, allowing us to run evolutionary searches for more generations given a fixed computation budget (i.e., a fixed number of total test case evaluations). We expected that subsampling improves lexicase's problem-solving success because it enables deeper searches. To test this hypothesis, we repeated the performance experiment (described previously in Sect. 1.3.3.1), except we evolved *all* populations (regardless of selection scheme and subsampling level) for 300 generations. We compared the number of successful replicates from each of down-sampled, cohort, and standard lexicase. If down-sampled and cohort lexicase lose their performance edge over standard lexicase, the distinction must come from the time after the 300 generation limit that they would have continued evolving. This finding would suggest that subsampling's improved problem-solving success results from its facilitation of deeper evolutionary searches.

---

[1] Evaluating a single program on a single test case is one test case evaluation.

1.3.3.3 Does Random Subsampling Reduce the Computational Effort Required to Solve Problems with Lexicase Selection?

Our previous work [13] shows that subsampling can improve lexicase selection's problem-solving success given a fixed computational budget. Here, we are interested in whether or not subsampling reduces the total computational effort required to find solutions; that is, do down-sampled and cohort lexicase generally find solutions using fewer total evaluations than standard lexicase selection? We evolved programs on the four program synthesis problems described previously (Sect. 1.3.2) using down-sampled, cohort, and standard lexicase (at a 10% subsampling level for down-sampled and cohort lexicase). For each condition, we ran 50 replicate populations. Because we wanted to compare how much computational effort it generally took for a particular selection scheme to solve a problem, we only used data from the first 25 replicates of each condition to solve the problem (i.e., the 25 replicates per condition that used the least computational effort). We also included truncated lexicase [28], another lexicase selection variant that works to reduce the rigidness in lexicase selection by limiting the number of test cases used in a selection event before a candidate solution is selected. Truncated lexicase also has the potential to reduce the computational effort needed to find solutions. For our truncated lexicase condition, we used a truncation level equal to 10% of the training set.

1.3.3.4 Does Subsampling Degrade Lexicase Selection's Diversity Maintenance?

Part of lexicase selection's success is known to be the result of its effectiveness at diversity maintenance [4, 9, 24]. Subsampling, however, is likely to degrade diversity maintenance because it both reduces the total number of niches available each generation (i.e., there are fewer possible orderings of test cases) *and* decreases niche stability from generation to generation (i.e., the set of possible test case permutations changes every generation). Thus, we expected populations evolved using down-sampled and cohort lexicase selection to have lower overall diversity and more frequent selective sweeps (coalescence events) than those evolved with standard lexicase selection. Additionally, cohort lexicase inherently buffers populations against selective sweeps, slowing down the rate at which a lineage can take over a population by limiting competition each generation to within cohorts. As such, we expected cohort lexicase to have fewer selective sweeps (and thus more phylogenetic diversity) than down-sampled lexicase.

To test our hypotheses, we replicated the experiment in Sect. 1.3.3.1, running both subsampling lexicase variants (at a range of subsampling levels) and standard lexicase for 30,000,000 total evaluations. In these runs, we collected data on genotypic, phenotypic, and phylogenetic diversity. We measured genotypic and phenotypic diversity with the Shannon diversity index. To assess phylogenetic diversity, we used a suite of phylogenetic diversity metrics (see [3] for a review). After all replicates terminated, we analyzed the results of each of these diversity measures *at the time solutions were found.*[2] Within each subsampling level, we compared cohort, down-sampled, and standard lexicase selection.

1.3.3.5 Does Subsampling Reduce Lexicase Selection's Capacity to Maintain *specialists*?

Recent work Helmuth et al. [10] demonstrates lexicase's tendency to select specialist individuals (i.e., individuals that have a low aggregate fitness but perform well on a subset of tests that the majority of the population fails). Helmuth et al. found that lexicase's ability to select specialists is a major driver behind its problem-solving success. Just as we expected subsampling to degrade lexicase selection's diversity maintenance, we also expected subsampling to inhibit specialist maintenance. Because specialists perform well on few test cases (and potentially poorly on the rest), a specialist's likelihood of being selected by lexicase selection is reduced if any of the test cases it passes are not sampled. Thus, we hypothesized that both down-sampled and cohort lexicase reduce lexicase selection's capacity to maintain specialist individuals.

To test our hypothesis, we investigated the extreme case of populations with a single specialist. We generated hypothetical populations, each containing a 'specialist' and many 'generalists'. In each generated population, the specialist individual was able to solve only one focal test case, and none of the generalists were allowed to solve the focal test case. We varied the probability at which generalists could solve each non-focal test case, ranging from 0.1 to 1.0 (where all generalists solved all non-focal test cases). We also varied population size and the total number of test cases. Table 1.1 shows all parameter values used in this experiment. We generated 100 populations for each combination of these parameters.

Table 1.1 Generated population configurations

Parameter	Values
Population size	10, 20, and 100
# test cases	10, 20
Generalist pass rate on non-focal tests	0.1, 0.2, 0.3, 0.4, 0.5, 0.6, 0.7, 0.8, 0.9, 1.0

We generated 100 populations for all combinations of the parameters given in this table

---

[2] Choosing when to measure diversity in evolutionary computation is an interesting problem. In evolutionary computation, diversity maintenance is often viewed as a mechanism to avoid premature convergence on suboptimal solutions. If our goal is to compare how well different selection schemes maintain diversity, *when* should we measure diversity? Measuring diversity *after* a global solution is found is not particularly meaningful, as finding the solution often causes the population to converge, decreasing diversity. We measured diversity at the time the solution is found to mitigate this problem. However, this solution only partially addresses the underlying problem: the process of evolution often involves many selective sweeps and subsequent divergences and we cannot know where in this cycle our measurements occurred.

For each population, we calculated the probability of each candidate solution being selected *at least once* to be a parent in the next generation under standard, down-sampled, and cohort lexicase selection. For standard lexicase selection, we calculated exact probabilities: we enumerated all possible orderings of test cases, counting the number of enumerations where each candidate solution is selected. This is intractable for the subsampled lexicase variants, so we took a sampling approach. To approximate the selection probability in the lexicase variants, we randomly subsampled the population according to the selection scheme being tested. After subsampling, down-sampled lexicase is equivalent to standard lexicase with fewer test cases, while cohort lexicase is equivalent to standard lexicase conducted separately on each cohort. Thus, we calculated the selection probabilities for each candidate solution with that particular random subsampling. This process was repeated 100,000 times to approximate the true selection probabilities under down-sampled and cohort lexicase. These calculations allowed us to compare the specialist's selection probability across configurations.

1.3.4   Statistical Analyses

All statistics were calculated using the R statistical computing language v3.6.0 [26], and all figures in this work were created using the ggplot2 R package [31]. We compared problem-solving success rates among different independent conditions using Fisher's exact tests, and we corrected for multiple comparisons using the Holm–Bonferroni method where appropriate. For measures of computational effort and diversity, we performed a Kruskal–Wallis test to look for statistically significant differences among independent conditions. For comparisons in which the Kruskal–Wallis test was significant (significance level of 0.05), we performed a post-hoc Mann–Whitney test between relevant conditions (with a Holm–Bonferonni correction for multiple comparisons where appropriate). Statistical analyses for the specialist experiment also used a Kruskal–Wallis test, but swapped the Mann–Whitney test for a Wilcoxon test because the data were paired. Analysis and visualizations scripts can all be found in the supplemental material [5].

1.4   Results and Discussion

1.4.1   Subsampling Improves Lexicase Selection's Problem-Solving Success

Figure 1.1 shows the fraction of replicates where a perfect solution evolved within 30,000,000 evaluations under each of down-sampled, cohort, reduced, and standard lexicase selection. For each program synthesis problem, we conducted a Fisher's

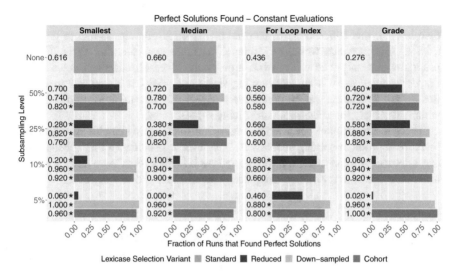

Fig. 1.1 Problem-solving success after 30,000,000 evaluations. Bars show the fraction of replicates that found a perfect solution. An asterisk (*) to the left of a bar denotes a significant difference compared to the standard lexicase results (using a Holm–Bonferroni correction for multiple comparisons). Results for standard lexicase (light purple) consist of 250 replicates per problem, while results for reduced lexicase (dark purple), down-sampled lexicase (yellow), and cohort lexicase (orange) consist of 50 replicates for each configuration

exact test (0.05 significance level) between the 250 standard lexicase replicates and the 50 subsampled replicates of each experimental condition; we corrected for multiple comparisons using the Holm–Bonferonni method.

Our data are largely consistent with previous work [13]. For three of the four problems (Smallest, Median, and Grade), statically reducing the training set beyond a critical threshold significantly decreased problem-solving success. For example, at 5 and 10% subsampling levels, reduced lexicase performs significantly worse than standard lexicase in each of the Smallest, Median, and Grade problems. Reduced lexicase rarely outperformed standard lexicase, only doing so in three cases: Grade at 25 and 50% subsampling, and For Loop Index at 10% subsampling. Statically reducing the size of the training set did not inhibit our capacity to solve the For Loop Index problem; we suspect this is because the training set (100 test cases) is much larger than necessary. The same trend is true for 50- and 25%-reduced lexicase on the Grade problem.

Both down-sampled and cohort lexicase performed significantly better than standard lexicase on at least one subsampling level for every problem. Specifically, down-sampled lexicase significantly outperformed standard lexicase on all problems at the 5 and 10% subsampling levels, while cohort lexicase also outperformed standard lexicase at 5 and 10% subsampling on all problems except For Loop Index at the 10% subsampling level. Neither down-sampled nor cohort lexicase performed significantly worse than standard lexicase in any experimental configuration.

These results achieved better performance on more extreme subsampling levels than in [13]; this is because we removed all selection pressure to reduce program size. In this previous work, we included a single test case that favored small programs that only took effect when a program solved all other test cases *it was evaluated against*. At high subsampling levels (e.g., 5%), it is easy for programs that do not generalize well to prematurely trigger this size-minimization test case, which negatively impacted problem-solving success rates.

These results support our previous claim that subsampling can improve lexicase selection's problem-solving success. Although there is evidence that subsampling can improve solution rates, a different approach is needed to tease apart *why* this difference exists, or how down-sampled and cohort lexicase actually differ.

1.4.2 Deeper Evolutionary Searches Contribute to Subsampling's Success

Figure 1.2 shows the fraction of replicates where a perfect solution evolved after 300 generations under each of down-sampled, cohort, and standard lexicase selection. After 300 generations, conditions with aggressive subsampling (e.g., 5%) have made fewer total evaluations than conditions with milder subsampling (e.g., 50%) or standard lexicase. To be exact, 50, 25, 10, and 5% subsampling complete 15,000,000,

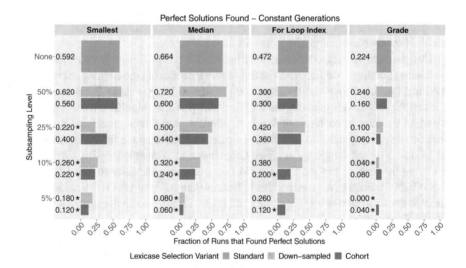

Fig. 1.2 Evolutionary results at the end of 300 generations. Bars show the fraction of replicates that found a perfect solution on or before 300 generations. An asterisk (*) to the left of a bar denotes significant difference compared to the standard lexicase results. Results for standard lexicase (light purple) consist of 250 replicates per problem, while results for down-sampled lexicase (yellow) and cohort lexicase (orange) consist of 50 replicates for each experimental configuration

7,500,000, 3,000,000, and 1,500,000 evaluations, respectively. We hypothesized that random subsampling improves lexicase selection because it allows evolutionary searches to run for more generations given a fixed evaluation budget. By terminating all replicates after 300 generations, we expected subsampling to lose its advantage over standard lexicase.

Given a fixed number of generations, neither down-sampled nor cohort lexicase significantly outperformed standard lexicase at any subsampling level. In fact, down-sampled and cohort lexicase performed significantly worse than standard lexicase on all problems with 5 and 10% subsampling rates except in three cases: cohort at 10% subsampling on Grade, down-sampled at 10 and 5% subsampling on For Loop Index.

As shown in Sect. 1.4.1, when given equivalent computational budgets (i.e., total number of training case evaluations), subsampling significantly improves lexicase's problem-solving success. However, this experiment shows that when we restrict down-sampled and cohort lexicase to the same number of *generations* as standard lexicase, they both have significantly diminished success on the same problems. These data support our hypothesis that deeper evolutionary searches contribute to the success of the subsampled variations on lexicase selection.

1.4.3   Subsampling Reduces Computational Effort

Next, we explored how subsampling affects the amount of computational effort required to solve problems in the context of lexicase selection. For this experiment, we removed all evaluation and generation termination criteria. Figure 1.3 shows the number of test case evaluations in each of the first 25 replicates for each condition in which a solution evolved (i.e., the 25 replicates that required the least computational effort to solve the problem). We performed a Kruskal–Wallis test (significance level 0.05) to look for significant differences among selection schemes for each program synthesis problem. For problems in which the Kruskal–Wallis test was significant,

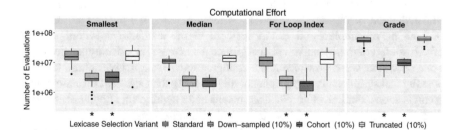

Fig. 1.3 The number of evaluations required for each treatment to solve the specified problems. The 25 replicates with the fewest evaluations for each treatment are shown. An asterisk (*) under a box denotes significant difference between that treatment and standard lexicase

we performed a post-hoc Mann–Whitney test between standard lexicase and each of the down-sampled, cohort, and truncated lexicase (with a Holm–Bonferonni correction for multiple comparisons).

Both down-sampled and cohort lexicase used significantly fewer evaluations than standard lexicase on all four problems. Across all problems, truncated lexicase did not use significantly fewer evaluations than standard lexicase; on the Median problem, truncated lexicase actually used significantly *more* evaluations than standard lexicase. The data show a clear trend that 10% subsampling, whether via down-sampling or cohorts, can significantly reduce the number of evaluations needed to solve these program synthesis problems. However, truncated lexicase (using 10% of the training cases per selection event) causes either no effect or a significant increase in required evaluations.

1.4.4   Subsampling Does Not Systematically Decrease Phenotypic Diversity in Lexicase Selection

Mutations to the binary tags used by the programs to reference modules and memory are often silent (i.e., the phenotype and fitness remain the same) allowing populations to endure high mutation rates that drive adaptive evolution. As a result, almost all replicates maximize genotypic diversity, rendering comparisons uninformative. Therefore, we examined the phenotypic diversity of lexicase and the two subsampled variants.

When evolution produced a candidate solution capable of solving all test cases in the training set, we immediately tested that solution on the cases in the reserved validation set as well. If this candidate solution continued to pass all test cases, we declared it a "perfect solution" and proceeded to measure the phenotypic diversity of the population it arose from. To do so, we tested all programs in the population on all test cases across both the training and validation sets. We designated each candidate solution's performances (in sequence) on all test cases as that solution's phenotype. Figure 1.4 shows the Shannon diversity of these results.

Minimal evidence was found to support our hypothesis that subsampling results in a reduction of phenotypic diversity. After comparing the phenotypic diversity of both down-sampled and cohort lexicase to the standard algorithm, only 2 of 32 configurations resulted in a significant decrease in phenotypic diversity, both of which were down-sampled configurations. Conversely, cohort lexicase actually had significantly *higher* phenotypic diversity than standard lexicase in two configurations. Further, cohort lexicase results had a significantly higher phenotypic diversity than down-sampled lexicase in 4 of 16 comparisons.

With only two configurations leading to decreased phenotypic diversity, we cannot conclude that there is a systematic decrease in phenotypic diversity due to subsampling for these program synthesis problems. However, these results hint

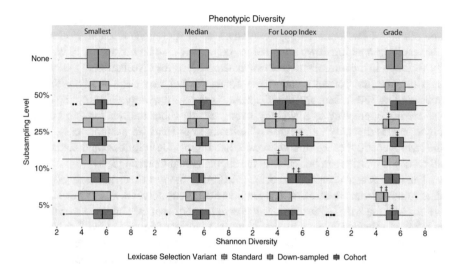

Fig. 1.4 Shannon diversity of candidate solution phenotypes at the first generation a perfect solution was found; individual phenotypes were measured as a program's performance on each test from the training and validation sets. A dagger (†) above a box denotes significant difference with standard lexicase. A double dagger (‡) denotes significant difference between cohort lexicase and down-sampled lexicase at that subsampling level. Results consist of replicates that found a perfect solution out of 250 replicates for standard lexicase on each problem (purple boxes) and 50 replicates for each combination of problem and subsampling level for down-sampled lexicase (yellow boxes) and cohort lexicase (orange boxes)

at a difference between diversity due to down-sampled lexicase and cohort lexicase; we plan to explore this difference in future work.

1.4.5 Cohort Lexicase Enables More Phylogenetic Diversity Than Down-Sampled Lexicase

As with phenotypic diversity, we recorded the phylogenetic diversity metrics at the time point when populations first found a perfect solution. This timing was necessary; the discovery of a perfect solution is likely to produce a selective sweep, radically altering the structure of the phylogeny. An unavoidable side effect is that the measurements are taken after different numbers of generations have elapsed in different replicates. This discrepancy is potentially concerning, as phylogenetic diversity measurements are sensitive to the number of generations represented within the phylogeny. Adding more generations will, in many cases, legitimately increase the diversity of evolutionary history that a population contains. However, the number of generations elapsed can have a disproportionately large effect on a phylogenetic diversity metric, swamping out other effects. In this case, it is these other effects that we are most interested in, as we have already analyzed the causes and effects of the number of generations a population goes through. Fortunately, our results comparing down-sampled vs. cohort lexicase do not appear to be driven by variation in the number of generations elapsed, as the distribution of generations at which the first perfect solution was found did not vary consistently *within* any subsampling level. Because this distribution did vary *among* subsampling levels, we are not attempting to make any strong claims about the relationship between phylogenetic diversity and degree of subsampling. Here we examine only two of the pyhlogenetic metrics that were calculated; plots, descriptions, and statistics of all recorded metrics can be found in the supplemental material [5].

The most recent common ancestor (MRCA) is the most recently evolved candidate solution from which all extant candidate solutions descend. For this experiment we tracked the MRCA throughout the evolutionary search, and we examined the number of selective sweeps by counting the number of times the MRCA changed (see Fig. 1.5). For all problems tested, cohort lexicase has significantly fewer MRCA changes than down-sampled lexicase for 5, 10, and 25% subsampling levels. This pattern suggests that cohort lexicase inhibits selective sweeps in a way that down-sampled lexicase does not. A likely mechanism for this behavior is that, by explicitly fragmenting the population into groups, cohort lexicase prevents any single candidate solution from sweeping more than one cohort per generation.

Another phylogenetic measure we examined was the phylogenetic divergence (i.e., how distinct the extant taxa are from each other) [3]. Here we quantify

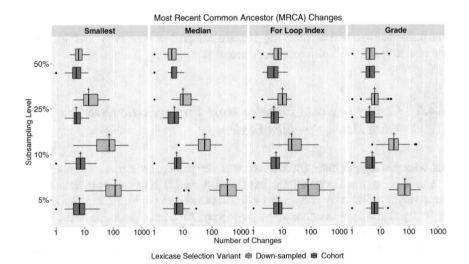

Fig. 1.5 Number of times the most recent common ancestor (MRCA) of all extant candidate solutions changed for each evolutionary run. Changes shown on a logarithmic scale. A dagger (†) above a box denotes significant difference between cohort lexicase and down-sampled lexicase at that subsampling level. All results shown are from the replicates that found a perfect solution out of 50 replicates per experimental condition

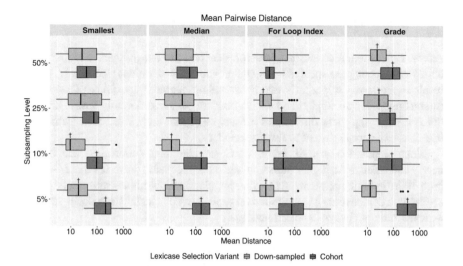

Fig. 1.6 Mean distance between all pairs of extant taxa in the phylogenetic tree for runs of both subsampled lexicase variants at different subsampling levels. A dagger (†) above a box denotes significant difference between cohort lexicase and down-sampled lexicase at that subsampling level. All results shown consist of the replicates that found a perfect solution out of 50 replicates per experimental condition

phylogenetic divergence via mean pairwise distance of the extant solutions in the phylogeny. This metric is calculated as the average distance in the phylogenetic tree between each pair of extant candidate solutions (see Fig. 1.6) [30]. Cohort lexicase has significantly higher mean pairwise distance than down-sampled lexicase for all problems at the 5 and 10% subsampling levels. This result indicates that cohort lexicase has significantly higher phylogenetic divergence than down-sampled lexicase, providing further evidence that cohort lexicase is better than down-sampled lexicase at maintaining phylogenetic diversity. Other phylogenetic diversity metrics were consistent with these results.

Because the differing generation counts prevent us from meaningfully comparing phylogenetic diversity across subsampling levels, all we can say conclusively is that subsampling does not appear to decrease phylogenetic diversity. That said, it may well be the case that greater phylogenetic diversity helps produce better candidate solutions. If so, this factor could explain why more generations (as opposed to more evaluation thoroughness) increases the computational efficiency of lexicase selection. A more targeted investigation will be required to determine how important phylogenetic diversity is to the success of lexicase selection variants.

1.4.6   Subsampling Degrades Specialist Maintenance

Across experimental conditions, lexicase selection has a significantly higher probability of selecting the specialist than either subsampled variant (see Fig. 1.7). This result supports our hypothesis that subsampling degrades specialist preservation. Interestingly, down-sampled and cohort lexicase behave differently across the conditions. Exploring these differences can help us better understand the mechanisms that cause a lexicase variant to favor specialists.

When population size is large, down-sampled and cohort lexicase behave nearly identically. At higher subsampling rates specialists have a higher survival probability in both treatments. At smaller population sizes, higher subsampling rates continue to demonstrate a higher survival probability of specialists in down-sampled lexicase, but not always in cohort lexicase.

At the extreme, when population size, subsampling rate, and generalist pass rate are all small, cohort lexicase has a drastically higher probability of specialist survival than down-sampled lexicase. In this case, the specialist benefits from the low generalist pass rate, since many non-specialists will fail to solve many of the test cases. Specifically, if *all* candidate solutions competing against the specialist fail a given test case, it will be non-discriminatory and effectively ignored. This effect is more pronounced in cohort lexicase, when the specialist is competing only within

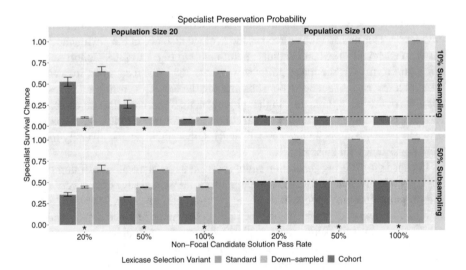

Fig. 1.7 Bars show the median probability that a focal specialist will be selected as a parent in the next generation *at least once*; data are aggregated over 100 experimental populations. Error bars show the minimum and maximum probabilities across all populations for that configuration. The dashed lines show the expected probability for both subsampled lexicase variants for configurations where population size is 100. An asterisk (*) denotes a significant difference between cohort lexicase and down-sampled lexicase; standard lexicase was always significantly different. All configurations shown are for 20 test cases

its cohort (e.g., a cohort of size 2 for a population size of 20 with 10% subsampling), rather than the full population. At a population size of 100, this benefit is lessened because cohorts still contain a relatively large number of candidate solutions. In the remaining configurations, down-sampled lexicase has a higher probability of specialist survival than cohort lexicase.

To better understand these probabilities, consider a situation with two constraints: (1) the specialist solves only its one assigned test case, and (2) every other candidate solution can solve all test cases but the specialist's (i.e., the generalist pass rate is 1.0). While the situation is improbable, it is the worst-case scenario for selecting the specialist; relaxing either constraint could only increase the chance of selecting the specialist. In this situation, the specialist's odds of selection *in a single selection event* under lexicase selection is $\frac{1}{T}$ where T is the number of test cases; that is, the probability of its focal test case being chosen first. The specialist's probability of selection for the entire next generation can be expressed as Eq. 1.1 where N is the total population size [4] (for further discussion of selection probabilities under full lexicase selection, see [15]).

$$P_{lexicase} = 1 - (1 - \frac{1}{T})^N \tag{1.1}$$

We can modify Eq. 1.1 to accommodate down-sampled lexicase by accounting for two cases. First, the specialist's sole test case can be included in the test cases used for this generation, in which case the specialist has a $\frac{D}{T}$ chance of being selected (recall D is the down-sample factor, which divides the number of training cases such that each organism sees $\frac{1}{D}$ of the full training set each generation). Otherwise the specialist's test case is not included, and the specialist has no chance of being selected. Thus, we arrive at Eq. 1.2.

$$P_{down\text{-}sampled} = \frac{1 - (1 - \frac{D}{T})^N}{D} \tag{1.2}$$

Finally, we can also account for cohort lexicase selection. Cohort lexicase also gives the specialist a $\frac{1}{D}$ chance of being evaluated against its sole test case. The only difference is in the number of selection events; cohort lexicase can be thought of as standard lexicase being conducted on each cohort. Thus, in the case where the specialist is in the same cohort as its test case, it does not have N selection events to be selected, but instead $\frac{N}{D}$. This gives us the final equation, Eq. 1.3.

$$P_{cohort} = \frac{1 - (1 - \frac{D}{T})^{\frac{N}{D}}}{D} \tag{1.3}$$

Plotting these equations, we can see both that down-sampled and cohort lexicase approach a maximum specialist survival probability of $\frac{1}{D}$, and that down-sampled approaches that limit at lower population sizes than cohort lexicase (see Fig. 1.8). The plots also show that increasing the number of training cases increases the

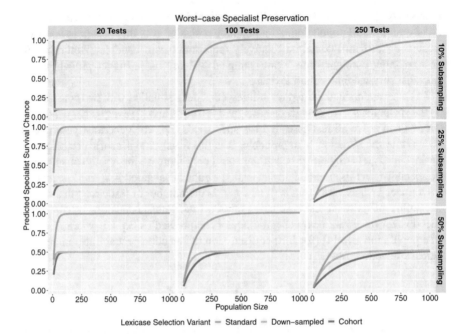

Fig. 1.8 Probabilities that the focal specialist will be selected to be a parent in the next generation *at least once* in the situation where there is one specialist, which solves only one test case, but is also the only candidate solution to solve that specific test case. Meanwhile, all other candidate solutions solve all other test cases. Note the special case of a population size of 10 with 10% subsampling. Here, each cohort has one solution, which guarantees selection exactly once with no selective pressure

required population size to reach the $\frac{1}{D}$ limit. Thus the two subsampled lexicase variants have the same maximum specialist selection probability, but smaller populations will see a lower value for cohort lexicase. These theoretical findings help explain our empirical results.

Again, this is the worst-case scenario for the specialist. Further work is needed to see how specialist preservation changes under different situations (e.g., more copies of the specialist, less elite generalists, specialists that solve more than one test case, etc.). Figure 1.8 shows only the lower bound on the specialist selection probability.

1.5 Conclusion

Here, we investigated the effects of random subsampling on lexicase selection. We replicated previous results [13], demonstrating that subsampling improves lexicase's problem-solving success, and we have shown that subsampling's success is a result of it enabling deeper evolutionary searches (i.e., running searches for more generations). Moreover, we have shown that subsampling reduces the total computational effort required to evolve solutions in the context of lexicase selection. We expected that applying subsampling to lexicase selection would degrade phenotypic diversity, but have found no evidence of systematic degradation. However, we did find evidence that cohort lexicase is better at generating and preserving phylogenetic diversity than down-sampled lexicase. Finally, we have shown that subsampling does reduce lexicase's capacity to maintain specialist individuals.

Overall, our results highlight the value of random subsampling in lexicase selection, showing that it can improve problem-solving success and save computational effort. However, we also demonstrate that subsampling degrades specialist preservation, and as such, for problems where maintaining specialists is especially important, subsampling might have an overall negative effect on problem-solving success. Future work should explore how subsampling affects both overall population diversity and specialist maintenance at a fine-grained scale and on a wider range of problem types.

Acknowledgements This research was supported by the National Science Foundation through the BEACON Center (Coop. Agreement No. DBI-0939454), a Graduate Research Fellowship to AL (Grant No. DGE-1424871), and Grant No. DEB-1655715 to CO. Michigan State University provided computational resources through the Institute for Cyber-Enabled Research.

References

1. Aenugu, S., Spector, L.: Lexicase selection in learning classifier systems. In: Proceedings of the Genetic and Evolutionary Computation Conference - GECCO 2019, pp. 356–364. ACM Press, Prague, Czech Republic (2019)
2. Curry, R., Heywood, M.: Towards efficient training on large datasets for genetic programming. In: A. Tawfik, S. Goodwin (eds.) Conference of the Canadian Society for Computational Studies of Intelligence, pp. 161–174. Springer (2004)
3. Dolson, E., Lalejini, A., Jorgensen, S., Ofria, C.: Quantifying the tape of life: Ancestry-based metrics provide insights and intuition about evolutionary dynamics. In: Artificial Life Conference Proceedings, pp. 75–82. MIT Press (2018)
4. Dolson, E.L., Banzhaf, W., Ofria, C.: Ecological theory provides insights about evolutionary computation. preprint, PeerJ Preprints (2018). URL https://peerj.com/preprints/27315
5. Ferguson, A.: FergusonAJ/gptp-2019-subsampled-lexicase: GPTP Chapter Companion (2020). https://doi.org/10.5281/zenodo.3679380, https://github.com/FergusonAJ/gptp-2019-subsampled-lexicase
6. Forstenlechner, S., Fagan, D., Nicolau, M., O'Neill, M.: Towards Understanding and Refining the General Program Synthesis Benchmark Suite with Genetic Programming. In: 2018 IEEE Congress on Evolutionary Computation (CEC), pp. 1–6. IEEE, Rio de Janeiro (2018)
7. Gathercole, C., Ross, P.: Dynamic training subset selection for supervised learning in Genetic Programming. In: Y. Davidor, H.P. Schwefel, R. Maenner (eds.) Parallel Problem Solving from Nature - PPSN III, vol. 866, pp. 312–321. Springer Berlin Heidelberg, Berlin, Heidelberg (1994)
8. Gonçalves, I., Silva, S., Melo, J.B., Carreiras, J.M.: Random sampling technique for overfitting control in genetic programming. In: A. Moraglio, S. Silva, K. Krawiec, P. Machado, C. Cotta (eds.) European Conference on Genetic Programming

9. Helmuth, T., McPhee, N.F., Spector, L.: Effects of lexicase and tournament selection on diversity recovery and maintenance. In: Proceedings of the 2016 on Genetic and Evolutionary Computation Conference Companion, pp. 983–990. ACM (2016)

10. Helmuth, T., Pantridge, E., Spector, L.: Lexicase selection of specialists. In: Proceedings of the Genetic and Evolutionary Computation Conference on - GECCO 2019, pp. 1030–1038. ACM Press, Prague, Czech Republic (2019)

11. Helmuth, T., Spector, L.: General program synthesis benchmark suite. In: Proceedings of the 2015 Annual Conference on Genetic and Evolutionary Computation, pp. 1039–1046. ACM (2015)

12. Helmuth, T., Spector, L., Matheson, J.: Solving uncompromising problems with lexicase selection. IEEE Transactions on Evolutionary Computation **19**(5), 630–643 (2015)

13. Hernandez, J.G., Lalejini, A., Dolson, E., Ofria, C.: Random Subsampling Improves Performance in Lexicase Selection. In: Proceedings of the Genetic and Evolutionary Computation Conference Companion, GECCO 2019, pp. 2028–2031. ACM, New York, NY, USA (2019). Event-place: Prague, Czech Republic

14. Hmida, H., Hamida, S.B., Borgi, A., Rukoz, M.: Sampling Methods in Genetic Programming Learners from Large Datasets: A Comparative Study. In: P. Angelov, Y. Manolopoulos, L. Iliadis, A. Roy, M. Vellasco (eds.) Advances in Big Data, vol. 529, pp. 50–60. Springer International Publishing, Cham (2017)

15. La Cava, W., Helmuth, T., Spector, L., Moore, J.H.: A Probabilistic and Multi-Objective Analysis of Lexicase Selection and ϵ-Lexicase Selection. Evolutionary Computation **27**, 377–402 (2018)

16. La Cava, W., Spector, L., Danai, K.: Epsilon-Lexicase Selection for Regression. In: Proceedings of the Genetic and Evolutionary Computation Conference 2016, GECCO 2016, pp. 741–748. ACM, New York, NY, USA (2016). Event-place: Denver, Colorado, USA

17. Lalejini, A., Ofria, C.: Evolving event-driven programs with SignalGP. In: Proceedings of the Genetic and Evolutionary Computation Conference on - GECCO 2018, pp. 1135–1142. ACM Press, Kyoto, Japan (2018)

18. Lalejini, A., Ofria, C.: Tag-accessed memory for genetic programming. In: Proceedings of the Genetic and Evolutionary Computation Conference Companion - GECCO 2019, pp. 346–347. ACM Press, Prague, Czech Republic (2019)

19. Lalejini, A., Wiser, M.J., Ofria, C.: Gene duplications drive the evolution of complex traits and regulation. In: Artificial Life Conference Proceedings 14, pp. 257–264. MIT Press (2017)

20. Martinez, Y., Naredo, E., Trujillo, L., Legrand, P., Lopez, U.: A comparison of fitness-case sampling methods for genetic programming. Journal of Experimental & Theoretical Artificial Intelligence **29**, 1203–1224 (2017)

21. Melo, V.V., Vargas, D.V., Banzhaf, W.: Batch Tournament Selection for Genetic Programming. In: Proceedings of the Genetic and Evolutionary Computation Conference Companion - GECCO 2019, pp. 994–1002. ACM Press, Prague, Czech Republic (2019)

22. Metevier, B., Saini, A.K., Spector, L.: Lexicase selection beyond genetic programming. In: W. Banzhaf, L. Spector, L. Sheneman (eds.) Genetic Programming Theory and Practice XVI, pp. 123–136. Springer International Publishing, Cham (2019)

23. Moore, J.M., Stanton, A.: Lexicase selection outperforms previous strategies for incremental evolution of virtual creature controllers. In: Proceedings of the 14th European Conference on Artificial Life ECAL 2017, pp. 290–297. MIT Press, Lyon, France (2017)

24. Moore, J.M., Stanton, A.: Tiebreaks and Diversity: Isolating Effects in Lexicase Selection. In: The 2018 Conference on Artificial Life, pp. 590–597. MIT Press, Tokyo, Japan (2018)

25. Moore, J.M., Stanton, A.: The Limits of Lexicase Selection in an Evolutionary Robotics Task. In: The 2019 Conference on Artificial Life, pp. 551–558. MIT Press, Newcastle, United Kingdom (2019)

26. R Core Team: R: A Language and Environment for Statistical Computing. R Foundation for Statistical Computing, Vienna, Austria (2019). URL https://www.R-project.org/

27. Spector, L.: Assessment of problem modality by differential performance of lexicase selection in genetic programming: a preliminary report. In: Proceedings of the 14th annual conference companion on Genetic and evolutionary computation, pp. 401–408. ACM (2012)
28. Spector, L., Cava, W.L., Shanabrook, S., Helmuth, T., Pantridge, E.: Relaxations of Lexicase Parent Selection. In: W. Banzhaf, R.S. Olson, W. Tozier, R. Riolo (eds.) Genetic Programming Theory and Practice XV, pp. 105–120. Springer International Publishing, Cham (2018)
29. Spector, L., Martin, B., Harrington, K., Helmuth, T.: Tag-based modules in genetic programming. In: Proceedings of the 13th annual conference on Genetic and evolutionary computation - GECCO 2011, p. 1419. ACM Press, Dublin, Ireland (2011)
30. Webb, C.O.: Exploring the phylogenetic structure of ecological communities: an example for rain forest trees. The American Naturalist **156**(2), 145–155 (2000)
31. Wickham, H.: ggplot2: Elegant Graphics for Data Analysis. Springer-Verlag New York (2016). URL https://ggplot2.tidyverse.org

Chapter 2
It Is Time for New Perspectives on How to Fight Bloat in GP

Francisco Fernández de Vega, Gustavo Olague, Francisco Chávez,
Daniel Lanza, Wolfgang Banzhaf, and Erik Goodman

2.1 Introduction

A well known phenomenon in GP is the inherent bloating behavior correlated with fitness improvement [4]. Many approaches to fix this problems have been described and applied although no perfect solution has yet been found.

This paper considers the problem from a new point of view. The main goal is to understand whether new paths are possible, so that in the future new bloat control methods can be produced. Instead of using more traditional size-based approaches (such as penalty functions associated to chromosome size), we are particularly interested in analyzing whether the influence of parallel and distributed computing models that reduce computing time may be also useful for reducing bloat. Although island models have been analyzed before in this context [16, 17], these previous approaches relied more on spatial structure of the models, while we are here more interested in the standard GP algorithm, when it is run in parallel using the standard fitness parallelization approach.

The analysis that we present is useful to understand the relationship between individual size and computing time, and therefore with the bloat phenomenon. Moreover, a new set of bloat-control mechanisms could be easily derived, that

F. Fernández de Vega (✉) · F. Chávez · D. Lanza
University of Extremadura, Badajoz, Spain
e-mail: fcofdez@unex.es; fchavez@unex.es

G. Olague
CICESE, Ensenada, BC, Mexico
e-mail: olague@cicese.mx

W. Banzhaf · E. Goodman
Beacon Center, Michigan State University, East Lansing, MI, USA
e-mail: banzhafw@msu.edu; goodman@msu.edu

© Springer Nature Switzerland AG 2020
W. Banzhaf et al. (eds.), *Genetic Programming Theory and Practice XVII*, Genetic and Evolutionary Computation, https://doi.org/10.1007/978-3-030-39958-0_2

makes use of parallel architectures that are nowadays present in every computer system.

We thus analyze the bloat phenomenon using execution time instead of memory consumption (size). To the best of our knowledge this is the first time such approach is considered and applied. As we describe below, preliminary tests with a well known benchmark problem shows the feasibility of the idea.

Thus, the main contribution of this chapter is providing a new perspective for bloat fighting in GP, and any variable-size chromosome based evolutionary approach; secondly, we describe how this perspective may inspire new bloat-control methods for GP; and finally, one of such control methods is described and tested with success. Although results are still preliminary, we are confident with results obtained, which will be confirmed in the future with a series of experiments in a wider set of benchmark problems.

The rest of the chapter is organized as follows: In Sect. 2.2 we contextualize the problem, and then, their relationship with parallel architectures and scheduling are described in Sect. 2.3. The methodology applied is presented in Sect. 2.4, while Sect. 2.5 shows the experiments performed and results obtained. Finally, we draw our conclusions in Sect. 2.6.

2.2   The Bloat Phenomenon

The bloat problem has been addressed frequently in the GP literature since first described by J. Koza in [3]. A good review of the topic may be found in [10]. We will refer here to ideas that have been described in the last decade and are more directly related with the approach we follow in this chapter.

Among the available techniques to control bloat, particularly relevant is a recent method called *the waiting room*, introduced by Luke and Panait [6]. The idea is for individuals to add a pre-birth phase to all newly created individuals. Children must wait for a period of time proportional to their size before they are allowed to enter the population and compete. Although the authors recognized that the idea was associated with the relationship between individual sizes and evaluation times, they maintained the emphasis on size-control mechanism and hence did not elaborate on the time concept, nor did they take into account the possibilities associated with parallel and distributed infrastructures available, given their influence on evaluation time when individuals are sent to available processors. Thus, they relied on the total number of nodes individuals feature, similarly to all of the other methods, although using a somewhat different approach.

Another technique of interest is *operator equalization* presented in [1] aimed at controlling the distribution of program sizes at each generation, defining a specific shape for the distribution. Some of the best results were achieved by using a uniform or flat distribution [9], and also by applying speciation, fitness sharing or elitism, see [11]. But again, difficulties in effectively applying the method include how to control the shape of the distribution without changing the nature of the search, and how to efficiently account for individuals' sizes and shapes.

Although plenty of size-related techniques can be found in the literature for bloat control in GP, few times, if any, a computing time analysis have been tackled. Here we come up with a deeper analysis of computing times that sheds light on the problem, in contrast to the standard approach of individuals' sizes. Moreover, we want to see if the relationship between size and evaluation time can be exploited in parallel systems not just to save time, but to address the bloat phenomenon in a more natural way.

We must also remember that in a series of papers, it was observed that the island model offers some possibilities for fighting bloat [2, 15, 16], and this observation was later exploited in a new proposal by Whigham [17] that considers spatial distribution of islands in GP. The connection between the dynamics of some of the parallel models for GP and the bloat phenomenon has been shown to be mainly due to the spatial structure of the model, which relies on islands of individuals. But there is still a second source of possible improvement in parallel EAs, as we described above: the number of computing resources employed to run the algorithm, which has not been studied yet from the point of view of its influence on algorithm's bloating behavior. Even when the simplest embarrassingly parallel model is used for running a GP experiment, a load balancing technique must decide how individuals are distributed among the available computing resources, and this may also have an influence on the bloat phenomenon, as we show below.

2.3 Load-Balancing and Parallel GP

Among the available parallel models for EAs and GP, the only one that does not change the algorithm behavior is the embarrassingly parallel model. All algorithm's steps are performed as in the sequential version, and the only change is introduced in the most expensive part of the algorithm: the fitness evaluation. Thus, instead of sequentially evaluating every individual in the population, they are distributed among the available processors, and fitness values are computed in parallel.

This model is frequently used in the parallel computing literature, being known as the client/server model. It requires some kind of load-balancing mechanism that allows to reduce latencies and distribute tasks efficiently among computing resources, so that *makespan* is reduced. Interested readers can find a taxonomy of load-balancing methods in [7], while [19] presents a comparison of different strategies.

If we focus on GP, given that a large number of individuals featuring different sizes must be managed across the available processors, which are typically smaller in number than the population size, some kind of load-balancing mechanism must be applied, in charge of sending individuals to idle processors, and this mechanism might provide new hidden properties: sometimes, a deeper analysis of the new version of a given algorithm allows us to discover some properties that were not noticed before. We are here interested in both the parallel model itself, and the load-balancing technique that can be used and considered as the basis for a new proposal that we will describe and analyze below.

Load-balancing techniques have already been considered as an implicit component of parallel versions of genetic programming. Since the nineties, static load-balancing mechanisms—the ones we will consider here—have been applied within parallel versions of GP, when facing difficult real-world problems. For instance, [8] describes a parallel version of GP that considers complexity of individuals as the basis for establishing the load-balancing policy.

Nevertheless, few papers since then have studied the importance of load-balancing techniques in GP. We may refer to [14], where several methods were tested. But again, no specific study on their relationship with the bloat phenomenon has already been described.

2.3.1   Structural Complexity of GP Individuals

When any load-balancing technique is to be employed, a prediction of computing time for the task must be applied, so that the method can properly decide when to launch the task. In GP, an important feature of GP individuals is their structural complexity [12]. Although this value is typically computed taking into account the number of nodes, such as the case of evaluating *computing effort* [13] or lexicographic complexity [5], both are approximate estimates of the real value required: in other words, the evaluation time of the individuals' fitness function.

But we can adopt a different point of view, as we do in the approach we present below: given that the estimation of the real complexity of an individual is measured when the individual has been evaluated, we can characterize individuals using that computing time, so that we can employ it in future decisions. In any case, that value will not be available when the load-balancing mechanism must decide when to launch the evaluation of a new individual; nevertheless, it will be available after the individual's evaluation. This could be useful, if not for that individual, given that it was already sent to be evaluated, then at least for its children, as a value to somehow approximate its evaluation time.

And this is basically the idea we will apply: Our approach takes into account an individual's computing time, as a value to decide how to distribute children among available computing resources, and ultimately, to reduce computing time while simultaneously reducing the bloat phenomenon. We thus need to keep a record of the time that each program spends during testing and use that information to create clusters of programs with similar durations that will be useful for load-balancing individuals: clusters will be send to different processors. Thus, the load-balancing mechanism relies on individuals' computing time, and this allows to return individuals after evaluation in an order that depends on their computing time. We use the computer's clock to give a value to runtime of a program; this, of course, is correlated with the size of the computer program and the number of instruction cycles required to execute it. All clusters are created without regard to the fitness function. We do not measure directly the size of the individuals, nor use any information about the complexity of breeding programs other than time.

As we show below, the above described load-balancing technique has an impact on the bloat phenomenon without increasing the computational complexity of the algorithm.

2.4   Methodology

As described above, we will consider execution time as a measure of an individual's complexity given that a correlation exists between size and running time, that can be more deeply investigated in future work.

This idea can be easily applied when individuals are evaluated: we just have to take elapsed time during an individual's evaluation as the complexity value required.

The idea is particularly useful when multicore or manycore computer architectures are to be employed: ideally, all of the individuals in the population could be evaluated simultaneously, and their evaluation time obtained simultaneously.

We simplify the measurements by directly using the elapsed evaluation time as the representation of the individual's complexity.

Once the individuals' evaluation times have been obtained, and with the hypothesis that individuals of similar size will produce offspring of similar size, our proposed method groups individuals by computing time, always understanding it as an indirect—and easier to compute—measure of an individual's size. We must again consider that in a parallel system with as many processors as individuals, individuals of similar size will finish their evaluation simultaneously and will be ready to reproduce. Therefore, an automatic grouping mechanism naturally arise from these parallel architectures (see Figs. 2.1, 2.2, 2.3, and 2.4). If the number of processors is smaller, then, the load balancing mechanism which is always in charge of distributing tasks among processors, will decide which individuals group together in single tasks, and may thus apply grouping according to the ending time of individuals evaluation.

After grouping, selection and breeding phases are performed within each group, so only individuals of similar size-time are allowed to crossover. Then, the load-balancing mechanism is in charge of creating tasks by grouping individuals of the same cardinality by evenly dividing the whole population.

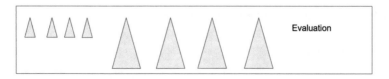

Fig. 2.1 First step: individuals are sent to be evaluated

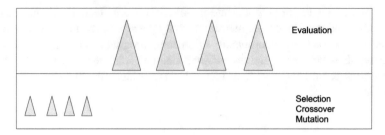

Fig. 2.2 Second step: smaller individuals fitness available first

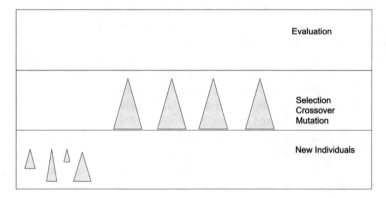

Fig. 2.3 Third step: smaller individuals produce children while larger ones are coming back from evaluation

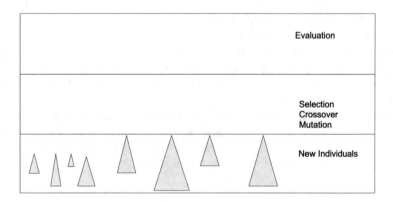

Fig. 2.4 Fourth step: larger individuals produce children

Our hypothesis states that individuals of similar size will produce offspring of similar size. This will not be the case if the crossover operation does not divide the individual into two similar-size parts. If different, crossover will produce small and big individuals, whose sizes do not follow our hypothesis. Nevertheless, considering that individuals are randomly divided, we expect a central tendency which results aim to be of a size in between both parents. This can be considered as a weak point of our offspring-size prediction, which can be improved in a future version. However, generally speaking, we expect that offspring will have similar sizes compared to their parents.

2.4.1   Implementation

Therefore, when designing a specific bloat control mechanism for GP that uses the new time-based perspective, the parallel version of the algorithms, in particular multi-thread based models, has been initially chosen, which are available today in some of the most popular EAs tools. Thus, all operations performed within each group are done in one thread, and the number of threads created corresponds to the number of groups. Each thread collects its corresponding individuals and performs the selection and breeding steps. In this way, all operations are isolated within each group, that can be naturally conformed when individuals return from evaluation according to evaluation time.

After the breeding phase, the mechanism takes advantage of the fact that each group of individuals is contained in a different thread and it performs the evaluation of all the corresponding individuals. Each group/thread contains the same amount of individuals, individuals of similar execution time. As a result, the evolutionary process is considerably speeded up by parallelizing the evaluation phase. Afterwards, once all individuals of the population are evaluated, and their computing times obtained, they are sent to the thread corresponding to their computing time (size surrogate) value, so that the next breeding operations can be performed.

2.4.1.1   Software Tool

With the goal of making the bloat method easily usable, it has been implemented based on a popular existing tool, ECJ [18]. Such system has been built in modules in order to facilitate the replacement of any part involved in the evolutionary process. In our case, we replace the module that carry out the breeding phase with the new time-based approach.

Our bloat control mechanism slightly modifies the way individuals are bred. In order to apply the bloat control mechanism, two new operations have been implemented.

- *GroupBreeder* orchestrates the breeding phase and starts the corresponding threads.

  - As the first step, individuals need to be grouped according to evaluation times. During their evaluations, elapsed time has been captured so each individual already contains it as a new feature. Before grouping them, all the individuals of the population are sorted by evaluation time.
  - Then, the same number of individuals goes to each group, so they are taken in order and sent to their groups. In case individuals cannot be equally split into groups, first groups will get one individual more.
  - Next, one thread is instantiated per group. A group of individuals is assigned to each of them.
  - Threads are started and the program continues until all threads have finished.

- *GroupBreederThread* represents the threads that actually perform the selection, breeding and evaluation of individuals.

  - A call to the module that performs the selection and breeding is done, specifying the group to which these operations need to be applied. Here, the only change that has been done to the original implementation is to apply these operations to only the individuals that correspond to the specified group—the group that corresponds to the thread.
  - Once selection and breeding phases have finished, evaluation of new individuals generated by this thread takes place.

Note that the evaluation step has not been modified. Therefore, it goes through all individuals and tries to evaluate them; however, it will not actually evaluate any, since they have been previously evaluated and marked as such.

2.4.2 Experiments

All experiments described below were run on an Intel(R) Xeon(R) CPU (E5530) that offers 8 cores at 2.4 GHz and 8 GB of memory. Default configuration parameters set in ECJ has been used for the benchmark problem. Generations were set to 50, and 30 runs were launched, for statistical purposes. The well known even parity problem from the GP literature was used with the basic configurations already available in the ECJ toolkit. The only change is the number of bits in the chromosome: 12 bits so that the problem is difficult enough for long runs. Regarding the load-balancing mechanism—number of groups to be used—several configurations were employed: 1 group, which corresponds with the standard GP algorithm, and also 2, 4, 8, 16, 32, 64, and 128 groups.

2.5   Results

We present and discuss below results obtained in the experiments. Average fitness and size are plotted on the figures included. The proposed method uses parallel execution and results are presented here: as many threads as possible are launched so that individuals of different sizes are evaluated in different processors; Yet, ideas extracted can be also adapted when running experiments in a sequential fashion. We also include an analysis of results in this context.

2.5.1   Parallel Model

In the parity problem, fitness is monotonically affected by the number of threads (groups), as can be observed in Fig. 2.5. Nevertheless, if we take into account the scale employed, differences are really narrow.

If we focus instead on size evolution, a dramatic reduction up to a third of the size as compared with a standard run (1 group) can be found in Fig. 2.6. The slightly

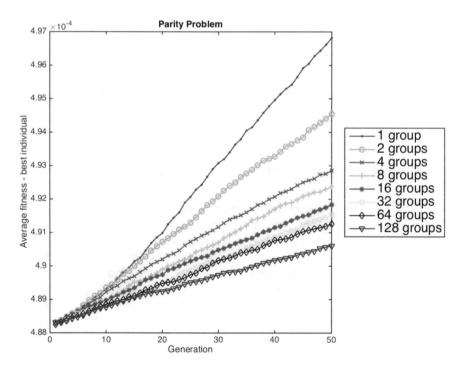

Fig. 2.5 Best-fitness evolution along generations (averaged over 30 runs) for the parity problem (maximizing fitness)

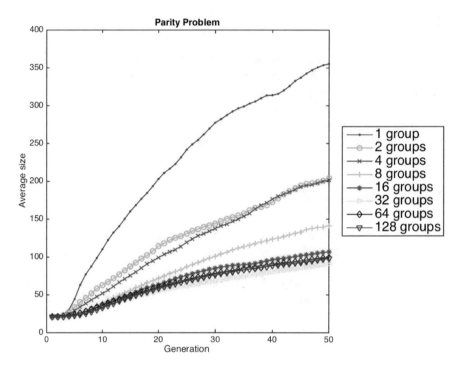

Fig. 2.6 Size-evolution along generations (averaged over 30 runs) for the parity problem

affected fitness may be acceptable, taking into account the considerable reduction in size.

2.5.2 Sequential Execution

Although the previously described idea was born considering how individuals can be run on parallel computing systems, given that individuals may naturally group when a number of them are launched to be evaluated on different processors, and return simultaneously when their running time is similar, the idea can be adapted to sequential environments with minimal changes: allowing individuals to group according to running time. Unfortunately delays are present in this latest approach, given that individuals can only be grouped when all of them has been evaluated, while in the parallel model this is not the case: they can breed with other individuals that has finish their evaluation process simultaneously. In any case, the idea for the sequential version is to simply emulate the parallel version. Therefore, no population structure support the model, although some resemblances with structured models may be seen in this sequential approach. But this distinction is pertinent to properly understand the new bloat control approach: while the method naturally fits parallel

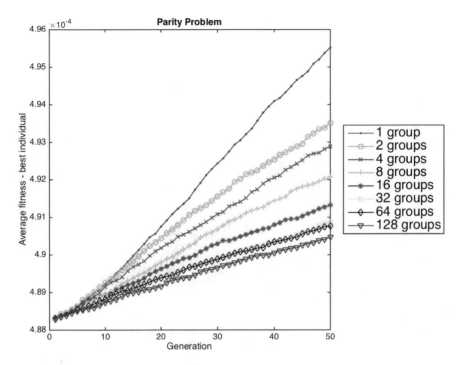

Fig. 2.7 Best-fitness evolution along generations (averaged over 30 runs) for the parity problem (sequential execution) (maximizing fitness)

computing environments, and can be applied without any additional effort, other structured-based approaches can only be applied when specific grouping tasks are added to the algorithm.

Results from sequentially executed runs are shown for the even parity problems in Figs. 2.7 and 2.8. Similarly to what is seen in the results from parallel execution, fitness quality is not strongly affected, while the individual size is notably reduced. This phenomenon is produced solely by the fact of grouping individuals by computing time before carrying out selection and crossover stages.

Summarizing, we have shown in this preliminary study that the proposed time and individual duration method points to new research avenues where GP could be studied to address the problem of size growth from a computing time perspective. Moreover, under the new approach this paper introduces a first method for controlling individual sizes. Its main idea was to group individuals according to evaluation time, and it naturally allows keeping individuals' size under control, while fitness quality remains high. The experiments allow us to see the interest of the approach in both sequential and parallel models, although it requires further analysis with a larger set of problems in future work to confirm the findings presented above.

Although we have focused here on the time-size relationship as well as on the specific method implemented under this perspective, we believe that new

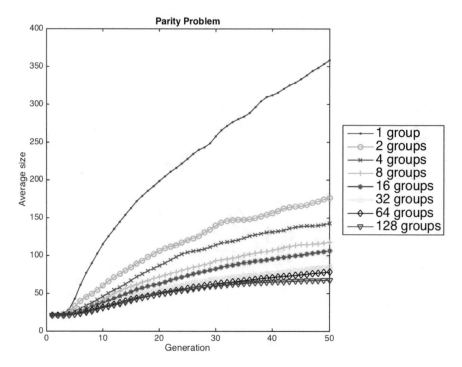

Fig. 2.8 Size-evolution along generations (averaged over 30 runs) for the parity problem (sequential execution)

approaches may be developed in the future considering time-space relationships in variable-size chromosome-based evolutionary techniques. Moreover, we think that specific improvements in the method presented here for GP will be attained when parallel environments are used and a proper load-balancing approach is applied.

2.6   Conclusions

This chapter presents a new approach to evaluating individuas complexity in variable-size chromosomes based evolutionary algorithms: using computing time instead of individuals size. Moreover, given that in parallel/distributed computing environment load-balancing methods may allow individuals to naturally group according to arrival time, this idea can shed light on the bloating phenomenon, providing clues for new control methods.

To demonstrate the usefulness of the approach, we present a first method which is based on an individual's computing time—which is automatically obtained when fitness is computed—as a trait employed for characterizing and grouping individuals together in a natural way, so that they can only bred within their groups. The reason for this idea is to keep computing time—and thus, indirectly, an individual's size growth—under control.

Based on the above described idea, we have run a set of experiments on a well known benchmark problem: parity, and results—both in parallel and sequential environments—show that the idea works, and a first specific method to prevent bloat has been presented, although other possible ones that relies in load-balancing techniques may be derived.

Acknowledgements We acknowledge support from Spanish Ministry of Economy and Competitiveness under project TIN2017-85727-C4-f2,4g-P, Regional Government of Extremadura, Department of Commerce and Economy, the European Regional Development Fund, a way to build Europe, under the project IB16035, Junta de Extremadura, project GR15068, and CICESE project 634-128.

References

1. Dignum, S., Poli, R.: Operator equalisation and bloat free GP. In: European Conference on Genetic Programming, *LNCS*, vol. 4971, pp. 110–121. Springer (2008)
2. Galeano, G., Fernández de Vega, F., Tomassini, M., Vanneschi, L.: Studying the influence of synchronous and asynchronous parallel GP on programs length evolution. In: Congress on Evolutionary Computation, vol. 2, pp. 1727–1732. IEEE (2002)
3. Koza, J.R.: Genetic programming: On the programming of computers by means of natural selection. MIT Press (1992)
4. Langdon, W.B., Poli, R.: Fitness causes bloat, pp. 13–22. Soft Computing in Engineering Design and Manufacturing. Springer (1998)
5. Luke, S., Panait, L.: Lexicographic parsimony pressure. In: Conference on Genetic and Evolutionary Computation, pp. 829–836 (2002)
6. Luke, S., Panait, L.: A comparison of bloat control methods for genetic programming. Evolutionary Computation **14**(3), 309–344 (2006)
7. Osman, A., Ammar, H.: Dynamic load balancing strategies for parallel computers. In: International Symposium on Parallel and Distributed Computing, vol. 11, pp. 110–120 (2002)
8. Oussaidène, M., Chopard, B., Pictet, O.V., Tomassini, M.: Parallel genetic programming and its application to trading model induction. Parallel Computing **23**(8), 1183–1198 (1997)
9. Silva, S.: Reassembling operator equalisation: A secret revealed. ACM SIGEVOlution **5**(3), 10–22 (2011)
10. Silva, S., Costa, E.: Dynamic limits for bloat control in genetic programming and a review of past and current bloat theories. Genetic Programming and Evolvable Machines **10**(2), 141–179 (2009)
11. Trujillo, L., Munoz, L., Galván-López, E., Silva, S.: Neat genetic programming: Controlling bloat naturally. Information Sciences (2016)
12. Vanneschi, L., Castelli, M., Silva, S.: Measuring bloat, overfitting and functional complexity in genetic programming. In: Conference on Genetic and Evolutionary Computation, pp. 877–884. ACM (2010)
13. Fernández de Vega, F.: Distributed genetic programming models with application to logic synthesis on FPGAs. Ph.D. thesis, University of Extremadura (2001)
14. Fernández de Vega, F., Abengózar Sánchez, J.G., Cotta, C.: A preliminary analysis and simulation of load balancing techniques applied to parallel genetic programming. In: International Work-Conference on Artificial Neural Networks, *LNCS*, vol. 6692, pp. 308–315. Springer (2011)

15. Fernández de Vega, F., Galeano, G., Gómez, J.A., and, J.M.S.: Efficient use of computational resources in genetic programming: Controlling the bloat phenomenon by means of the island model. In: Conference of the Industrial Electronics Society, vol. 3, pp. 2520–2524. IEEE (2002)

16. Fernández de Vega, F., Gil, G.G., Gómez Pulido, J.A., Guisado, J.L.: Control of bloat in genetic programming by means of the island model. In: Parallel Problem Solving from Nature-PPSN VIII, *LNCS*, vol. 3242, pp. 263–271. Springer (2004)

17. Whigham, P.A., Dick, G.: Implicitly controlling bloat in genetic programming. IEEE Transactions on Evolutionary Computation **14**(2), 173–190 (2010)

18. White, D.R.: Software review: The ECJ toolkit. Genetic Programming and Evolvable Machines **13**(1), 65–67 (2012)

19. Zaki, M.J., Li, W., Parthasarathy, S.: Customized dynamic load balancing for a network of workstations. Journal of Parallel and Distributed Computing (1997)

Chapter 3
Explorations of the Semantic Learning Machine Neuroevolution Algorithm: Dynamic Training Data Use, Ensemble Construction Methods, and Deep Learning Perspectives

Ivo Gonçalves, Marta Seca, and Mauro Castelli

3.1 Introduction

The success of artificial intelligence can be partially attributed to Artificial Neural Networks (ANNs), a machine learning algorithm that was invented in the late 1950s [74]. Inspired by the anatomy of the human brain, classic ANNs consist of neurons: atomic operators that receive a set of inputs and generate one output, determined by their activation function. To create networks, neurons are connected over synapses, so that the output of one neuron serves as input for the other. ANNs were used for a wide variety of tasks in many different fields [62], showing their suitability in addressing both classification and regression problems. Several properties of ANNs make them a suitable approach for addressing forecasting tasks: (1) ANNs are data-driven self-adaptive methods and there are few a priori assumptions about the models for the problems under study [79]. Thus, ANNs are a valuable technique for problems whose solutions require knowledge that is difficult to specify, but for which there are enough data or observations; (2) ANNs are universal functional approximators. In particular, it was proven that ANNs can approximate any continuous function to any desired accuracy [28, 29, 31].

Despite this result, in the literature there is no established procedure to determine the correct number of neurons for a given application. Another critical aspect relates with the topology of the network, that is, how the neurons are connected among them. For ANNs to perform well at a certain task, it is critical to find suitable connection weights. For this purpose, weights are adjusted in a learning

I. Gonçalves (✉)
INESC Coimbra, DEEC, University of Coimbra, Coimbra, Portugal

M. Seca · M. Castelli
NOVA IMS, Universidade Nova de Lisboa, Lisboa, Portugal
e-mail: M20170451@novaims.unl.pt; mcastelli@novaims.unl.pt

© Springer Nature Switzerland AG 2020
W. Banzhaf et al. (eds.), *Genetic Programming Theory and Practice XVII*, Genetic and Evolutionary Computation, https://doi.org/10.1007/978-3-030-39958-0_3

process, based on provided training data. The most prevalent approach is backprop-agation [60], where the error between prediction and ground truth is distributed back recursively through adjacent connections. However, backpropagation fails to answer the question of how to define the general topology of neurons and synapses. Devising suitable topologies is crucial, since it directly affects the speed and accuracy of the subsequent learning process.

Neuroevolution addresses these issues by applying Evolutionary Computation (EC) methods with the goal of evolving ANNs. Within neuroevolution, some approaches are specifically designed to automatically discover suitable combina-tions of topology and weights. More recently, a neuroevolution algorithm called Semantic Learning Machine (SLM) was proposed by Gonçalves et al. [22]. The most interesting characteristic of SLM is that it searches over unimodal error landscapes in any supervised learning problem where the error is measured as a distance to the known targets. It was empirically verified that this characteristic allows SLM to outperform other neuroevolution methods over a considerable set of supervised learning problems [32]. This work continues the investigation of SLM by empirically studying: (1) different methods of dynamic training data use; (2) different ensemble constructions methods. Furthermore, the extension of SLM to Convolutional Neural Networks is also discussed.

This chapter is organized as follows: Sect. 3.2 overviews the field of neu-roevolution; Sect. 3.3 describes the SLM neuroevolution algorithm and presents its distinctive features; Sect. 3.4 outlines the experimental methodology; Sect. 3.5 reports and discusses the experimental results; and Sect. 3.6 presents some initial results with SLM and Convolutional Neural Networks, and discusses future Deep Learning perspectives.

3.2 Neuroevolution Overview

One of the most challenging problems when using an ANN is the choice of its architecture or topology. In this context, the terms architecture and topology are used as synonyms to indicate some specific hyperparameters of the ANN, namely the number of hidden layers, the number of hidden neurons, and how the neurons are connected among them. Despite the vast number of applications where ANNs were used [1], the literature lacks an established procedure to determine the most suitable topology of a network when addressing a given task. Consequently, a lot of effort is currently put into automating the process of finding good ANN architectures. Solving this requires addressing several topics, such as:

1. how to design the components of the architecture
2. how to put them together
3. how to set the hyperparameters

There are two main approaches followed when approaching these topics, namely (a) using search methods based on artificial intelligence, and (b) using evolutionary techniques to generate networks. The first approach uses the gradient descent algorithm to optimize the weights of the network and to dynamically modify the hyperparameters of ANNs, while the second approach is characterized by the use of evolution to optimize the network's topology.

Neuroevolution techniques were successfully applied in different domains [14] and several attempts were proposed for using EC techniques for the optimization of ANNs [76]. The existing works can be categorized into three main approaches: (1) use of EC to train the ANN; (2) use of EC to optimize the network of an ANN; (3) use of EC to optimize the ANN's topology and to train the ANN. These different approaches are briefly discussed to present the reader the evolution of this research field in the last three decades.

The initial works aimed at using EC techniques for optimizing the weights of the connections, with a fixed topology [10, 52, 53, 68, 73]. The main idea of these works was to counteract the limitations of the backpropagation algorithm [7, 30], that due to the use of gradient descent [71] can get trapped in a local minimum of the error function and it is not capable of determining a global minimum if the error function is multimodal and/or non-differentiable [76]. These works replace the backpropagation algorithm with an evolutionary technique for learning the weights of the ANN's connections. The use of EC techniques for optimizing the weights of an ANN is attractive because they can handle the global search problem better in a vast, complex, multimodal, and non-differentiable surface [76]. Additionally, it does not depend on gradient information of the error function and thus is particularly appealing when this information is difficult to obtain. These two advantages allowed the use of evolutionary-based methods to train different kind of ANNs, including recurrent ANNs [26, 42] and higher order ANNs [12].

Subsequently, a second strand of research investigated the design of ANN architectures, namely the number of neurons, the number of hidden layers, and the connectivity among the neurons. Differently with respect to the previous studies, where it was assumed that the architecture of an ANN is predefined and fixed during the evolution of connection weights, the main aim here is to use EC techniques for evolving the topology of the network. The problem can be formulated as an optimization problem in which the objective is to determine the global optimum into a search space where each point represents an architecture. Given an objective function to be optimized (i.e., a function that quantifies the quality or performance of each architecture), the design of the optimal topology corresponds in determining the highest point on the surface induced by the objective function on the search space of the architectures. This surface (or fitness landscape), presents some properties that make EC a suitable approach for finding the sought architecture [52]. The following features were identified and discussed by Miller et al. [52]: (1) the number of possible neurons and connections is unbounded, thus the fitness landscape is infinitely large; (2) since changes in the number of neurons or connections must be discrete, the surface is non-differentiable, thus making gradient-based approaches impossible to be used; (3) the mapping from network design to network performance after learning, is indirect, strongly epistatic, and dependent on initial conditions (e.g., initial weights), so the surface is complex and noisy; (4) structurally similar networks can show very different information processing capabilities, thus making the surface highly deceptive; and (5) structurally dissimilar networks can provide similar performance, thus the surface is multimodal.

For all these reasons, EC techniques seem to be a natural choice for addressing the problem at hand, and they represent an alternative approach, with respect to constructive and destructive algorithms, toward the automatic design of architectures. A constructive algorithm [13, 15] is a hill climbing method that starts with a minimal network (an ANN with a minimal number of hidden layers, nodes, and connections) and adds new layers, nodes, and connections during the training phase when deemed necessary and based on some criterion. On the other hand, destructive algorithms [8, 56, 66] search for the optimal topology starting with a maximal network and by subsequently removing unnecessary layers, nodes, and connections during the training phase [76]. While these approaches are simpler to implement with respect to EC-based methods, they are susceptible to becoming trapped at structural local optima and they are able to explore only a small fraction of the possible ANN topologies [4].

Several works based on EC for optimizing the topology of an ANN appeared in the literature. One of the first works was proposed by Miller et al. [52], where a method based on a genetic algorithm is used for evolving neural network architectures for specific tasks. Each network architecture is represented as a connection constraint matrix mapped directly into a bit-string genotype. Modified standard genetic operators act on populations of these genotypes to produce network architectures with higher fitnesses over successive generations. Architecture fitness is assessed by training particular network instantiations and recording their final performance error. Schaffer et al. [63] demonstrated, using a genetic algorithm, that an evolved network architecture performs better than a large network using backpropagation learning alone when the criterion is correct generalization from a set of examples. In the same line of research, Wilson [75] showed that when genetic search is applied, a set of perceptrons can learn more complex tasks than initially apparent. Schiffmann et al. [64] presented a crossover operator for a genetic algorithm specifically created for automatic topology-optimization. In contrast to competing approaches, it allows that two parent networks with a different number of units can mate and produce a (valid) child network, which inherits genes from both of the parents. Similarly, Alba et al. [3] relied on a genetic algorithm to address the connectivity and structure definition problems, in order to accomplish a full genetic ANN design. Nikolopoulos and Fellrath [57] proposed the use of genetic algorithms and classifier systems to optimize the architecture of an ANN to be used for investment advising.

In the methods previously discussed, only the architecture of the ANN is evolved, but it is assumed that the activation function of each node in the architecture is fixed and predefined a priori. Despite the simplicity of this assumption, some studies demonstrated that the choice of the activation function plays a important role in determining the performance of an ANN [9, 48].

An important attempt to evolve the architecture of an ANN, as well as the activation functions, was proposed by Schoenauer and Ronald [65], where authors investigated the tuning of the slopes of the transfer functions of the individual neurons in the ANN. White and Ligomenides [72] adopted a simpler approach to the evolution of both topological structures and node transfer functions. The initial population contained ANNs with an 80% of the neurons using the sigmoid function and a 20% of the neurons using a Gaussian function. The evolutionary process was used to determine the optimal blend of these two functions in an automatic fashion. The idea of evolving the activation functions is nowadays investigated given the popularity of deep learning. For instance, the rectified linear activation (ReLU) function [33] has simplified the training of deep neural networks by counteracting the problems related to weight initialization and the vanishing gradient. As summarized by Manessi and Rozza [47], variations of ReLU have been proposed over the years, such as leaky ReLU (LReLU) [46], which addresses dead neuron issues in ReLU networks, thresholded ReLU [38], which tackles the problem of large negative biases in autoencoders, and parametric ReLU (PReLU) [27], which treats the leakage parameter of LReLU as a per-filter learnable weight. While these works introduced new and useful activation functions, other works used more advanced strategies to learn the most suitable activation function for the particular architecture at hand. Agostinelli et al. [2] designed a novel form of piecewise linear activation function that is learned independently for each neuron using gradient descent. With this adaptive activation function, they were able to improve upon deep neural network architectures composed of static rectified linear units, achieving state-of-the-art performance on CIFAR-10, CIFAR-100, and a benchmark from high-energy physics involving Higgs boson decay modes. Manessi and Rozza [47] introduced two approaches to automatically learn different combinations of base activation functions (such as the identity function, ReLU, and hyperbolic tangent) during the training phase. They presented a thorough comparison of their novel approaches with well-known architectures on three standard datasets showing substantial improvements in the overall performance. Thus, the evolution of the activation functions is nowadays deemed as important as the evolution of the architectures of the ANNs [47].

The evolutionary methods just discussed only evolve the architecture of ANNs, without any connection weights. That is, connection weights have to be learned in a subsequent step. While this approach reduces the complexity of evolving both the topology and the weights, there is a major problem with the evolution of architectures without connection weights as pointed out by Yao and Liu [77]. In particular it is possible to identify two critical issues: (1) different random initial weights may produce different training results. Thus, the same genotype may have different fitness due to different random initial weights used in training; and (2) different training algorithms may produce different training results even from the same set of initial weights. This is especially true for multimodal error functions.

Thus, the remaining part of this section recalls contributions where EC-based techniques were used to optimize the weights and the topology of an ANNs simultaneously. The idea behind this approach is that each individual in a population is a fully specified ANN with complete weight information. As a consequence, there is a one-to-one mapping between a genotype and its phenotype, thus allowing the search process to overcome the issues related to the fitness evaluation. Srinivas and Patnaik [68] presented a technique for reducing the search space of the genetic algorithm to improve its performance in searching for the globally optimal set of connection weights. They used the notion of equivalent solutions in the search space, and included in the reduced search-space only one solution, called the base solution, from each set of equivalent solutions. The iteration of the genetic algorithm consisted of an additional step where the solutions are mapped to the respective base solutions. A genetic algorithm based method was also proposed by Bornholdt and Grauden [5] for evolving a network that represented a model for a brain with sensory and motor neurons. Oliker et al. [58] proposed a distributed genetic algorithm for designing and training neural networks. The method sets the neural network's architecture and weights for a given task where the network is comprised of binary linear threshold units. White and Ligomenides [72] introduced a new algorithm which uses a genetic algorithm to determine the topology and link weights of a neural network. If the genetic algorithm fails to find a satisfactory solution network, the best network developed by the genetic algorithm is used to try to find a solution via back-propagation. In this way, each algorithm is used to its greatest advantage: the genetic algorithm (with its global search) determines a (sub-optimal) topology and weights to solve the problem, and back-propagation (with its local search) seeks the best solution in the neighborhood of the weight and topology spaces found by the genetic algorithm.

Besides genetic algorithms, other EC methods were used to address the optimization problem at hand. Koza and Rice [39] showed how to use genetic programming to find both the weights and architecture for a neural network, including the number of layers, the number of processing elements per layer, and the connectivity between processing elements. Jian and Yugeng [34] presented a new design method for the structure and weights of static neural networks based on evolutionary programming. The method is further extended to design recurrent neural networks through introducing delayed links into networks. Particle Swarm Optimization (PSO) has also been used to evolve both the weights and the topology of the networks [17, 37, 78]. In particular, Kiranyaz et al. [37] presented a Multi-Dimensional Particle Swarm Optimization (MD-PSO) technique for the automatic design of Artificial Neural Networks by evolving to the optimal network configuration (connections, weights, and biases) within the architecture space. Similarly, Garro and Vázquez [17] explored the simultaneous evolution of the three principal components of an ANN: the set of synaptic weights, the connections or architecture, and the transfer function for each neuron. The main topic of this contribution was the evaluation of eight different proposed fitness functions used to evaluate the quality of each solution and find the best design. Zhang et al. [78] introduced a new evolutionary system for evolving Feed-Forward ANNs, which is constrained to the use of PSO. Both the architecture and the weights of ANNs were adaptively adjusted according to the quality of the network. One of the most popular and broadly used approaches in neuroevolution is the NeuroEvolution of Augmenting Topologies (NEAT) algorithm [69]. Recently, NEAT has been evolved to CoDeepNEAT [51], an algorithm capable of covering more complex areas such as vision, speech and language.

To conclude, among the many existing references reporting on the use of EC to optimize ANNs, the reader is particularly referred to [61, 70, 76] for a comprehensive overview of this research field.

3.3   Semantic Learning Machine

3.3.1   Algorithm

In the proposal of Geometric Semantic Genetic Programming (GSGP), Moraglio et al. [54] showed that any supervised learning problem where the error is measured as a distance to the known targets has a unimodal error landscape. This property can be exploited by constructing specific variation operators. These operators are known as geometric semantic operators. In this context, the term semantics is used to refer to the outputs of any supervised learning model (e.g., a neural network) over a set of data instances. In GSGP, geometric semantic operators were defined for some domains: boolean, arithmetic, and conditional rules. GSGP was shown to outperform the traditional syntactic genetic programming approach in several datasets [21, 54].

The reasoning behind these geometric semantic operators can be used to create equivalent operators for other representations or computational models. With the proposal of the Semantic Learning Machine (SLM) neuroevolution algorithm [22, 24], it is achievable to perform semantic search for the space of Neural Networks (NNs). This was made possible by deriving the arithmetic mutation from GSGP to the space of NNs, therefore defining a geometric semantic mutation for NNs. This allows SLM to effectively and efficiently explore the space of NNs by exploiting the underlying unimodal error landscape. Given that these error landscapes are unimodal, no local optima exist. In the case of SLM this means that, with the exception of the global optimum, every point in the search space has at least one neighbor with better fitness, and that neighbor is reachable through the application of the mutation operator. The direct consequence is that a hill climbing strategy can effectively advance the search.

SLM is essentially a geometric semantic hill climber for NNs that follows a $(1 + \lambda)$ strategy. Without local optima, the search can be focused around the current best NN without incurring in any particular disadvantage. SLM can be summarized in the following steps:

1. Generate N initial random NNs
2. Choose the best NN (B) from the initial random NNs, according to the selected performance criterion
3. Repeat the following steps until a given stopping criterion is met:

(a) Apply the geometric semantic mutation to the current best (B) N times to generate N new NNs (known as children or neighbors)
(b) Update B as being the NN with the best performance according to the selected criterion, considering the current B and the N newly generated NNs

4. Return B as the best performing NN according to the selected performance criterion

The initial random NNs can be generated without any particular restriction. They can have any number of layers and neurons, with any activation functions, while the weights in the connections between the neurons can be freely selected. The networks do not have to be fully-connected and can be as sparsely connected as desired. The crucial aspect of SLM is the geometric semantic mutation which takes a parent NN and produces a child NN. This mutation works by adding new hidden neurons while ensuring that the semantics of the parent's hidden neurons are not affected by these new hidden neurons. To ensure this fundamental aspect of the geometric semantic mutation, the new hidden neurons do not feed their computations to the parent's hidden neurons, with the exception of the output neurons. The weights of connections from the new hidden neurons in the last hidden layer to the output neurons are defined by the learning step. The learning step can be computed optimally with the Moore–Penrose pseudoinverse (similarly to the case of GSGP [21, 23, 55]), or it can be defined as a parameter to be tuned. Each new hidden neuron added can select from which neurons it receives incoming connections. This means that the sparseness level can be easily controlled by defining how many incoming connections each new neuron will receive. The weights of each connection can be freely selected as in the initialization step. As is common in neuroevolution algorithms, SLM does not rely on backpropagation to adjust the weights of the NNs. For further SLM details the reader is referred to Gonçalves et al. [22] and Gonçalves [24].

3.3.2 Previous Comparisons with Other Neuroevolution Methods

Jagusch et al. [32] explored several SLM variants and performed a comparison with other neuroevolution methods as well as other well-established supervised machine learning techniques. Regarding the neuroevolution methods, NEAT and a fixed-topology neuroevolution approach were used as points of comparison. NEAT was one the focus of the comparison given its popularity. The comparisons were performed on a total of nine real-world datasets freely available from the UCI Machine Learning Repository [43]: four binary classification datasets and five regression datasets. The results showed that, in terms of learning the training data, SLM was superior to the other neuroevolution methods in all the nine datasets considered (all with statistically significant differences). In this comparison the best

SLM variant was, naturally, always the one that computed the optimal learning step. Focusing on the NEAT comparison and on the generalization performance, SLM was found to be superior to NEAT, with statistically significant differences, in eight out of the nine datasets considered. No statistically significant difference was found in the remaining dataset. Furthermore, particularly the SLM variant that computed the optimal learning step and used a semantic stopping criterion [25] (further details on Sect. 3.4.2) also resulted in much smaller neural networks and achieved speed-ups of various orders of magnitude over NEAT on several datasets.

3.4 Experimental Methodology

3.4.1 Datasets and Parameter Tuning

In the experimental phase, four real-world binary classifications datasets are considered: Cancer, Credit, Diabetes, and Sonar. In Credit, the objective is to classify the individuals as either good or bad credit whereas in Diabetes and Cancer, the goal is to predict whether an individual has diabetes or cancer, respectively. The Sonar task aims at classifying sonar signals as they either bounced off a metal cylinder or a roughly cylindrical rock. All of these datasets are freely available from the UCI Machine Learning Repository [43]. Table 3.1 presents the number of features (input variables), the number of instances (observations), and the % of class 1 instances in each of the four datasets under consideration.

Different SLM variants are compared with the Multi-layer Perceptron (MLP) trained with backpropagation. A nested k-fold cross-validation (CV) methodology is followed. A 30-fold outer CV is used to obtain 30 final generalization values (test set values) to assess the statistical significance of the results. For each outer training fold, a twofold inner CV is conducted to perform parameter tuning for each algorithm. Both algorithms are allowed to explore a total of 72 random parameter combinations during parameter tuning. Four SLM and two MLP variants are tested (detailed on Sects. 3.4.2 and 3.4.3). To ensure fairness, SLM tests 18 parameter combinations for each of the four variants considered, while MLP tests 36 parameter combinations for each of the two variants considered.

Table 3.1 Binary classification datasets considered

Dataset	Features	Instances	% of class 1 instances
Cancer	30	569	≈37%
Credit	24	1000	30%
Diabetes	8	768	≈35%
Sonar	60	208	≈47%

3.4.2 SLM Variants

The base SLM configuration is the following:

- In the initial population each NN is generated with a random number of hidden layers selected between 1 and 5
- In the initial population each NN randomly selects the number of neurons for each hidden layer between 1 and 5
- Each hidden neuron randomly selects its activation function from the following options: Logistic, Relu, and Tanh
- Each hidden neuron randomly selects the weight of each incoming connection from values in the interval $[-mncw, mncw]$, where $mncw$ represents the maximum neuron connection weight parameter (subject to parameter tuning)
- Each hidden neuron randomly selects the weight of its bias from values in the interval $[-mbw, mbw]$, where mbw represents the maximum bias weight parameter (subject to parameter tuning)
- Each time a new NN is created by the mutation operator, the number of new neurons to be added to each layer is randomly selected between 1 and 3

The three main differences between the SLM variants under study are the following: (1) the strategy regarding the learning step; (2) the type of training data use (static or dynamic); (3) the stopping criterion to decide the termination of the training process.

Regarding the learning step, two variants are considered: computing the Optimal Learning Step (OLS) for each application of the mutation operator; and using a Bounded Learning Step (BLS). The SLM-BLS variants introduce an additional parameter that defines the maximum learning step (mls) that bounds the learning step. At each application of the mutation operator, the effective learning step is randomly selected from values in the interval $[-mls, mls]$.

In terms of dynamic training data use, two approaches are considered: randomly selecting a subset of the training data at each iteration and computing the quality of each solution with this subset; and always using the complete training data but randomly weighting each instance (between 0 and 1) and changing these weights at each iteration. The first approach is referred to as Random Sampling Technique (RST) following Gonçalves et al. [18–20], while the second approach is referred to as Random Weighting Technique (RWT). In genetic programming, RST successfully contributed to avoid overfitting and improve generalization on high-dimensional datasets [19, 20]. Other studies using dynamic training data in genetic programming have followed [16, 49, 50, 67].

Finally, regarding the termination of the training process, the following approaches are considered: termination based on a given number of iterations; and termination based on a semantic stopping criterion . The Semantic Stopping Criteria (SSC) proposed by Gonçalves et al. [25] use information gathered from the semantic neighborhood (the set of new models generated by the mutation) to decide when to stop the search. These are named the Error Deviation Variation

(EDV) criterion and the Training Improvement Effectiveness (TIE) criterion. EDV stops the search when a considerable majority of the neighbors are improving the training performance at the expense of larger error deviations. TIE stops the search when training error improvements become harder to find within the semantic neighborhood. This can signal that the training error improvements are being forced at the expense of the resulting generalization. These SSC can be used to avoid setting a maximum number of iterations and to avoid setting data aside to use as a validation set to decide when to stop.

With these different aspects into consideration, the following SLM variants are grouped and named as follows:

1. BLS variants: SLM-BLS, SLM-BLS + RST, and SLM-BLS + RWT
2. OLS variants: SLM-OLS, SLM-OLS + RST, and SLM-OLS + RWT
3. BLS + TIE/EDV: SLM-BLS + TIE/EDV
4. OLS + EDV: SLM-OLS + EDV

When SLM-BLS is mentioned by itself it refers to SLM-BLS without using RST and RWT. Similarly, when SLM-OLS is mentioned by itself it refers to SLM-OLS without using RST and RWT.

All SLM variants can tune the maximum neuron connection weight ($mncw$) and the maximum bias weight (mbw) in the range [0.1, 0.5]. The BLS variants and BLS + TIE/EDV can tune the maximum learning step (mls) in the range [0.1, 2], and the number of iterations in the range [1, 100]. The BLS and the OLS variants select with equal probability the use of RST, RWT, or none. BLS + TIE/EDV selects with equal probability the use of EDV or TIE as the semantic stopping criterion. Whenever RST is used, the parameter that defines the ratio of the total training data to be used (the subset ratio) is selected from the range [0.01, 0.99].

3.4.3 MLP Variants

Two MLP variants are considered: the most common stochastic gradient descent (SGD) [35, 59] variant, and the Adam SGD variant [36]. For SGD and Adam, the following parameters are tuned:

- The number of iterations in the range [1, 100]
- The batch size between 50 and the maximum number of training instances available
- The activation function to be used in the hidden layers: Logistic, Relu, and Tanh
- The number of hidden layers in the range [1, 5]
- The number of hidden neurons per layer in the range [1, 200]
- The learning rate in the range [0.1, 2]
- The L2 penalty in the range [0.1, 10]

SGD can also select the momentum in the range [0.0000001, 1] and decide to use or not the Nesterov's momentum. Adam can also select the beta 1 and beta 2 parameters in the range [0, 1[.

3.5 Results and Analysis

This section analyzes the results obtained in the experimental phase. Section 3.5.1 presents the results achieved by the different variants of the SLM algorithm taken into account, analyzing the performance obtained on the validation set and discussing some aspects related to the choice of the parameters. Subsequently, Sect. 3.5.2 presents the results produced by the MLP over the same benchmark problems and discusses the main performance differences between MLP and SLM. Section 3.5.3 compares SLM and MLP after their best configuration are found and explores the generalization ability of SLM under different ensemble construction methods.

3.5.1 SLM

This section presents the results obtained by considering different groups of SLM variants. The first analysis refers to the validation Area Under Receiver Operating Characteristic (AUROC) curve values produced by the considered SLM variants, and the results are summarized in Table 3.2. For each benchmark problem and for each technique, this table reports the mean and the standard deviation of the validation AUROC. These values were obtained from the nested cross-validation procedure previously described. Thus they are the mean and standard deviation values achieved by the best models obtained in the inner cross-validation procedure (that was performed to determine the most suitable values of the hyperparameters). According to the results reported in this table and complementing them with the ones in Table 3.3, it is possible to state that OLS variants are the best performer, outperforming the other variants taken into account. In particular, the OLS variants outperformed the other competitors 23 times on both the Cancer and the Credit datasets, 20 times on the Diabetes dataset, and 26 times on the Sonar dataset. The

Table 3.2 Validation AUROC for each SLM variant considered

Dataset	BLS variants	OLS variants	BLS + TIE/EDV	OLS + EDV
Cancer	0.951 ± 0.095	0.937 ± 0.124	0.896 ± 0.185	0.959 ± 0.061
Credit	0.679 ± 0.134	0.733 ± 0.120	0.564 ± 0.166	0.688 ± 0.108
Diabetes	0.680 ± 0.169	0.784 ± 0.110	0.626 ± 0.153	0.738 ± 0.131
Sonar	0.648 ± 0.282	0.816 ± 0.213	0.636 ± 0.277	0.724 ± 0.220

Table 3.3 Best SLM configuration by variant

Dataset	BLS variants	OLS variants	BLS + TIE/EDV	OLS + EDV
Cancer	5	23	0	2
Credit	0	23	0	7
Diabetes	1	20	0	9
Sonar	2	26	1	1

Table 3.4 Number of iterations for each SLM variant considered

Dataset	BLS variants	OLS variants	BLS + TIE/EDV	OLS + EDV
Cancer	79.567 ± 16.332	64.067 ± 23.712	3.067 ± 2.741	1.133 ± 0.434
Credit	68.267 ± 20.793	63.433 ± 26.165	3.967 ± 2.399	2.233 ± 1.406
Diabetes	76.000 ± 18.819	60.467 ± 22.508	5.567 ± 8.336	1.967 ± 1.217
Sonar	75.033 ± 22.172	58.667 ± 24.288	3.533 ± 3.391	3.167 ± 4.900

second-best performer when considering the Cancer and the Sonar datasets is the BLS variants group, while OLS + EDV outperforms the other competitors 7 times on the Credit dataset and 9 times on the Diabetes dataset. BLS + TIE/EDV performs poorly in relative terms. A possible explanation might be that, for the datasets considered, the maximum number of iterations is not high enough for the semantic stopping criterion to take effect under a bounded learning step. Overall, OLS families (OLS variants and OLS + EDV) seem to provide higher AUROC values with respect to the BLS families. A global view on the average AUROC values reported in Table 3.2 suggests that: (1) within the BLS groups, BLS + TIE/EDV always achieved a lower average validation AUROC than the BLS variants; (2) within the OLS groups, OLS + EDV is the best performer on the Cancer dataset, while the OLS variants group is the best performer over the remaining benchmarks taken into account; (3) the OLS variants represent the most suitable choice for the classification problems at hand.

The subsequent analysis considers the average number of iterations performed by each SLM variant. Results of this analysis are reported in Table 3.4. The BLS variants require a larger number of iterations with respect to the other competitors. This behavior was expected considering that the BLS variants do not use any semantic stopping criterion and optimal learning step. Focusing on the other competitors, the OLS variants (the best performer over these problems) ran for a significantly larger number of iterations with respect to BLS + TIE/EDV. When comparing OLS variants with OLS + EDV, it is possible to see that also in this case the number of iterations performed by OLS + EDV is significantly lower than the one of the OLS variants. The use of the optimal learning step allows OLS + EDV to reach satisfactory performance in all of the considered benchmarks and to outperform the OLS variants over the Cancer dataset. To summarize the results of this analysis, it seems that the use of a semantic stopping criterion with the OLS and the BLS is effective at reducing the computational effort needed to perform the

Table 3.5 EDV and TIE use
in SLM-BLS

Dataset	EDV	TIE
Cancer	27	3
Credit	26	4
Diabetes	25	5
Sonar	25	5

Table 3.6 RST and RWT
use in the BLS and the OLS
variants

	BLS variants			OLS variants		
Dataset	None	RST	RWT	None	RST	RWT
Cancer	8	9	13	15	6	9
Credit	18	4	8	15	4	11
Diabetes	11	6	13	12	4	14
Sonar	15	4	11	15	6	9

training process, but according to the problem at hand, it may not result in the best overall performance.

With respect to the semantic stopping criterion, Table 3.5 compares the usage of EDV and TIE in SLM-BLS. According to these values, it is clear that EDV is more effective than the TIE stopping criterion in the classification problems considered. An additional analysis performed during the experimental phase aimed at understanding whether the random weighting/sampling techniques are beneficial when coupled with the SLM variants. Results of this analysis are reported in Table 3.6, where the use of RST, RWT, and the whole original set of observations (None) are compared in the context of BLS variants and OLS variants. According to these values, it seems that RWT is more effective than RST in both of the considered SLM variants. Comparing the use of RST and RWT with the use of the whole original set of observations, the results of Table 3.6 suggest that random sampling and random weighting techniques can improve the performance of SLM. This shows that the dynamic use of training data can indeed be beneficial within SLM. This is particularly clear when RWT is used in conjunction with the optimal learning step computation.

3.5.2 MLP

This section presents the results obtained by both MLP variants: Adam and SGD. The first part of this discussion considers the performance on the validation set and the corresponding results are reported in Table 3.7. According to these values, Adam is the best performer over the Cancer dataset, while SGD is the best performer over the Credit dataset and the Sonar dataset. The two MLP variants produce the same performance on the Diabetes dataset. According to these results it is difficult to draw a general conclusion, and it seems that the choice between Adam and SGD must be evaluated according to the particular problem at hand. To complement this analysis,

Table 3.7 Validation
AUROC for each MLP
variant considered

Dataset	Adam	SGD
Cancer	0.542 ± 0.110	0.500 ± 0.000
Credit	0.509 ± 0.031	0.525 ± 0.048
Diabetes	0.498 ± 0.016	0.498 ± 0.011
Sonar	0.496 ± 0.023	0.581 ± 0.109

Table 3.8 Best MLP
configuration by variant

Dataset	Adam	SGD
Cancer	28	2
Credit	9	21
Diabetes	24	6
Sonar	7	23

Table 3.9 Number of
iterations for each MLP
variant considered

Dataset	Adam	SGD
Cancer	56.200 ± 25.183	55.500 ± 26.046
Credit	56.400 ± 24.210	57.167 ± 26.592
Diabetes	42.433 ± 27.320	56.700 ± 30.326
Sonar	50.267 ± 24.237	49.667 ± 29.352

Table 3.8 shows the best MLP configurations by variant. According to these values, SGD produces the best performance most of the times over the Credit and Sonar dataset, while Adam returns the best performance on the remaining datasets in the vast majority of the runs considered.

At this stage, it is important to compare the results of Table 3.7 (obtained with MLP) with the ones reported in Table 3.2 (obtained with SLM). According to these results, all the SLM variants are able to outperform the best MLP variant over all the classification problems under exam. This comparison clearly demonstrates the superiority of SLM (with respect to MLP) in creating models characterized by a greater validation AUROC. Overall, the SLM algorithm is a competitive option to consider in these classification problems given that its performance (independently of the selected variant) is significantly better than the best MLP variant.

Another important aspect to analyze in the comparison between MLP and SLM, is the number of iterations required to produce the final model. While this analysis was performed for the SLM algorithm (see Table 3.4), Table 3.9 reports the same information for MLP-based models. In particular, the SLM variants that use a semantic stopping criterion are able to build a classification model in a considerably lower amount of iterations with respect to an MLP variant based on the backpropagation algorithm. This smaller number of iterations does not negatively affect the performance of the final models, as these SLM variants are able to outperform MLP over all the considered benchmarks.

To further understand the different MLP variants, Table 3.10 reports the activation function used by each one of them. From these values it is interesting to point out that the Adam variant has a clear preference for the Relu function, disregarding the problem under analysis. SGD shows a preference for the Relu function when

Table 3.10 Activation functions use by MLP variant

Dataset	Adam			SGD		
	Logistic	Relu	Tanh	Logistic	Relu	Tanh
Cancer	8	15	7	7	16	7
Credit	9	13	8	9	8	13
Diabetes	4	16	10	7	15	8
Sonar	8	17	5	0	14	16

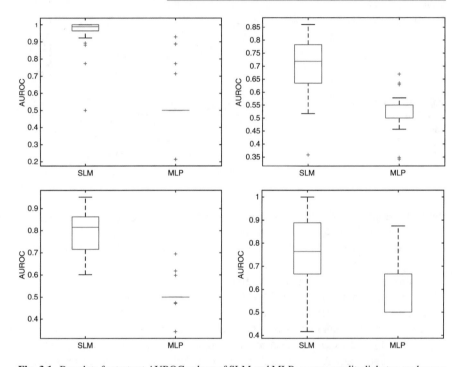

Fig. 3.1 Boxplots for test set AUROC values of SLM and MLP: cancer, credit, diabetes, and sonar

considering the Cancer and Diabetes datasets, while Tanh is the function used most of the times over the Credit dataset. For the Sonar dataset the Logistic function was never selected and Relu and Tanh were selected 14 and 16 times respectively.

3.5.3 Generalization and Ensemble Analysis

This section starts by assessing the generalization (i.e., the test set performance over the 30 outer folds) of SLM and MLP considering the best parameter configurations found after parameter tuning. Figure 3.1 presents the boxplots for the AUROC values of both algorithms. On each box, the central mark is the median, the edges

Table 3.11 p-values of
Mann–Whitney U-tests over
test set AUROC values of
SLM and MLP

Dataset	p-value
Cancer	3.486×10^{-11}
Credit	1.054×10^{-8}
Diabetes	1.036×10^{-11}
Sonar	9.894×10^{-5}

of the box are the 25th and 75th percentiles, and the whiskers extend to the most extreme data points that are not considered outliers.

These boxplots show that SLM consistently achieves better AUROC values than MLP across all datasets. A set of statistical tests is performed to assess the statistical significance of these results. Firstly, a Kolmogorov-Smirnov test is applied to assess if these values come from a normal distribution. The result of this test suggests that the alternative hypothesis (i.e., the data do not come from a normal distribution) cannot be rejected considering a significance level (α) of 0.05. Given this outcome, a rank-based statistic is selected for the next step. A Mann-Whitney U-test is performed with the null hypothesis that the samples have equal medians. As in the previous test, a significance level of 0.05 is considered. The outcomes suggest that SLM outperforms MLP in all datasets with statistically significant differences. The p-values for these comparisons can be found in Table 3.11.

The final part of the analysis studies the outcomes of different ensemble construction methods when using SLM as a base learner. Bagging [6] and Boosting [11] methods are compared with a common simple averaging construction method that trains the base learner N times without changing the training instances provided. This simpler ensemble construction method can be effective if the base learner already has an inherent diversity within its search process. This might be the case of SLM given that it is a stochastic algorithm. These three ensemble construction methods are used to create ensembles of 30 NNs using SLM as the base learner. In the Boosting case, four variations of AdaBoost.R2 are studied and labeled as follows:

- Boosting-1: weighted median prediction and fixed learning rate of 1
- Boosting-2: weighted median prediction and variable learning rate selected randomly in the interval [0, 1] for each new NN added to the ensemble
- Boosting-3: weighted mean prediction and fixed learning rate of 1
- Boosting-4: weighted mean prediction and variable learning rate selected randomly in the interval [0, 1] for each new NN added to the ensemble

Figure 3.2 presents the boxplots for the test set AUROC values of each ensemble construction method considered: Simple (averaging), Bagging, and the four Boosting variations.

These ensemble results show that the simple averaging method performs similarly to Bagging and Boosting. In terms of the median AUROC value, the simple averaging method even achieves the highest value in three of the four datasets: Credit, Diabetes, and Sonar. Overall, these different ensemble construction methods

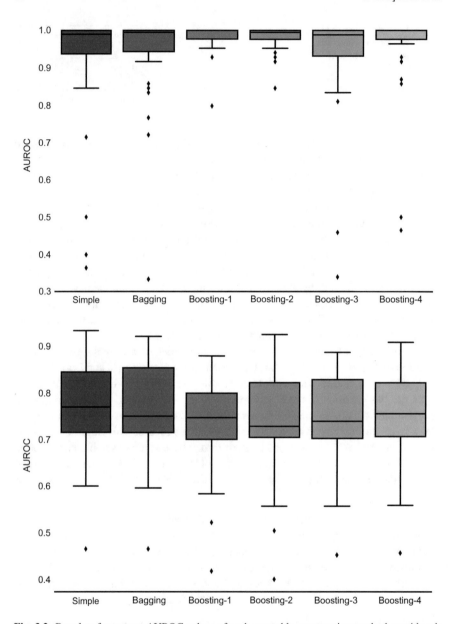

Fig. 3.2 Boxplots for test set AUROC values of each ensemble construction method considered: cancer, credit, diabetes, and sonar

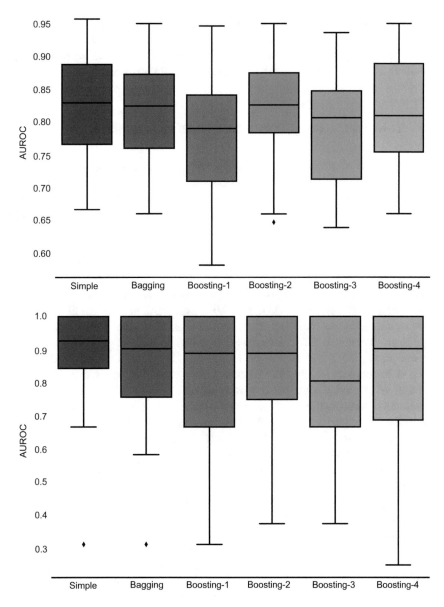

Fig. 3.2 (continued)

perform similarly in terms of the distribution of the values. These results suggest that the stochastic nature of SLM allows the simple averaging method to perform well without having to explicitly confer more diversity to the base learner, e.g., by providing different training instances to each ensemble member.

3.6   Toward the Deep Semantic Learning Machine

Recently, SLM was used in conjunction with Convolutional Neural Networks (CNNs) for the first time [40, 41]. In these contributions, the task of discriminating between benign and malignant prostate cancer lesions given multiparametric magnetic resonance imaging was addressed. This image classification task was proposed in the context of the PROSTATEx competition [44]. SLM was tested as a backpropagation replacement for the training of the last fully-connected layers of CNNs. In this approach, the outputs from the convolutional layers of a given CNN are passed as inputs (without pre-training) to SLM. The empirical comparison is performed with XmasNet [45], a state-of-the-art CNN specifically developed to address the PROSTATEx 2017 competition. The results show that SLM achieves higher AUROC curve values than XmasNet with a statistically significant difference. This performance is achieved without neither pre-training the underlying CNN nor relying on backpropagation. Furthermore, SLM is also much more computationally efficient in the training phase. SLM achieves an average speed-up of around 14 over training with backpropagation. This is also important as neuroevolution methods are sometimes perceived as slow. Additionally, it is important to emphasize that SLM was only run on CPU (whereas XmasNet was trained using a GPU) and without any explicit parallelization. This further reinforces the results obtained, given that each network evaluation could be suitably parallelized, thus achieving a higher speed-up. Furthermore, inside each network evaluation, the new nodes of a given layer can also be evaluated in parallel. SLM can be further extended to include the convolutional layers within the search process. This would remove the need for a fixed CNN topology to be provided and it would also eliminate the burden of assessing several CNN topologies in order to find a suitable topology for the task at hand. Adapting SLM to include the convolutional layers within the search would result in a unimodal search over the space of CNNs. Such a development could result in considerable improvements in the field of deep learning and computer vision. This is something that is currently under study.

Acknowledgements This work was partially supported by projects UID/MULTI/00308/2019 and by the European Regional Development Fund through the COMPETE 2020 Programme, FCT— Portuguese Foundation for Science and Technology and Regional Operational Program of the Center Region (CENTRO2020) within project MAnAGER (POCI-01-0145-FEDER-028040). This work was also partially supported by national funds through FCT (Fundação para a Ciência e a Tecnologia) under project DSAIPA/DS/0022/2018 (GADgET).

References

1. Abiodun, O.I., Jantan, A., Omolara, A.E., Dada, K.V., Mohamed, N.A., Arshad, H.: State-of-the-art in artificial neural network applications: A survey. Heliyon **4**(11), e00,938 (2018)
2. Agostinelli, F., Hoffman, M., Sadowski, P., Baldi, P.: Learning activation functions to improve deep neural networks. arXiv preprint arXiv:1412.6830 (2014)
3. Alba, E., Aldana, J., Troya, J.M.: Full automatic ANN design: A genetic approach. In: International Workshop on Artificial Neural Networks, pp. 399–404. Springer (1993)
4. Angeline, P.J., Saunders, G.M., Pollack, J.B.: An evolutionary algorithm that constructs recurrent neural networks. IEEE Transactions on Neural Networks **5**(1), 54–65 (1994)
5. Bornholdt, S., Graudenz, D.: General asymmetric neural networks and structure design by genetic algorithms. Neural Networks **5**(2), 327–334 (1992)
6. Breiman, L.: Bagging predictors. Machine Learning **24**(2), 123–140 (1996).
7. Chauvin, Y., Rumelhart, D.E.: Backpropagation: Theory, architectures, and applications. Psychology Press (2013)
8. Cun, Y.L., Denker, J.S., Solla, S.A.: Advances in neural information processing systems 2. chap. Optimal Brain Damage, pp. 598–605. Morgan Kaufmann Publishers Inc., San Francisco, CA, USA (1990)
9. DasGupta, B., Schnitger, G.: Efficient approximation with neural networks: A comparison of gate functions. Pennsylvania State University, Department of Computer Science (1992)
10. Dill, F.A., Deer, B.C.: An exploration of genetic algorithms for the selection of connection weights in dynamical neural networks. In: Proceedings of the IEEE 1991 National Aerospace and Electronics Conference NAECON 1991, vol. 3, pp. 1111–1115 (1991)
11. Drucker, H.: Improving regressors using boosting techniques. In: Proceedings of the Fourteenth International Conference on Machine Learning, ICML '97, pp. 107–115. Morgan Kaufmann Publishers Inc., San Francisco, CA, USA (1997).
12. Epitropakis, M.G., Plagianakos, V.P., Vrahatis, M.N.: Evolutionary Algorithm Training of Higher-Order Neural Networks. IGI Global (2009)
13. Fahlman, S.E., Lebiere, C.: Advances in neural information processing systems 2. chap. The Cascade-correlation Learning Architecture, pp. 524–532. Morgan Kaufmann Publishers Inc., San Francisco, CA, USA (1990)
14. Floreano, D., Dürr, P., Mattiussi, C.: Neuroevolution: From architectures to learning. Evolutionary Intelligence **1**(1), 47–62 (2008)
15. Frean, M.: The upstart algorithm: A method for constructing and training feedforward neural networks. Neural Computation **2**(2), 198–209 (1990)
16. Galván-López, E., Vázquez-Mendoza, L., Schoenauer, M., Trujillo, L.: On the use of dynamic GP fitness cases in static and dynamic optimisation problems. In: International Conference on Artificial Evolution (Evolution Artificielle), pp. 72–87. Springer (2017)
17. Garro, B.A., Vázquez, R.A.: Designing artificial neural networks using particle swarm optimization algorithms. Computational Intelligence and Neuroscience **2015**, 61 (2015)
18. Gonçalves, I., Silva, S.: Experiments on controlling overfitting in genetic programming. In: Local proceedings of the 15th Portuguese Conference on Artificial Intelligence, EPIA 2011 (2011)
19. Gonçalves, I., Silva, S., Melo, J.B., Carreiras, J.M.B.: Random sampling technique for overfitting control in genetic programming. In: Genetic Programming, pp. 218–229. Springer (2012)
20. Gonçalves, I., Silva, S.: Balancing learning and overfitting in genetic programming with interleaved sampling of training data. In: Genetic Programming, pp. 73–84. Springer (2013)
21. Gonçalves, I., Silva, S., Fonseca, C.M.: On the generalization ability of geometric semantic genetic programming. In: Genetic Programming, pp. 41–52. Springer (2015)
22. Gonçalves, I., Silva, S., Fonseca, C.M.: Semantic learning machine: A feedforward neural network construction algorithm inspired by geometric semantic genetic programming. In: Progress in Artificial Intelligence, *Lecture Notes in Computer Science*, vol. 9273, pp. 280–285. Springer (2015)

23. Gonçalves, I., Silva, S., Fonseca, C.M., Castelli, M.: Arbitrarily close alignments in the error space: A geometric semantic genetic programming approach. In: Proceedings of the 2016 on Genetic and Evolutionary Computation Conference Companion, pp. 99–100. ACM (2016)

24. Gonçalves, I.: An exploration of generalization and overfitting in genetic programming: Standard and geometric semantic approaches. Ph.D. thesis, Department of Informatics Engineering, University of Coimbra, Portugal (2017)

25. Gonçalves, I., Silva, S., Fonseca, C.M., Castelli, M.: Unsure when to stop? ask your semantic neighbors. In: Proceedings of the Genetic and Evolutionary Computation Conference, GECCO'17, pp. 929–936. ACM, New York, NY, USA (2017).

26. Greenwood, G.W.: Training partially recurrent neural networks using evolutionary strategies. IEEE Transactions on Speech and Audio Processing 5(2), 192–194 (1997)

27. He, K., Zhang, X., Ren, S., Sun, J.: Delving deep into rectifiers: Surpassing human-level performance on imagenet classification. In: Proceedings of the IEEE International Conference on Computer Vision, pp. 1026–1034 (2015)

28. Hornik, K., Stinchcombe, M., White, H.: Multilayer feedforward networks are universal approximators. Neural Networks 2(5), 359–366 (1989)

29. Hornik, K.: Approximation capabilities of multilayer feedforward networks. Neural Networks 4(2), 251–257 (1991)

30. Hush, D.R., Horne, B.G.: Progress in supervised neural networks. IEEE Signal Processing Magazine 10(1), 8–39 (1993)

31. Irie, B., Miyake, S.: Capabilities of three-layered perceptrons. In: IEEE International Conference on Neural Networks, vol. 1, p. 218 (1988)

32. Jagusch, J.B., Gonçalves, I., Castelli, M.: Neuroevolution under unimodal error landscapes: An exploration of the semantic learning machine algorithm. In: Proceedings of the Genetic and Evolutionary Computation Conference Companion, pp. 159–160. ACM (2018)

33. Jarrett, K., Kavukcuoglu, K., LeCun, Y., et al.: What is the best multi-stage architecture for object recognition? In: 2009 IEEE 12th International Conference on Computer Vision, pp. 2146–2153. IEEE (2009)

34. Jian, F., Yugeng, X.: Neural network design based on evolutionary programming. Artificial Intelligence in Engineering 11(2), 155–161 (1997)

35. Kiefer, J., Wolfowitz, J.: Stochastic estimation of the maximum of a regression function. Ann. Math. Statist. 23(3), 462–466 (1952).

36. Kingma, D.P., Ba, J.: Adam: A method for stochastic optimization. CoRR **abs/1412.6980** (2014).

37. Kiranyaz, S., Ince, T., Yildirim, A., Gabbouj, M.: Evolutionary artificial neural networks by multi-dimensional particle swarm optimization. Neural Networks 22(10), 1448–1462 (2009)

38. Konda, K., Memisevic, R., Krueger, D.: Zero-bias autoencoders and the benefits of co-adapting features. arXiv preprint arXiv:1402.3337 (2014)

39. Koza, J.R., Rice, J.P.: Genetic generation of both the weights and architecture for a neural network. In: IJCNN-91-Seattle International Joint Conference on Neural Networks, vol. 2, pp. 397–404. IEEE (1991)

40. Lapa, P., Gonçalves, I., Rundo, L., Castelli, M.: Enhancing classification performance of convolutional neural networks for prostate cancer detection on magnetic resonance images: A study with the semantic learning machine. In: Proceedings of the Genetic and Evolutionary Computation Conference Companion, GECCO '19. ACM, New York, NY, USA (2019).

41. Lapa, P., Gonçalves, I., Rundo, L., Castelli, M.: Semantic learning machine improves the cnn-based detection of prostate cancer in non-contrast-enhanced mri. In: Proceedings of the Genetic and Evolutionary Computation Conference Companion, GECCO '19. ACM, New York, NY, USA (2019).

42. Lei, J., He, G., Jiang, J.P.: The state estimation of the cstr system based on a recurrent neural network trained by HGAs. In: Proceedings of International Conference on Neural Networks (ICNN'97), vol. 2, pp. 779–782 (1997)

43. Lichman, M.: UCI Machine Learning Repository (2013).

44. Litjens, G., Debats, O., Barentsz, J., Karssemeijer, N., Huisman, H.: "PROSTATEx Challenge data", The Cancer Imaging Archive. https://wiki.cancerimagingarchive.net/display/Public/SPIE-AAPM-NCI+PROSTATEx+Challenges (2017). Online; Accessed on January 25, 2019
45. Liu, S., Zheng, H., Feng, Y., Li, W.: Prostate cancer diagnosis using deep learning with 3D multiparametric MRI. In: Medical Imaging 2017: Computer-Aided Diagnosis, *Proceedings SPIE*, vol. 10134, p. 1013428. International Society for Optics and Photonics (2017).
46. Maas, A.L., Hannun, A.Y., Ng, A.Y.: Rectifier nonlinearities improve neural network acoustic models. In: Proc. ICML, vol. 30, p. 3 (2013)
47. Manessi, F., Rozza, A.: Learning combinations of activation functions. In: 2018 24th International Conference on Pattern Recognition (ICPR), pp. 61–66. IEEE (2018)
48. Mani, G.: Learning by gradient descent in function space. In: 1990 IEEE International Conference on Systems, Man, and Cybernetics Conference Proceedings, pp. 242–247 (1990)
49. Martinez, Y., Trujillo, L., Naredo, E., Legrand, P.: A comparison of fitness-case sampling methods for symbolic regression with genetic programming. In: EVOLVE-A Bridge between Probability, Set Oriented Numerics, and Evolutionary Computation V, pp. 201–212. Springer (2014)
50. Martínez, Y., Naredo, E., Trujillo, L., Legrand, P., López, U.: A comparison of fitness-case sampling methods for genetic programming. Journal of Experimental & Theoretical Artificial Intelligence **29**(6), 1203–1224 (2017)
51. Miikkulainen, R., Liang, J., Meyerson, E., Rawal, A., Fink, D., Francon, O., Raju, B., Shahrzad, H., Navruzyan, A., Duffy, N., Hodjat, B.: Evolving deep neural networks. In: R. Kozma, C. Alippi, Y. Choe, F.C. Morabito (eds.) Artificial Intelligence in the Age of Neural Networks and Brain Computing. Amsterdam: Elsevier (2018).
52. Miller, G.F., Todd, P.M., Hegde, S.U.: Designing neural networks using genetic algorithms. In: Proceedings of the Third International Conference on Genetic Algorithms, pp. 379–384. Morgan Kaufmann Publishers Inc., San Francisco, CA, USA (1989)
53. Montana, D.J., Davis, L.: Training feedforward neural networks using genetic algorithms. In: IJCAI, vol. 89, pp. 762–767 (1989)
54. Moraglio, A., Krawiec, K., Johnson, C.G.: Geometric semantic genetic programming. In: Parallel Problem Solving from Nature-PPSN XII, pp. 21–31. Springer (2012)
55. Moraglio, A., Mambrini, A.: Runtime analysis of mutation-based geometric semantic genetic programming for basis functions regression. In: Proceedings of the 15th annual conference on Genetic and Evolutionary Computation, pp. 989–996. ACM (2013)
56. Mozer, M.C., Smolensky, P.: Advances in neural information processing systems 1. chap. Skeletonization: A Technique for Trimming the Fat from a Network via Relevance Assessment, pp. 107–115. Morgan Kaufmann Publishers Inc., San Francisco, CA, USA (1989)
57. Nikolopoulos, C., Fellrath, P.: A hybrid expert system for investment advising. Expert Systems **11**(4), 245–250 (1994)
58. Oliker, S., Furst, M., Maimon, O.: Design architectures and training of neural networks with a distributed genetic algorithm. In: IEEE International Conference on Neural Networks, pp. 199–202. IEEE (1993)
59. Robbins, H., Monro, S.: A stochastic approximation method. Ann. Math. Statist. **22**(3), 400–407 (1951).
60. Rumelhart, D.E., Hinton, G.E., Williams, R.J.: Learning representations by back-propagating errors. Nature **323**(6088), 533 (1986)
61. S. Ding H. Li, C.S.J.Y.F.J.: Evolutionary artificial neural networks: A review. Artificial Intelligence Review **39**, 251–260 (2013).
62. Samarasinghe, S.: Neural networks for applied sciences and engineering: From fundamentals to complex pattern recognition. Auerbach Publications (2016)
63. Schaffer, J.D., Caruana, R.A., Eshelman, L.J.: Using genetic search to exploit the emergent behavior of neural networks. Physica D: Nonlinear Phenomena **42**(1–3), 244–248 (1990)
64. Schiffmann, W., Joost, M., Werner, R.: Synthesis and performance analysis of multilayer neural network architectures (1992)

65. Schoenauer, M., Ronald, E.: Genetic extensions of neural net learning: Transfer functions and renormalisation coefficients
66. Sietsma, J., Dow, R.J.: Creating artificial neural networks that generalize. Neural Networks **4**(1), 67–79 (1991)
67. Silva, S., Ingalalli, V., Vinga, S., Carreiras, J.M., Melo, J.B., Castelli, M., Vanneschi, L., Gonçalves, I., Caldas, J.: Prediction of forest aboveground biomass: An exercise on avoiding overfitting. In: European Conference on the Applications of Evolutionary Computation, pp. 407–417. Springer (2013)
68. Srinivas, M., Patnaik, L.M.: Learning neural network weights using genetic algorithms-improving performance by search-space reduction. In: [Proceedings] 1991 IEEE International Joint Conference on Neural Networks, vol. 3, pp. 2331–2336 (1991)
69. Stanley, K.O., Miikkulainen, R.: Evolving neural networks through augmenting topologies. Evolutionary Computation **10**(2), 99–127 (2002)
70. Stanley, K.O., Clune, J., Lehman, J., Miikkulainen, R.: Designing neural networks through neuroevolution. Nature Machine Intelligence **1**(1), 24–35 (2019)
71. Sutton, R.S.: Two problems with backpropagation and other steepest-descent learning procedures for networks. In: Proceedings of the Eighth Annual Conference of the Cognitive Science Society. Hillsdale, NJ: Erlbaum (1986)
72. White, D., Ligomenides, P.: Gannet: A genetic algorithm for optimizing topology and weights in neural network design. In: International Workshop on Artificial Neural Networks, pp. 322–327. Springer (1993)
73. Whitley, D., Starkweather, T., Bogart, C.: Genetic algorithms and neural networks: Optimizing connections and connectivity. Parallel Computing **14**(3), 347–361 (1990)
74. Widrow, B., Lehr, M.A.: 30 years of adaptive neural networks: Perceptron, madaline, and backpropagation. Proceedings of the IEEE **78**(9), 1415–1442 (1990)
75. Wilson, S.W.: Perception redux: Emergence of structure. Physica D: Nonlinear Phenomena **42**(1–3), 249–256 (1990)
76. Yao, X.: Evolving artificial neural networks. Proceedings of the IEEE **87**(9), 1423–1447 (1999)
77. Yao, X., Liu, Y.: A new evolutionary system for evolving artificial neural networks. IEEE Transactions on Neural Networks **8**(3), 694–713 (1997)
78. Zhang, C., Shao, H., Li, Y.: Particle swarm optimisation for evolving artificial neural network. In: Systems, Man, and Cybernetics, 2000 IEEE International Conference on, vol. 4, pp. 2487–2490. IEEE (2000)
79. Zhang, G., Patuwo, B.E., Hu, M.Y.: Forecasting with artificial neural networks:: The state of the art. International Journal of Forecasting **14**(1), 35–62 (1998)

Chapter 4
Can Genetic Programming Perform Explainable Machine Learning for Bioinformatics?

Ting Hu

4.1 Introduction

In recent years, with the increasing availability of computational power, machine learning has seen a rapid growth of applications in a variety of research fields and industries [6, 9, 11, 15, 20]. However, questions with regard to the explainability of machine learning have been raised. How trustworthy is the learned predictive model? What is the mechanism of the prediction? Given a specific testing sample, how can we explain the prediction result? Such an explainability issue is in an urgent need to be addressed in order to allow wider applications and further developments of machine learning [7, 12, 26].

Some initial attempts have been put forward to enable machine learning to be more explainable and transparent. For instance, Ribeiro et al. proposed the local interpretable model-agnostic explanations (LIME) algorithm to identify a local linear model to explain the prediction on a specific instance [24]. Such a local linear model is faithfully derived from the globally learned highly non-linear model. Thus it is able to provide an explainable linear relationship of the predictors and the response while retaining the prediction accuracy. In a bioinformatics application study of machine learning, Yu et al. designed a deep learning algorithm DCell, a neural network embedded in the hierarchical structure of more than two thousand subsystems comprising a eukaryotic cell [22]. DCell simulated cellular growth and was trained to encode complex genotypes for the prediction of diseases.

The research on explainable machine learning is still at its infancy since the most popular methods for interpretation and explanation are either problem dependent,

T. Hu (✉)
School of Computing, Queen's University, Kingston, ON, Canada

Department of Computer Science, Memorial University, St. John's, NL, Canada
e-mail: ting.hu@queensu.ca

© Springer Nature Switzerland AG 2020   63
W. Banzhaf et al. (eds.), *Genetic Programming Theory and Practice XVII*, Genetic and Evolutionary Computation, https://doi.org/10.1007/978-3-030-39958-0_4

i.e., domain knowledge is needed to design the architecture of the predictive model, or question dependent, i.e., which aspect is needed to explain the prediction model.

Genetic programing (GP), a powerful automatic learning and evolutionary algorithm, may be a good candidate for solving the explainability problem. First, the representation of a predictive model can be flexible in GP. Second, feature selection is intrinsic and co-evolved with the predictive models. Third, allowing multiple objectives, GP may facilitate the evolution of both compact and accurate predictive models.

In this study, we explored some initial ideas of using GP for explainable machine learning by designing a linear GP algorithm and feature importance evaluation methods for the bioinformatics application problem of predicting disease risk using metabolite abundance levels in blood samples. Among multiple issues related to explainability, we aimed to address the questions as follows: Which features are more influential on the prediction of the disease risk? How do these features influence the prediction?

4.2 Methods

4.2.1 Metabolomics Data for Osteoarthritis

In the metabolomics data for osteoarthritis (OA) used in the current study, knee OA patients were selected from the Newfoundland Osteoarthritis Study (NFOAS) initiated in 2011 [17, 27]. The NFOAS aimed at identifying novel genetic, epigenetic, and biochemical markers for OA. The NFOAS recruited OA patients who underwent a total knee replacement surgery due to primary OA between November 2011 and December 2013 at the St. Clare's Mercy Hospital and Health Science Centre General Hospital in St. John's, the capital city of Newfoundland and Labrador (NL), Canada. Healthy controls were selected from the CODING study (The Complex Diseases in the Newfoundland population: Environment and Genetics), where participants were adult volunteers [10].

Both cases and controls were from the same source population of Newfoundland and Labrador. Knee OA diagnosis was made based on the American College of Rheumatology clinical criteria for the classification of idiopathic OA of the knee [2] and the judgment of the attending orthopedic surgeons. Controls were individuals without self-reported family doctor diagnosed knee OA based on their medical information collected by a self-administered questionnaire. A total of 153 OA cases and 236 healthy controls were collected.

Blood samples were collected after at least 8 h of fasting and plasma was separated from blood using the standard protocol. Metabolic profiling was performed on plasma using the Waters XEVO TQ MS system (Waters Limited, Mississauga, Ontario, Canada) coupled with Biocrates AbsoluteIDQ p180 kit, which measures 186 metabolites including 90 glycerophospholipids, 40 acylcarnitines (1 free carnitine), 21 amino acids, 19 biogenic amines, 15 sphingolipids and 1 hexose (above 90% is glucose). The details of the 186 metabolites and the metabolic profiling method were described in our previous publication [29]. Over 90% of the metabolites (167/186) were successfully determined in each sample.

Prior to performing the informatics analyses, several steps of preprocessing were applied to the dataset. Batch correction was performed by multiplying each metabolite concentration value by the ratio of the overall mean and the batch mean for that metabolite. Then, covariate adjustment was performed to remove the variation due to individual's age, gender, and body mass index (BMI). The samples were randomly assigned to either a *discovery* or *replication* dataset, such that cases and controls were divided evenly between the two datasets. The purpose of the split was to allow two sets of independent analysis and ensure the generalization of the results. Finally, each metabolite concentration value was normalized to zero mean and unit variance across the population.

4.2.2 Linear Genetic Programming Algorithm

In this study, a Linear Genetic Programming (LGP) algorithm [5] was developed to evolve symbolic models that predict the disease risk using the metabolite concentrations in the blood samples. A population of diverse candidate prediction models is generated randomly in the step of initialization and will evolve to improve prediction accuracy gradually through a number of generations. After evolution halts, the best model of the population in the final generation will be the output.

Each candidate prediction model takes the form of a symbolic computer program comprised of a set of sequential instructions. An instruction can be an assignment statement or a conditional statement. The conditional if instructions affect the program flow such that the instruction immediately following the if instruction is not executed if the condition is false. In the case of nested if instructions, each of the successive conditions needs to be true in order for the instruction following the chain of if instructions to be executed.

A register r stores the value of a feature, a calculation variable, or a constant. A feature can be a predictor or an attribute used to make a prediction of the outcome. In the context of the current study, features are concentration levels of metabolites in the samples. A calculation variable serves as a temporary buffer that enhances the computation capacity. In an assignment instruction, only registers storing calculation variables can serve as the return on the left side of the assignment symbol "=", but any register can serve as an operand on the right-hand side. This is to prevent overwriting the feature values. When a prediction model is evaluated on a given sample, feature registers take all the values of the sample, and the set of instructions are executed sequentially. The sigmoid transformation of the final value stored in the designated calculation register r[0] is used to predict the outcome of the sample, i.e., if $S(r[0])$ is greater than or equal to 0.5, the sample is predicted as diseased (class one), otherwise the sample is predicted as healthy (class zero).

An example of classification model with eight instructions is given below. Here, the output register r[0] and calculation registers r[4] and r[5] are all initialized with ones. Feature registers r[1-3] take input values from three metabolite concentration levels m[1-3] respectively. For instance, when a sample with m[1-3] values as {0.2, 0.01, 0.085} is input to this classification model, the conditional statement r[1] > r[3] in instruction I1 becomes true, so in instruction I2, r[0] changes its value to 0.51. The rest of the instructions are executed sequentially, and the final value of r[0] is set to 1.0039. Its sigmoid transformation $S(1.0039)$ is greater than 0.5, so this sample will be classified by this model as class one, i.e., diseased.

```
I1:   if r[1] > r[3]
I2:       then r[0] = r[2] + 0.5
I3:   r[4] = r[2] / r[0]
I4:   if r[0] > 4
I5:       then if r[3] < 10
I6:               then r[5] = r[3] - r[4]
I7:   r[4] = r[4] * r[1]
I8:   r[0] = r[5] + r[4]
```

At the initial generation, a population of diverse linear genetic programs (classification models) was generated randomly. The fitness of each model was evaluated using mean classification error (MCE), computed as the average number of incorrectly classified training samples. A set of models were chosen as parents based on their fitness, and variation operators, including mutation and recombination, were applied to them. A mutation alters an element of a randomly picked instruction, i.e., replacing a return or an operand register by a randomly generated one or replacing the operator. Recombination swaps segments of instructions of two parent models. Survival selection picks fitter models to form the population for the next generation. Such an evolution process iterates for a certain number of generations, and the model with the lowest MCE at the end is output as the final best model of a run.

This LGP algorithm was implemented using the Julia programming language [4]. The main parameters used in the implementation are shown in Table 4.1. A fivefold cross-validation was used so that each run of the LGP algorithm produced *five* best classification models as its output. We first ran the LGP algorithm 200 times using the discovery dataset and collected 1000 evolved best predictive models. We investigated the resulting classification models by calculating various statistics of the fitness (MCE) values, sensitivity, specificity and area under the curve (AUC) as computed on the testing fold for each run.

Table 4.1 Parameter
configuration of the LGP
algorithm

Fitness function	Mean classification error (MCE)
Program initialization	Random
Program length	[1, 500]
Population size	500
Number of parents	500
Parent selection	Tournament with size 16
Survival selection	Truncation
Number of generations	500
Operator set	$\{+, -, \times, \div, x^y, \text{if} <, \text{if} >\}$
Constant set	$\{1, 2, 3, 4, 5, 6, 7, 8, 9, 10\}$
Calculation registers	150
Mutation operator	Effective instructions only

4.2.3   Training Using the Full and the Focused Feature Sets

Although all 167 features (metabolites) in the metabolomics data were provided as input variables to the LGP algorithm, not all of them will be picked and influence the output of a final best genetic program (predictive model). For a linear genetic program, we define its *effective features* as the ones that effectively influence the value of its output register. Note that this definition excludes features that are present in the program but are structurally or semantically ineffective in modifying the final value of the output register. This is also a powerful feature of GP algorithms where feature selection is intrinsic and co-evolved with the model accuracy.

Since we had 1000 evolved best predictive models, we counted the occurrence frequency of each metabolite and used it as a quantitative measure of the feature importance of a metabolite. The intuition is that if a metabolite appeared more often as an effective feature in the best predictive models, it was regarded as more important. Therefore, we performed two rounds of training by running the LGP algorithm provided with (1) the 167 full feature set and (2) the top-ranked focused feature subset.

4.2.4   Feature Synergy Analysis

In addition to looking at the individual occurrence of single metabolites in the best classification models, the co-occurrence of metabolites in the models was studied by counting the number of times each metabolite pair appeared together in the same model. The top 1% of the resulting metabolite pairs, ranked by decreasing frequency, were used to construct a metabolite synergy network. Network science has seen increasing applications in biomedical research [1, 3, 8, 13, 14, 16, 18], where biological entities are represented as vertices and their relationships can be modeled using edges linking pairs of vertices. Network modeling is a powerful tool to study interconnections among a large number of biological entities. In this study, vertices represent metabolites and an edge links two metabolites if they have a co-occurrence frequency in the set of 1000 best prediction models greater than the given cutoff. The network was rendered and analyzed using the Cytoscape software [25].

For the second round of more focused analysis, only the subset of metabolites appearing in the metabolite synergy network was used as a restricted feature set in a repeated model learning implementation, allowing the evolutionary algorithm to only use these more important metabolites to construct the classification models. The analysis was performed on both the discovery and replication datasets, each resulting in another set of 1000 best classification models. The intersection of the top 20 most common metabolites from the discovery and replication runs was reported, and such metabolites are regarded as interacting metabolites with high potential associations to the disease of OA.

4.3 Results and Discussion

4.3.1 Best Genetic Programs Evolved on the Full Feature Set

The initial training used the discovery data and the full set of 167 metabolite features. Recall that we used fivefold cross-validation and ran the LGP algorithm 200 times for each fold, therefore, we collected 1000 best evolved linear genetic programs. Table 4.2 shows the evaluation statistics of the 1000 best programs including the fitness (MCE), sensitivity, specificity, and the AUC. The statistics were computed using the prediction results on the testing samples. We see that the prediction performance of the 1000 evolved best programs varies and the minimal error rate can be as low as 0.067 while the best sensitivity, specificity, and the AUC can be as high as 1.000, 0.933, and 0.947 respectively.

To take a closer look, Fig. 4.1 shows the distributions of the fitness values and the number of effective features in the best 1000 evolved genetic programs. We see that the majority of the evolved programs achieved an error rate between 0.3 and 0.5, and only used 20–45 out of the total 167 features for the prediction.

Table 4.2 Statistics of the testing results on the full feature set (discovery)

	MCE	Sensitivity	Specificity	AUC
Mean	0.367	0.684	0.584	0.663
Median	0.367	0.667	0.600	0.667
Min	0.067	0.200	0.200	0.320
Max	0.667	1.000	0.933	0.947
Std dev	0.095	0.146	0.142	0.110
5% confidence	0.181	0.398	0.305	0.447
95% confidence	0.553	0.970	0.862	0.879

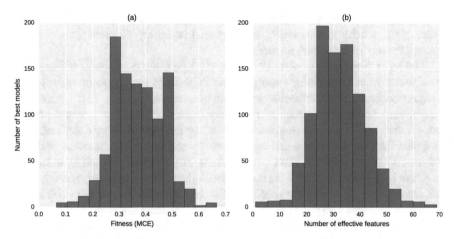

Fig. 4.1 The distributions of (**a**) the fitness and (**b**) the number of effective features in the best 1000 evolved predictive models

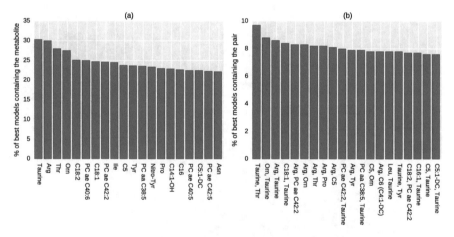

Fig. 4.2 The occurrence frequencies of the top 20 (**a**) individual and (**b**) pairs of features/metabolites using the full feature set and the discovery data

4.3.2 Identification of Important Features

Out of the 1000 best evolved programs, we counted the occurrence frequency of each metabolite feature, as well as the co-occurrence frequency of metabolite pairs being effective in a best program. We ranked the individual and pairs of metabolite features based on the frequencies to study their importance in making the prediction. Figure 4.2 shows the top 20 individual metabolites and co-occurring metabolite pairs. We see that the top individual metabolites including *taurine*, *arginine* (Arg), *tyrosine* (Thr), and *ornithine* (Orn) also appear in pairs most frequently. Metabolites such as *C6*, *leucine* (Leu), and *C16:1* appeared in the top 20 pairs without being among the top 20 individual features.

The top 1% metabolite pairs out of all $\binom{167}{2}$ possible combinations were then used to construct a *metabolite synergy network* (Fig. 4.3), in order to show the global connection structure of metabolite pairs with the strongest synergy. In the visualized graph, vertices are metabolites and two metabolites are directly connected by an edge if their co-occurrence frequency was among the top 1%. Vertex size and edge weight reflect the occurrence and co-occurrence frequencies. The network includes 70 metabolites as vertices. The two most frequent metabolite features, *taurine* and *arginine* locate in the center of the network with the highest vertex degrees of 44 and 43, respectively. The other metabolites with a vertex degree greater than 10 include *tyrosine* (a degree of 27), *ornithine* (19), *C18:1* (14), and *PC ae C40:6* (12).

We also looked at the distributions of metabolite concentrations comparing diseased cases and healthy controls in order to see if the LGP algorithm was able to identify metabolites that influence the disease risk through synergistically

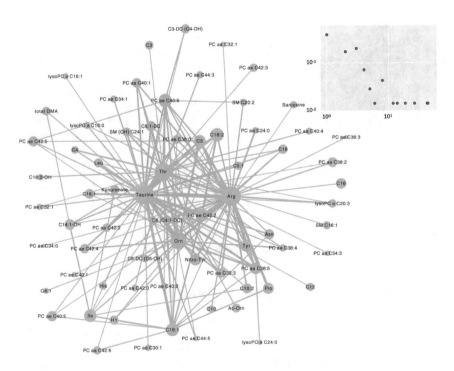

Fig. 4.3 The feature synergy network. Each vertex represents a feature/metabolite and an edge links to features if the pair appears among the top 1% the most frequent of all possible pairs in the best 1000 evolved predictive models. Vertex size and edge width are proportional to the individual and pairwise occurrence frequencies. The upper right inset shows the distribution of vertex degrees in a log-log scale

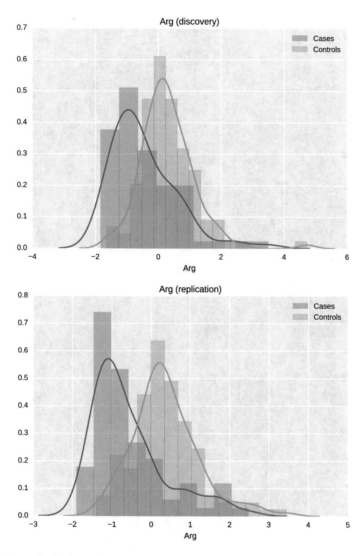

Fig. 4.4 The distributions of *arginine* (Arg) comparing case and control samples in the (**a**) discovery and (**b**) replication data

interacting with others rather than through individual and separate effects. We investigated two metabolites, *arginine* (Arg) and *C18:1*, both of which had a vertex degree greater than 10. Figures 4.4 and 4.5 show the comparison of metabolite concentration distributions in diseased cases (in pink color) v.s. healthy controls (in green color). We see that in both discovery and replication data, the concentration distribution of *arginine* differentiates significantly in the two populations. This indicates that using conventional uni-variable statistical methods could also detect

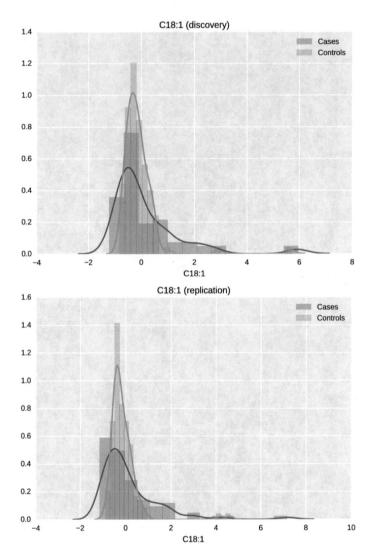

Fig. 4.5 The distributions of *C18:1* comparing case and control samples in the (**a**) discovery and (**b**) replication data

this metabolite feature. However, when we looked at the distribution comparison for metabolite *C18:1* (Oleic Acid) (Fig. 4.5), in both the discovery and replication data, the distributions heavily overlapped in the two populations. This indicates that conventional uni-variable methods would likely overlook this metabolite, however, the LGP algorithm was able to identify Oleic Acid as an important feature through interacting with other metabolites to influence disease risk.

4.3.3   Best Genetic Programs Evolved on the Focused Feature Subset

The second round of training repeated the LGP algorithm implementation using only the 70 metabolites included in the top 1% metabolite pairs as shown in Fig. 4.3. Tables 4.3 and 4.4 show the prediction measurements of the best 1000 evolved linear genetic programs using the reduced and focused feature subset on both the discovery and replication data. It can be seen that the prediction performance improved significantly comparing to the initial training using the full feature set (Table 4.2). This suggests that the feature selection of the LGP algorithm was effective.

We also investigated the feature importance ranking in the secondary training using only the focused feature subset. Figures 4.6 and 4.7 show the top 20 individual and pairs of metabolites ranked by the LGP algorithm by counting their occurrence or co-occurrence frequencies in the 1000 best evolved programs, using the discovery and replication data, respectively. Comparing to the frequencies reported in Fig. 4.2, the occurrence and co-occurrence frequencies increased by a large margin when a focused feature subset was used. This indicates that the model search was more effective and focused with a reduced and selected feature set.

Moreover, using the separate discovery and replication data was able to validate the results of the identification of the most influential metabolite features. Nine metabolites, *arginine*, *C16*, *C18:1*, *isoleucine*, *nitrotyrosine*, *ornithine*, *taurine*, *threonine* and *tyrosine*, were found most frequently appearing in the best models in the discovery dataset as the result of both rounds of analyses and were successfully replicated using the replication dataset, including the previous four key metabolites identified in the network.

Table 4.3 Statistics of the testing results on the focused feature subset (discovery)

	MCE	Sensitivity	Specificity	AUC
Mean	0.325	0.723	0.628	0.704
Median	0.333	0.733	0.600	0.709
Min	0.100	0.267	0.200	0.362
Max	0.600	1.000	1.000	1.000
Std dev	0.089	0.135	0.141	0.103
5% confidence	0.151	0.459	0.353	0.503
95% confidence	0.498	0.987	0.904	0.906

Table 4.4 Statistics of the testing results on the focused feature subset (replication)

	MCE	Sensitivity	Specificity	AUC
Mean	0.295	0.733	0.678	0.725
Median	0.300	0.733	0.667	0.739
Min	0.033	0.267	0.200	0.380
Max	0.600	1.000	1.000	1.000
Std dev	0.098	0.139	0.162	0.120
5% confidence	0.102	0.460	0.361	0.491
95% confidence	0.488	1.000	0.995	0.959

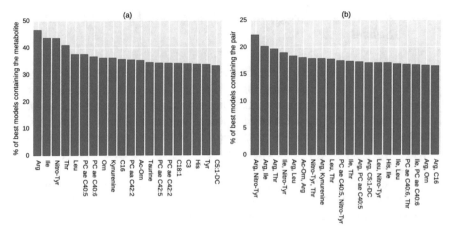

Fig. 4.6 The occurrence frequencies of the top 20 (**a**) individual and (**b**) pairs of features/metabolites using the focused feature set and the discovery data

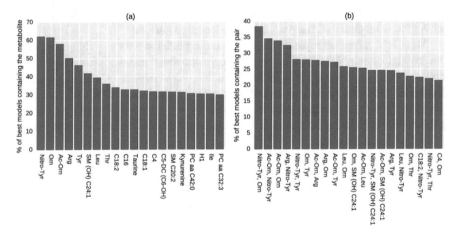

Fig. 4.7 The occurrence frequencies of the top 20 (**a**) individual and (**b**) pairs of features/metabolites using the focused feature set and the replication data

The results are interesting as *arginine* and its pathway related metabolites, such as *ornithine*, have been identified as being associated with OA in previous analysis using traditional methods including pairwise comparison and regression technique [30]. Similarly, *isoleucine* was also previously identified as OA-associated metabolite [28]. The current analyses applied a novel analytic method, the evolutionary algorithm, which confirmed our previous findings and also identified additional novel metabolic markers for OA. These included four amino acids and two acylcarnitines, which could have potential utility in the clinical management of OA. For example, *taurine* is the most abundant free amino acid in humans, and may play an important role in inflammation associated with oxidative stress [23]. It has been reported to be associated with rheumatoid arthritis [19]. *Nitrotyrosine* is also associated with oxidative damage and has been found associated with aging and the development of OA in cartilage samples from both monkeys and humans [21]. The findings in the current study certainly warrant further investigation of the role of those novel metabolic markers in OA.

4.4 Conclusion

In this study, we proposed to use an LGP algorithm to evolve predictive models for metabolomics study of human diseases. We aimed to provide explainable learning results to the application problem by showing which metabolite features were the most influential in the prediction through either individual effects or synergistic effects combining with other metabolite features.

The results showed that the LGP algorithm was able to evolve highly accurate classification models that took the metabolite concentrations as inputs and predicted the disease risk for unseen testing data. We highlighted that the feature selection was intrinsic to the LGP algorithm and was performed automatically through co-evolving with the genetic programs. It was shown that using the subset of only the most influential metabolite features, the LGP algorithm was able to further improve the prediction accuracy. In addition, we used a network presentation to depict the pairwise interaction patterns of the most influential metabolite pairs. Such a metabolite synergy network allowed us to identify the highly central metabolites that interacted with a large number of other metabolites and collectively contributed to the disease risk prediction.

The predictive model learning and feature importance analysis developed in this study open up more research directions for the field of genetic programming. As artificial intelligence and machine learning revolutionizing many research areas and industries, the explainability issue becomes more predominant. Questions like the following are often raised while applying machine learning algorithms, especially to bioinformatics problems. How trustworthy is the learned predictive model? What is the mechanism underlying the highly accurate prediction? Which features play a more important role in the prediction? In this study, we were only able to partially answer some of these questions and are leaving more explorations for future research.

Acknowledgements This research is supported by the Canadian Natural Sciences and Engineering Research Council (NSERC) Discovery grant RGPIN-04699-2016 to Ting Hu.

References

1. Almasi, S.M., Hu, T.: Measuring the importance of vertices in the weighted human disease network. PLoS ONE **14**(3), e0205,936 (2019)
2. Altman, R., Alarcon, G., Appelrouth, D., Bloch, D., Borenstein, D., Brandt, K., Brown, C., Cooke, T.D., et al.: The american college of rheumatology criteria for the classification and reporting of osteoarthritis of the hip. Arthritis and Rheumatology **34**(5), 505–514 (1991)
3. Barabasi, A.L., Oltvai, Z.N.: Network biology: Understanding the cell's functional organization. Nature Reviews Genetics **5**, 101–113 (2004)
4. Bezanson, J., Edelman, A., Karpinski, S., Shah, V.B.: Julia: A fresh approach to numerical computing. CoRR **abs/1411.1607** (2014). URL http://arxiv.org/abs/1411.1607
5. Brameier, M.F., Banzhaf, W.: Linear Genetic Programming. Springer (2007)
6. Camacho, D.M., Collins, K.M., Powers, R.K., Costello, J.C., Collins, J.J.: Next-generation machine learning for biological networks. Cell **173**(7), 1581–1592 (2018)
7. Caruana, R., Lou, Y., Gehrke, J., Koch, P., Sturm, M., Elhadad, N.: Intelligible models for healthcare: predicting pneumonia risk and hospital 30-day readmission. In: Proceedings of the 21th ACM SIGKDD International Conference on Knowledge Discovery and Data Mining, pp. 1721–1730 (2015)
8. Cho, D.Y., Kim, Y.A., Przytycka, T.M.: Network biology approach to complex diseases. PLoS Computational Biology **8**(12), e1002,820 (2012)
9. Dorani, F., Hu, T., Woods, M.O., Zhai, G.: Ensemble learning for detecting gene-gene interactions in colorectal cancer. PeerJ **6**, e5854 (2018)
10. Fontaine-Bisson, B., Thorburn, J., Gregory, A., Zhang, H., Sun, G.: Melanin-concentrating hormone receptor 1 polymorphisms are associated with components of energy balance in the complex diseases in the newfoundland population: Environment and genetics (coding) study. The American Journal of Clinical Nutrition **99**(2), 384–391 (2014)
11. Ghahramani, Z.: Probabilistic machine learning and artificial intelligence. Nature **521**, 452–459 (2015)
12. Gilpin, L.H., Bau, D., Yuan, B.Z., Bajwa, A., Specter, M., Kagal, L.: Explaining explanations: an overview of interpretability of machine learning. In: Proceedings of the 5th IEEE International Conference on Data Science and Advanced Analytics (DSAA), pp. 80–89 (2018)
13. Hu, T., Chen, Y., Kiralis, J.W., Moore, J.H.: ViSEN: Methodology and software for visualization of statistical epistasis networks. Genetic Epidemiology **37**, 283–285 (2013)
14. Hu, T., Moore, J.H.: Network modeling of statistical epistasis. In: M. Elloumi, A.Y. Zomaya (eds.) Biological Knowledge Discovery Handbook: Preprocessing, Mining, and Postprocessing of Biological Data, chap. 8, pp. 175–190. Wiley (2013)
15. Hu, T., Oksanen, K., Zhang, W., Randell, E., Furey, A., Sun, G., Zhai, G.: An evolutioanry learning and network approach to identifying key metabolites for osteoarthritis. PLoS Computational Biology **14**(3), e1005,986 (2018)
16. Hu, T., Sinnott-Armstrong, N.A., Kiralis, J.W., Andrew, A.S., Karagas, M.R., Moore, J.H.: Characterizing genetic interactions in human disease association studies using statistical epistasis networks. BMC Bioinformatics **12**, 364 (2011)
17. Hu, T., Zhang, W., Fan, Z., Sun, G., Likhodi, S., Randell, E., Zhai, G.: Metabolomics differential correlation network analysis of osteoarthritis. Pacific Symposium on Biocomputing **21**, 120–131 (2016)
18. Kafaie, S., Chen, Y., Hu, T.: A network approach to prioritizing susceptibility genes for genome-wide association studies. Genetic Epidemiology **43**(5), 477–491 (2019)
19. Kontny, E., Wojtecka-ŁUkasik, E., Rell-Bakalarska, K., Dziewczopolski, W., Maśliński, W., Maślinski, S.: Impaired generation of taurine chloramine by synovial fluid neutrophils of rheumatoid arthritis patients. Amino Acids **23**(4), 415–418 (2002)
20. Lee, M., Hu, T.: Computational methods for the discovery of metabolic markers of complex traits. Metabolites **9**(4), 66 (2019)

21. Loeser, R.F., Carlson, C.S., Carlo, M.D., Cole, A.: Detection of nitrotyrosine in aging and osteoarthritic cartilage: Correlation of oxidative damage with the presence of interleukin-1β and with chondrocyte resistance to insulin-like growth factor 1. Arthritis and Rheumatology **46**(9), 2349–2357 (2002)
22. Ma, J., Yu, M.K., Fong, S., Ono, K., Sage, E., Demchak, B., Sharan, R., Ideker, T.: Using deep learning to model the hierarchical structure and function of a cell. Nature Methods **15**(4), 290–298 (2018)
23. Marcinkiewicz, J., Kontny, E.: Taurine and inflammatory diseases. Amino Acids **46**(1), 7–20 (2014)
24. Ribeiro, M.T., Singh, S., Guestrin, C.: "why should I trust you?": Explaining the predictions of any classifier. In: Proceedings of the 22nd ACM SIGKDD International Conference on Knowledge Discovery and Data Mining, pp. 1135–1144 (2016)
25. Shannon, P., Markiel, A., Ozier, O., Baliga, N.S., Wang, J.T., Ramage, D., Amin, N., Schwikowski, B., Ideker, T.: Cytoscape: A software environment for integrated models of biomolecular interaction networks. Genome Research **13**, 2498–2504 (2003)
26. Yu, M.K., Ma, J., Fisher, J., Kreisberg, J.F., Raphael, B.J., Ideker, T.: Visible machine learning for biomedicine. Cell **173**(7), 1562–1565 (2018)
27. Zhai, G., Aref-Eshghi, E., Rahman, P., Zhang, H., Martin, G., Furey, A., Green, R.C., Sun, G.: Attempt to replicate the published osteoarthritis-associated genetic variants in the newfoundland & labrador population. Journal of Orthopedics and Rheumatology **1**(3), 5 (2014)
28. Zhai, G., Wang-Sattler, R., Hart, D.J., Arden, N.K., Hakim, A.J., Illig, T., Spector, T.D.: Serum branched-chain amino acid to histidine ratio: a novel metabolomic biomarker of knee osteoarthritis. Annals of the Rheumatic Diseases p. 120857 (2010)
29. Zhang, W., Likhodii, S., Aref-Eshghi, E., Zhang, Y., Harper, P.E., Randell, E., Green, R., Martin, G., Furey, A., Sun, G., Rahman, P., Zhai, G.: Relationship between blood plasma and synovial fluid metabolite concentrations in patients with osteoarthritis. The Journal of Rheumatology **42**(5), 859–865 (2015)
30. Zhang, W., Sun, G., Likhodii, S., Liu, M., Aref-Eshghi, E., Harper, P.E., Martin, G., Furey, A., Green, R., Randell, E., Rahman, P., Zhai, G.: Metabolomic analysis of human plasma reveals that arginine is depleted in knee osteoarthritis patients. Osteoarthritis and Cartilage **24**, 827–834 (2016)

Chapter 5
Symbolic Regression by Exhaustive Search: Reducing the Search Space Using Syntactical Constraints and Efficient Semantic Structure Deduplication

Lukas Kammerer, Gabriel Kronberger, Bogdan Burlacu, Stephan M. Winkler, Michael Kommenda, and Michael Affenzeller

5.1 Introduction

Symbolic regression is a task that we can solve with genetic programming (GP) and a common example where GP is particularly effective in practical applications. Symbolic regression is a machine learning task whereby we try to find a mathematical model represented as a closed-form expression that captures dependencies of variables from a dataset. Genetic programming has been proven to be well-suited for this task especially when there is little knowledge about the data-generating process. Even when we have a good understanding of the underlying process, GP can identify counterintuitive or unexpected solutions.

L. Kammerer (✉)
Heuristic and Evolutionary Algorithms Laboratory (HEAL), University of Applied Sciences Upper Austria, Hagenberg, Austria

Department of Computer Science, Johannes Kepler University, Linz, Austria

Josef Ressel Center for Symbolic Regression, University of Applied Sciences Upper Austria, Hagenberg, Austria
e-mail: lukas.kammerer@fh-hagenberg.at

G. Kronberger · B. Burlacu · M. Kommenda
Heuristic and Evolutionary Algorithms Laboratory (HEAL), University of Applied Sciences Upper Austria, Hagenberg, Austria

Josef Ressel Center for Symbolic Regression, University of Applied Sciences Upper Austria, Hagenberg, Austria

S. M. Winkler · M. Affenzeller
Heuristic and Evolutionary Algorithms Laboratory (HEAL), University of Applied Sciences Upper Austria, Hagenberg, Austria

Department of Computer Science, Johannes Kepler University, Linz, Austria

© Springer Nature Switzerland AG 2020
W. Banzhaf et al. (eds.), *Genetic Programming Theory and Practice XVII*, Genetic and Evolutionary Computation, https://doi.org/10.1007/978-3-030-39958-0_5

5.1.1 Motivation

GP has some practical limitations when used for symbolic regression. One limitation is that—as a stochastic process—it might produce highly dissimilar solutions even for the same input data. This can be very helpful to produce new "creative" solutions. However, it is problematic when we try to integrate symbolic regression in carefully engineered solutions (e.g. for automatic control of production plants). In such situations we would hope that there is an optimal solution and the solution method guarantees to identify the optimum. Intuitively, if the data changes only slightly, we expect that the optimal regression solution also changes only slightly. If this is the case we know that the solution method is trustworthy (cf. [15, 31]) and we can rely on the fact that the solutions are optimal at least with respect to the objective function that we specified. Of course this is only wishful thinking because of three fundamental reasons: (1) the symbolic regression search space is huge and contains many different expressions which are algebraically equivalent, (2) GP has no guarantee to explore the whole search space with reasonable computational resources and (3) the "optimal solution" might not be expressible as a closed-form mathematical expressions using the given building blocks. Therefore, the goal is to find an approximately optimal solution.

5.1.2 Prior Work

Different methods have been developed with the aim to improve the reliability of symbolic regression. Currently, there are several off-the-shelf software solutions which use enhanced variants of GP and are noteworthy in this context: the DataModeler package[1] [16] provides extensive capabilities for symbolic regression on top of Mathematica™. Eureqa™ is a commercial software tool[2] for symbolic regression based on research described in [27–29]. The open-source framework HeuristicLab[3] [36] is a general software environment for heuristic and evolutionary algorithms with extensive functionality for symbolic regression and white-box modeling.

In other prior work, several researchers have presented non-evolutionary solution methods for symbolic regression. Fast function extraction (FFX) [22] is a deterministic method that uses elastic-net regression [39] to produce symbolic regression solutions orders of magnitudes faster than GP for many real-world problems. The work by Korns toward "extremely accurate" symbolic regression [12–14] highlights the issue that baseline GP does not guarantee to find the optimal solution even for

---

[1] http://www.evolved-analytics.com/.

[2] https://www.nutonian.com/products/eureqa/.

[3] https://dev.heuristiclab.com.

rather limited search spaces. They give a useful systematic definition of increasingly larger symbolic regression search spaces using abstract expression grammars [10] and describes enhancements to GP to improve it's reliability. The work by Worm and Chiu on prioritized grammar enumeration [38] is closely related. They use a restricted set of grammar rules for deriving increasingly complex expressions and describe a deterministic search algorithm, which enumerates the search space for limited symbolic regression problems.

5.1.3 Organization of This Chapter

Our contribution is conceptually an extension of prioritized grammar enumeration [38], although our implementation of the method deviates significantly. The most relevant extensions are that we cut out large parts of the search space and provide a general framework for integrating heuristics in order to improve the search efficiency. Section 5.2 describes how we reduce the size of the search space which is defined by a context-free grammar:

1. We restrict the structure of solution to prevent too complicated solutions.
2. We use grammar restrictions to prevent semantic duplicates—solutions with different syntax but same semantics, such as algebraic transformations. With these restrictions, most solutions can only be generated in exactly one way.
3. We efficiently identify remaining duplicates with semantic hashing, so that (nearly) all solutions in the search space are semantically unique.

In Sect. 5.3, we explain the algorithm that iterates all these semantically unique solutions. The algorithm sequentially generates solutions from the grammar and keeps track of the most accurate one. For very small problems, it is even feasible to iterate the whole search space [19]. However, our goal in larger problems is to find accurate and concise solutions early during the search and to stop the algorithm after a reasonable time. The search order is determined with heuristics, which estimate the quality of solutions and prioritize promising ones in the search. A simple heuristic is proposed in Sect. 5.4. Modeling results in Sect. 5.5 show that this first version of our algorithm can already solve several difficult noiseless benchmark problems.

5.2 Definition of the Search Space

The search space of our deterministic symbolic regression algorithm is defined by a context-free grammar. Production rules in the grammar define the mathematical expressions that can be explored by the algorithm. The grammar only specifies possible model structures whereby placeholders are used for numeric coefficients. These are optimized separately by a curve-fitting algorithm (e.g. optimizing least squares with an gradient-based optimization algorithm) using the available training data.

In a general grammar for mathematical expressions—as it is common in symbolic regression with GP for example—the same formula can be derived in several forms. These duplicates inflate the search space. To reduce them, our grammar is deliberately restricted regarding the possible structure of expressions. Remaining duplicates that cannot be prevented by a context-free grammar are eliminated via a hashing algorithm. Using both this grammar and hashing, we can generate a search space with only semantically unique expressions.

5.2.1 Grammar for Mathematical Expressions

In this work we consider mathematical expressions as list of symbols which we call *phrases* or *sentences*. A phrase can contain both *terminal* and *non-terminal* symbols and a sentence only terminal symbols. Non-terminal symbols can be replaced by other symbols as defined by a grammar's *production rules* while terminal symbols represent parts of the final expression like functions or variables in our case.

Our grammar is very similar to the one by Kronberger et al. [19]. It produces only rational polynomials which may contain linear and nonlinear terms, as outlined conceptually in Eq. (5.1). The basic building blocks of terms are linear and non-linear functions $\{+, \times, \text{inv}, \exp, \log, \sin, \text{square root}, \text{cube root}\}$. Recursion in the production rules represents a strategy for generating increasingly complex solutions by repeated nesting of expressions and terms.

$$Expr = c_1 Term_1 + c_2 Term_2 + \ldots + c_n$$

$$Term = Factor_0 \times Factor_1 \times \ldots \tag{5.1}$$

$$Factor \in \{variable, \log(variable), \exp(variable), \sin(variable)\}$$

We explicitly disallow nested non-linear functions, as we consider such solutions too complex for real-world applications. Otherwise, we allow as many different structures as possible to keep accurate and concise models in the search space. We prevent semantic duplicates by generating just one side of mathematical equality relations in our grammar, e.g. we allow $xy + xz$ but not $x(y + z)$. Since each function has different mathematical identities, many different production rules are necessary to cover all special cases. Because we scale every term including function arguments, we also end up with many placeholders for coefficients in the structures. All production rules are detailed in Fig. 5.1 and described in the following.

We use a polynomial structure as outlined in Eq. (5.1) to prevent a factored form of solutions. The polynomial structure is enforced with the production rules Expr and Term. We restrict the occurrence of the multiplicative inverse $(= \frac{1}{\ldots})$, the square root and cube root function to prevent a factored form such as $\frac{1}{x+y}\frac{1}{x+z}$. This is necessary since we want to allow sums of simple terms as function arguments (see

```
G(Expr):
// Expressions and terms for polynomial structure
    Expr        -> "const" "*" Term "+" Expr    |
                   "const" "*" Term "+" "const"

    Term        -> RecurringFactors "*" Term    |
                   RecurringFactors             |
                   OneTimeFactors

    RecurringFactors -> VarFactor | LogFactor |
                        ExpFactor | SinFactor

    VarFactor   -> <variable>
    LogFactor   -> "log" "(" SimpleExpr ")"
    ExpFactor   -> "exp" "(" "const" "*" SimpleTerm ")"
    SinFactor   -> "sin" "(" SimpleExpr ")"

// Factors which can occur at most once per term
    OneTimeFactors -> InvFactor "*" SqrtFactor "*" CbrtFactor |
                      InvFactor "*" SqrtFactor                |
                      InvFactor "*"               CbrtFactor  |
                               SqrtFactor "*" CbrtFactor      |
                      InvFactor                                |
                               SqrtFactor                      |
                                          CbrtFactor

    InvFactor   -> "1/" "(" InvExpr ")"
    SqrtFactor  -> "sqrt" "(" SimpleExpr ")"
    CbrtFactor  -> "cbrt" "(" SimpleExpr ")"

// Function arguments
    SimpleExpr  -> "const" "*" SimpleTerm "+" SimpleExpr  |
                   "const" "*" SimpleTerm "+" "const"

    SimpleTerm  -> VarFactor "*" SimpleTerm | VarFactor

    InvExpr -> "const" "*" InvTerm "+" InvExpr |
               "const" "*" InvTerm "+" "const"

    InvTerm -> RecurringFactors "*" InvTerm    |
               RecurringFactors "*" SqrtFactor "*" CbrtFactor |
               RecurringFactors "*" SqrtFactor                |
               RecurringFactors "*"               CbrtFactor  |
                               SqrtFactor "*" CbrtFactor      |
               RecurringFactors                                |
                               SqrtFactor                      |
                                          CbrtFactor
```

Fig. 5.1 Context-free grammar for generating mathematical expressions

non-terminal symbol `SimpleExpr`). Therefore, these three functions can occur at most once time per term. This is defined with symbol `OneTimeFactors` and one production rule for each combination. The only function in which we do not allow sums as argument is exponentiation (see `ExpFactor`), since this form is substituted by the overall polynomial structure (e.g. we allow $e^x e^y$ but not e^{x+y}). Equation (5.2) shows some example identities and which forms are supported.

in the search space: not in the search space:

$$c_1 xy + c_2 xz + c_3 \equiv x(c_4 y + c_5 z) + c_6$$

$$c_1 \frac{1}{c_2 x + c_3 xx + c_4 xy + c_5 y + c_6} + c_7 \equiv c_8 \frac{1}{c_9 x + c_{10}} \frac{1}{c_{11}x + c_{12}y + c_{13}} + c_{14}$$

$$c_1 \exp(c_2 x) \exp(c_3 y) + c_4 \equiv c_5 \exp(c_6 x + c_7 y) + c_8$$

$$(5.2)$$

We only allow (sums of) terms of variables as function arguments, which we express with the production rules `SimpleExpr` and `SimpleTerm`. An exception is the multiplicative inverse, in which we want to include the same structures as in ordinary terms. However, we disallow compound fractions like in Eq. (5.3). Again, we introduce separate grammar rules `InvExpr` and `InvTerm` which cover the same rules as `Term` except the multiplicative inverse.

in the search space: not in the search space:

$$c_1 \frac{1}{c_2 \log(c_3 x + c_4) + c_5} + c_6 \equiv c_7 \frac{1}{c_8 \frac{1}{c_9 \log(c_{10}x + c_{11}) + c_{12}} + c_{13}} + c_{14} \qquad (5.3)$$

In the simplest case, the grammar produces an expression $E_0 = c_0 x + c_1$, where x is a variable and c_0 and c_1 are coefficients corresponding to the slope and intercept. This expression is obtained by considering the simplest possible `Term` which corresponds to the derivation chain `Expr` → `Term` → `RecurringFactors` → `VarFactor` → x. Further derivations could lead for example to the expression $E_1 = c_0 x + (c_1 x + c_2)$, produced by nesting E_0 into the first part of the production rule for `Expr`, where the `Term` is again substituted with the variable x.

However, duplicate derivations can still occur due to algebraic properties like associativity and commutativity. These issues cannot be prevented with a context-free grammar because a context-free grammar does not consider surrounding symbols of the derived non-terminal symbol in its production rules. For example the expression $E_1 = c_0 x + (c_1 x + c_2)$ contains two coefficients c_0 and c_1 for variable x which could be folded into a new coefficient $c_{new} = c_0 + c_1$. This type of redundancy becomes even more pronounced when `VarFactor` has multiple productions (corresponding to multiple input variables), as it becomes possible for multiple derivation paths to produce different expressions which are algebraically equivalent, such as $c_1 x + c_2 y$, $c_3 x + c_4 x + c_5 y$, $c_6 y + c_7 x$ for corresponding values of $c_1 \ldots c_7$. Another example are $c_1 x y$ and $c_2 y x$ which are both equivalent but derivable from the grammar.

To avoid re-visiting already explored regions of the search space, we implement a caching strategy based on expression hashing for detecting algebraically equivalent expressions. The computed hash values are the same for algebraically equivalent expressions. In the search algorithm we keep the hash values of all visited expressions and prevent re-evaluations of expressions with identical hash values.

5.2.2 Expression Hashing

We employ expression hashing by Burlacu et al. [3] to assign hash values to subexpressions within phrases and sentences. Hash values for parent expressions are aggregated in a bottom-up manner from the hash values of their children using any general-purpose hash function. We then simplify such expressions according to arithmetic properties such as commutativity, associativity, and applicable mathematical identities. The resulting canonical minimal form and associated hash value are then cached in order to prevent duplicated search effort.

Expression hashing builds on the idea of Merkle trees [23]. Figure 5.2 shows how hash values propagate towards the tree root (the topmost symbol of the expression) using hash function $\oplus$ to aggregate child and parent hash values. Expression hashing considers an internal node's own symbol, as well as associativity and commutativity properties. To account for these properties, each hashing step must be accompanied by a corresponding sorting step, where child subexpressions are reordered according to their type and hash value. Algorithm 1 ensures that child nodes are sorted and hashed before parent nodes, such that calculated hash values are consistent towards the root symbol.

An expression's hash value is then given by the hash value of its root symbol. After sorting, sub-expressions with the same hash value are considered isomorphic

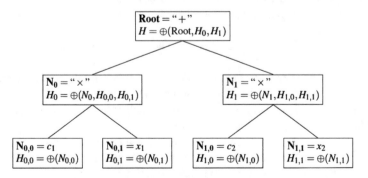

Fig. 5.2 Hash tree example, in which the hash values of all nodes are calculated from both their own node content and the has value of their children [3]

Algorithm 1 Expression hashing [3]

Input: An expression E
Output: The corresponding sequence of hash values

  1: hashes $\leftarrow$ empty list of hash values
  2: symbols $\leftarrow$ list of symbols in E
  3: **for all** symbol s in symbols **do**
  4:      $H(s) \leftarrow$ an initial hash value
  5:      **if** s is a terminal function symbol **then**
  6:          **if** s is commutative **then**
  7:              Sort the child nodes of s
  8:          **end if**
  9:          child hashes $\leftarrow$ hash values of s's children
10:          $H(n) \leftarrow \oplus$(child hashes, $H(s)$)
11:      **end if**
12:      hashes.append($H(n)$)
13: **end for**
14: **return** hashes

---

and are simplified according to arithmetic rules. The simplification procedure is illustrated in Fig. 5.3 and consists of the following steps:

1. **Fold**: Apply associativity to eliminate nested symbols of the same type. For example, postfix expression a b + c + consists of two nested additions where each addition symbol has arity 2. Folding flattens this expression to the equivalent form a b c + where the addition symbol has arity 3.
2. **Simplify**: Apply arithmetic rules and mathematical identities to further simplify the expressions. Since expression already include placeholders for numerical coefficients, we eliminate redundant subexpressions such as a a b + which becomes a b +, or a a + which becomes a.
3. Repeat steps 1 and 2 until no further simplification is possible.

Nested + and × symbols in Fig. 5.3 are folded in the first step, simplifying the tree structure of the expression. Arithmetic rules are then applied for further simplification. In this example, the product of exponentials

$$\exp(c_1 \times x_1) \times \exp(c_2 \times x_1) \equiv \exp((c_1 + c_2) \times x_1)$$

is simplified since from a local optimization perspective, optimizing the coefficients of the expression yields better results for a single coefficient $c_3 = c_1 + c_2$, thus it makes no sense to keep both original factors. Finally, the sum $c_4 x_1 + c_5 x_1$ is also simplified since one term in the sum is redundant.

After simplification, the hash value of the simplified tree is returned as the hash value of the original expression. Based on this computation we are able to identify already explored search paths and avoid duplicated effort.

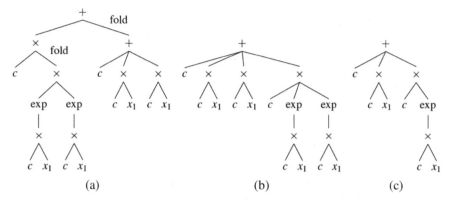

Fig. 5.3 Simplification to canonical minimal form during hashing (**a**) Original expression. (**b**) Folded expression. (**c**) Minimal form

5.3   Exploring the Search Space

By limiting the size of expressions, the grammar and the hashing scheme produce a large but finite search space of semantically unique expressions. In an exhaustive search, we iterate all these expressions and search for the best fitting one. Thereby, we derive sentences with every possible derivation path. An expression is rejected if another expression with the same semantic—according to hashing—has already been generated during the search. When a new, previously unseen sentence is derived, the placeholders for coefficients are replaced with real values and optimized separately. The best fitting sentence is stored.

Algorithm 2 outlines how all unique expressions are derived: We store unfinished phrases—expressions with non-terminal symbols—in a data structure such as a stack or queue. We fetch phrases from this data structure one after another, derive new phrases, calculate their hash values and compare these hash values to previously seen ones. To derive new phrases, we always replace the *leftmost* non-terminal symbol in the old phrase with the production rules of this non-terminal symbol. If a derived phrase becomes a sentence with only terminal symbols, its coefficients are optimized and its fitness is evaluated. Otherwise, if it still contains derivable non-terminal symbols, it is put back on the data structure.

We restrict the length of a phrase by its number of variable references—e.g. xx and $\log(x) + x$ have two variable references. Phrases that exceed this limit are discarded in the search. Since every non-terminal symbol is eventually derived to at least one variable reference, non-terminal symbols count as variable references. In our experiments, a limit on the complexity has been found to be the most intuitive way to estimate an appropriate search space limit. Other measures, e.g. the number of symbols are harder to estimate since coefficients, function symbols and the non-factorized representation of expression quickly inflate the number of symbols in a phrase.

Algorithm 2 Iterating the search space

Input: Data set ds, max. number of variable references $maxVariableRefs$
Output: Best fitting expression

 1: $openPhrases \leftarrow$ empty data structure
 2: $seenHashes \leftarrow$ empty set
 3: Add $StartSymbol$ to $openPhrases$
 4: $bestExpression \leftarrow$ constant symbol
 5:
 6: **while** $openPhrases$ is not empty **do**
 7:      $oldPhrase \leftarrow$ fetch and remove from $openPhrases$
 8:      $nonTerminalSymbol \leftarrow$ leftmost nonterminal symbol in $oldPhrase$
 9:      **for all** production $prod$ of $nonTerminalSymbol$ **do**
10:          $newPhrase \leftarrow$ apply $prod$ on copy of $oldPhrase$
11:          **if** VariableRefs($newPhrase$) $\leq maxVariableRefs$ **then**
12:              $hash \leftarrow$ Hash($newPhrase$)
13:              **if** $seenHashes$ not contains $hash$ **then**
14:                  Add $hash$ to $seenHashes$
15:                  **if** $newPhrase$ is sentence **then**
16:                      Fit coefficients of $newPhrase$ to ds
17:                      Evaluate $newPhrase$ on ds
18:                      **if** $newPhrase$ is better than $bestExpression$ **then**
19:                          $bestExpression \leftarrow newPhrase$
20:                      **end if**
21:                  **end if**
22:              **else**
23:                  Add $newPhrase$ to $openPhrases$
24:              **end if**
25:          **end if**
26:      **end for**
27: **end while**
28: **return** $bestExpression$

5.3.1 Symbolic Regression as Graph Search Problem

Without considering the semantics of an expression, we would end up exploring a search tree like in Fig. 5.4, in which semantically equivalent expressions are derived multiple times (e.g. $c_1x + c_2x$ and $c_1x + c_2x + c_3x$). However, hashing turns the search tree into a directed search graph in which nodes (derived phrases) are reachable via one or more paths, as shown in Fig. 5.5. Thus, hashing prevents the search in a graph region that was already visited. From this point of view, Algorithm 2 is very similar to simple graph search algorithms such as depth-first or breadth-first search.

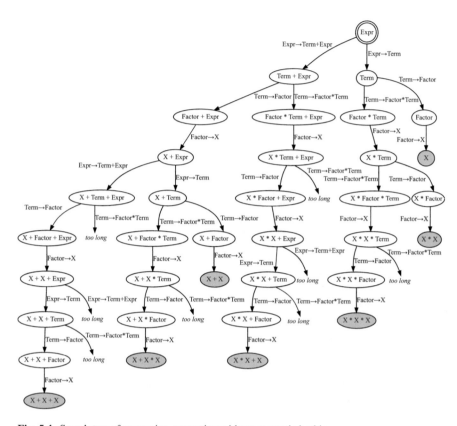

Fig. 5.4 Search tree of expression generation without semantic hashing

5.3.2 Guiding the Search

In Algorithm 2, the order in which expressions are generated is determined by the data structure used. A stack or a queue would result in a depth-first or a breadth-first search respectively. However, as the goal is to find well-fitting expressions quickly and efficiently, we need to guide the traversal of a search graph towards promising phrases.

Our general framework for guiding the search is very similar to the idea used in the A* algorithm [5]. We use a priority queue as data structure and assign a priority value to each phrase, indicating the expected quality of sentences which are derivable from that phrase. Phrases with high priority are derived first in order to discover well-fitting sentences, steering the algorithm towards good solutions.

Similar to the A* algorithm, we cannot make a definite statement about a phrase's priority before actually deriving all possible sentences from it. Therefore, we need to estimate this value with problem-specific heuristics. The calculation of phrase

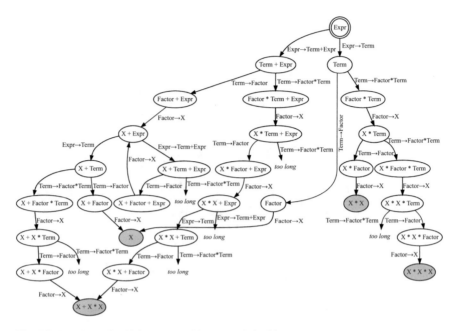

Fig. 5.5 Search graph with loops caused by semantic hashing

priorities provides us a generic point for integrating heuristics for improving the search efficiency and extending the algorithm's capabilities in future work.

5.4   Steering the Search

We introduce a simple heuristic for guiding the search and leave more complex and efficient heuristics for future work. The proposed heuristic makes a pessimistic estimation of the quality of a phrase's derivable sentences. This is done by evaluating phrases before they are derived to sentences. With the goal of finding short and accurate sentences quickly, the priority value considers both the expected quality and the length of a phrase.

5.4.1   Quality Estimation

Estimating the expected quality of an unfinished phrase is possible due to the polynomial structure of sentences and the derivation of the leftmost non-terminal symbol in every phrase. Since expressions are sums of terms ($c_1 Term_1 + c_2 Term_2 + \ldots$), repeated expansion of the leftmost non-terminal symbol derives one term after another. This results in phrases such as in Eq. (5.4), in which the first two terms $c_1 \log(c_2 x + c_3)$ and $c_4 x x$ contain only terminal symbols and the last non-terminal symbol is *Expr*.

$$
\underbrace{c_1 \log(c_2 x + c_3)}_{\textit{finishedTerm}_1} \quad + \quad \underbrace{c_4 x x}_{\textit{finishedTerm}_2} \quad + \quad \underbrace{\textit{Expr}}_{\text{Treat as coefficient}} \tag{5.4}
$$

Phrases where the only non-terminal symbol is Expr are evaluated as if they were full sentences by treating Expr as a coefficient during the local optimization phase. We get a pessimistic estimate of the quality of derivable sentences, since derived sentences with more terms can only have better quality. The quality can only improve with more terms because of separate coefficient optimization and one scaling coefficient per term, as shown in Eq. (5.5). If a term which does not improve the quality is derived, the optimization of coefficients will cancel it out by setting the corresponding scaling coefficient to zero (e.g. c_5 in Eq. (5.5)).

$$
\textit{finishedTerm}_1 \quad + \quad \textit{finishedTerm}_2 \quad + \quad \underbrace{c_5 \textit{Term}}_{\text{Can only improve quality}} \tag{5.5}
$$

This heuristic works only for phrases in which *Expr* is the only non-terminal symbol. For sentences with different non-terminal symbols, we reuse the estimated quality from the last evaluated parent phrase. The estimate is updated when a new term with only terminal symbols is derived and again only one *Expr* remains. For now, we do not have a reliable estimation method for terms that contain non-terminal symbols and leave this topic for future work.

5.4.2 Priority Calculation

To prevent arbitrary adding of badly-fitting terms that are eventually scaled down to zero, our priority measure considers both a phrase's length and its expected accuracy. To balance these two factors, these two measures need to be in the same scale. We use the normalized mean squared error (NMSE) as quality measure which is in the range [0, 1] for properly scaled solutions. This measure corresponds to $1 - R^2$ (coefficient of determination). As length measure we use the number of symbols relative to the maximum sentence length.

Since we limit the search space to a maximum number of variable references of a phrase, we cannot exactly calculate the maximum possible length of a phrase. Therefore, we estimate this maximum length with a greedy procedure: Starting with the grammar's start symbol *Expr*, we iteratively derive a new phrase using the longest production rule. If two production rules have the same length, we take the one with least non-terminal symbols and variable references.

Phrases with *lower* priority values are expanded first during the search. The priority for steering the search from Sect. 5.3 is the phrase's NMSE value minus its weighted relative length, as shown in Eq. (5.6). The weight w controls the greediness and allows corrections of over- or underestimations of the maximum length. However, in practice this value is not critical.

$$\text{priority}(p) = \text{NMSE}(p) + w\frac{len(p)}{length_{max}} \tag{5.6}$$

5.5 Experiments

We run our algorithm on several synthetic benchmark datasets to show that the search space defined by our restricted grammar is powerful enough to solve many problems in feasible time. As benchmark datasets, we use noiseless datasets from physical domains [4] and Nguyen-, Vladislavleva- and Keijzer-datasets [37] as defined and implemented in the HeuristicLab framework.

The search space was restricted in the experiments to include only sentences with at most 20 variable references. We evaluate at most 200,000 sentences. Coefficients are randomly initialized and then fitted with the iterative gradient-based Levenberg–Marquardt algorithm [20, 21] with at most 100 iterations. For each model structure, we repeat the coefficient fitting process ten times with differently initialized values to reduce the chance of finding bad local optima.

As a baseline, we also run symbolic regression with GP on the same benchmark problems. Therefore, we execute GP with strict offspring selection (OSGP) [1] and explicit optimization of coefficients [9]. The OSGP settings are listed in Table 5.1. The OSGP experiments were executed with the *HeuristicLab* software framework[4]

Table 5.1 OSGP experiment settings

Parameter	Setting
Population size	500
Max. selection pressure	300
Max. evaluated solutions	200,000
Mutation probability	15%
Selection	Gender-specific selection (random and proportional)
Crossover operator	Subtree swapping
Mutation operator	Point mutation, tree shaking, changing single symbols, replacing/removing branches
Max. tree size	Number of nodes: 30, depth: 50
Function set	$+, -, \times, \div$, exp, log, sin, cos, square, sqrt, cbrt

---

[4]https://dev.heuristiclab.com.

[36]. Since this comparison focuses only on particular weaknesses and strengths of our proposed algorithm over state of the art-techniques, we use the same OSGP settings for all experiments and leave out problem-specific hyper parameter-tuning.

5.5.1 Results

Both the exhaustive search and OSGP were repeated ten times on each dataset. All repetitions of the exhaustive search algorithm led to the exact same results. This underlines the determinism of the proposed methods, even though we rely on stochasticity when optimizing coefficients. Also the OSGP results do not differ much. Tables 5.2, 5.3, 5.4, and 5.5 show the achieved NMSE values for the exhaustive search and the median NMSE values of all OSGP repetitions. NMSE values in the Tables 5.2, 5.3, 5.4, and 5.5 smaller than 10^{-8} are considered as exact or good-enough approximations and emphasized in bold. The exhaustive search found a good solution (NMSE $< 10^{-8}$) within ten minutes for all datasets. If no such solution was found, the algorithm runs until it reaches the max. number of evaluated solutions, which can take days for larger datasets.

The experimental results show, that our algorithm struggles with problems with complex terms—for example with Keijzer data sets 4, 5 and 11 in Table 5.2. This is probably because our heuristic works "term-wise"—our algorithm searches completely broad without any guidance within terms which still contain non-terminal symbols. This issue becomes even more pronounced when we have to find

Table 5.2 Median NMSE results for Keijzer instances

	Problem	Exhaustive search		OSGP	
		Train.	Test	Train.	Test
1 [6, 7]	$0.3x\sin(2\pi x); x \in [-1, 1]$	3e−27	**2e−27**	1e−30	**8e−31**
2 [6]	$0.3x\sin(2\pi x); x \in [-2, 2]$	5e−22	**5e−22**	5e−18	**4e−18**
3 [6]	$0.3x\sin(2\pi x); x \in [-3, 3]$	6e−32	**3e−31**	4e−30	**3e−30**
4 [6, 26]	$x^3\exp(-x)\cos(x)\sin(x)(\sin(x)^2\cos(x)-1)$	1e−04	2e−04	1e−06	1e−06
5 [6]	$(30xz)/((x-10)y^2)$	3e−08	3e−08	3e−20	**3e−20**
6 [6, 32]	$\sum_{i=1}^{x}\frac{1}{i}$	8e−13	**6e−09**	5e−14	**5e−13**
7 [6, 32]	$\ln(x)$	2e−31	**3e−31**	1e−30	**2e−30**
8 [6, 32]	$\sqrt{x}$	2e−14	**8e−10**	5e−21	**1e−21**
9 [6, 32]	$\operatorname{arcsinh}(x)$ i.e. $\ln(x+\sqrt{x^2+1})$	5e−14	1e−05	5e−17	**6e−16**
10 [6, 32]	x^y	4e−04	1e−01	6e−32	2e−04
11 [6, 33]	$xy+\sin((x-1)(y-1))$	7e−04	7e−01	2e−22	9e−02
12 [6, 33]	$x^4-x^3+y^2/2-y$	5e−32	**1e−31**	7e−22	**8e−18**
13 [6, 33]	$6\sin(x)\cos(y)$	2e−32	**2e−31**	3e−32	**3e−32**
14 [6, 33]	$8/(2+x^2+y^2)$	4e−32	**2e−31**	1e−17	**1e−17**
15 [6, 33]	$x^3/5+y^3/2-y-x$	1e−22	**2e−21**	2e−11	**6e−10**

Table 5.3 Median NMSE results for Nguyen instances

	Problem	Exhaustive search		OSGP	
		Train.	Test	Train.	Test
1 [34]	$x^3 + x^2 + x$	5e−34	**3e−33**	8e−30	**2e−29**
2 [34]	$x^4 + x^3 + x^2 + x$	3e−33	**4e−33**	5e−30	**1e−28**
3 [34]	$x^5 + x^4 + x^3 + x^2 + x$	1e−33	**7e−33**	2e−16	**2e−15**
4 [34]	$x^6 + x^5 + x^4 + x^3 + x^2 + x$	6e−12	**6e−11**	2e−12	3e−08
5 [34]	$\sin(x^2)\cos(x) - 1$	9e−14	**3e−13**	3e−18	**4e−18**
6 [34]	$\sin(x) + \sin(x + x^2)$	2e−17	**2e−12**	6e−14	6e−08
7 [34]	$\log(x + 1) + \log(x^2 + 1)$	4e−13	**5e−12**	5e−13	**1e−09**
8 [34]	$\sqrt{x}$	6e−32	**2e−31**	7e−32	**1e−31**
9 [34]	$\sin(x) + \sin(y^2)$	2e−13	**2e−12**	8e−31	**8e−31**
10 [34]	$2\sin(x)\cos(y)$	5e−32	**1e−31**	1e−28	**8e−29**
11 [34]	x^y	2e−06	1e−02	6e−30	**3e−30**
12 [34]	$x^4 - x^3 + y^2/2 - y$	2e−31	**2e−31**	7e−18	**5e−17**

Table 5.4 Median NMSE results for Vladislavleva instances

	Problem	Exhaustive search		OSGP	
		Train.	Test	Train.	Test
1 [30]	$\exp(-(x_1 - 1)^2)/(1.2 + (x_2 - 2.5)^2)$	3e−03	3e−01	1e−09	9e−07
2 [26]	$\exp(-x)x^3\cos(x)\sin(x)(\cos(x)\sin(x)^2 - 1)$	3e−04	1e−02	3e−06	2e−03
3 [35]	$f_2(x_1)(x_2 - 5)$	1e−02	2e−01	3e−05	6e−04
4 [35]	$10/(5 + \sum_{i=1}^{5}(x_i - 3)^2)$	1e−01	2e−01	7e−03	1e−02
5 [35]	$30((x_1 - 1)(x_3 - 1))/(x_2^2(x_1 - 10))$	2e−03	9e−03	8e−16	**9e−15**
6 [35]	$6\sin(x_1)\cos(x_2)$	8e−32	**4e−31**	6e−31	**3e−19**
7 [35]	$(x_1 - 3)(x_2 - 3) + 2\sin((x_1 - 4)(x_2 - 4))$	1e−30	**9e−31**	5e−29	**4e−29**
8 [35]	$((x_1 - 3)^4 + (x_2 - 3)^3 - (x_2 - 3))/((x_2 - 2)^4 + 10)$	1e−03	2e−01	5e−05	2e−02

long and complex function arguments. It should also be noted that our algorithm only finds non-factorized representations of such arguments, which are even longer and therefore even harder to find in a broad search.

For the Nguyen datasets in Table 5.3 and the Keijzer datasets 12–15 in Table 5.2, we find the exact or good approximations in most cases with our exhaustive search. Especially for simpler datasets, the results of our algorithm surpasses the one of OSGP. This is likely due to the datasets' low number of training instances, which makes it harder for OSGP to find good approximations.

Some problems are not contained in the search space, thus we do not find any good solution for them. This is the case for Keijzer 6, 9 and 10 in Table 5.2, for which we do not support the required function symbols in our grammar. Also all Vladislavleva datasets except 6 and 7 in Table 5.4 and the problems "Fluid Flow" and "Pagie-1" in Table 5.5 are not in the hypothesis space as they are too complex.

Table 5.5 Median NMSE results for other instances

	Exhaustive search		OSGP	
Problem	Train.	Test	Train.	Test
Poly-10 [25] $x_1 x_2 + x_3 x_4 + x_5 x_6 + x_1 x_7 x_9 + x_3 x_6 x_{10}$	2e−32	**1e−32**	7e−02	1e−01
Pagie-1 (inverse dynamics) [24] $1/(1 + x^{-4}) + 1/(1 + y^{-4})$	1e−03	6e−01	9e−07	5e−05
Aircraft lift coefficient [4] $C_{L\alpha}(\alpha - \alpha_0) + C_{L\delta_e}\delta_e S_{HT}/S_{ref}$	3e−31	**3e−31**	2e−17	**2e−17**
Fluid flow [4] $V_\infty r \sin(\theta)(1 - R^2/r^2) + \Gamma/(2\pi)\ln(r/R)$	3e−04	4e−04	9e−06	2e−05
Rocket fuel flow [4] $p_0 A^\star/\sqrt{T_0}\sqrt{\gamma/R(2/(\gamma + 1))^{(\gamma+1)/(\gamma-1)}}$	3e−31	**3e−31**	1e−19	**1e−19**

Another issue is the optimization of coefficients. Although several problems have a simple structure and are in the search space, we do not find the right coefficients for arguments of non-linear functions, for example in Nguyen 5–7. The issue hereby is that we iterate over the actually searched model structure but determine bad coefficients. As we do never look again at the same model structure, we can only find an approximation. This is a big difference to symbolic regression with genetic programming, in which we might find the same structure again in next generations.

5.6 Discussion

Among the nonlinear system identification techniques, symbolic regression is characterized by its ability to identify complex nonlinear relationships in structured numerical data in the form of interpretable models. The combination of the power of nonlinear system identification without a priori assumptions about the model structure with the white-box ability of mathematical formulas represents the unique selling point of symbolic regression. If tree-based GP is used as search method, the ability to interpret the found models is limited due to the stochasticity of the GP search. Thus, at the end of the modeling phase, several similarly complex models of approximately the same quality can be produced, which have completely different structures and use completely different subsets of features. These last-mentioned limitations due to ambiguity can be countered using a deterministic approach in which only semantically unique models may be used. This approach, however, requires a lot of restrictions regarding search space complexity in order to specify a subspace in which an exhaustive search is feasible. On the other hand, the exhaustive claim enables the approach to generate extensive model libraries already in the offline phase, through which as soon as a concrete task is given in the online phase, it is only necessary to navigate in a suitable way.

In a very reduced summary, one could characterize the classical tree-based symbolic regression using GP and the approach of deterministically and exhaustively generating models in such a way that the latter enables a complete search in an incomplete search space while the classical approach performs an incomplete search in a rather complete search space.

5.6.1 Limitations

The approach we have described in this contribution also has several limitations. For the identification of optimal coefficient values we rely on the Levenberg–Marquardt method for least squares, which is a local search routine using gradient information. Therefore, we can only hope to find global optima for coefficient values. Finding bad local optima for coefficients is less of a concern when using GP variants with a similar local improvement scheme because there are implicitly many restarts through the evolutionary operations of recombination and mutation. In the proposed method we visit each structure only once and therefore risk to discard a good solution when we are unlucky to find good coefficients.

We have worked only with noiseless problem instances yet. We observed in first experiments with noisy problems instances that the algorithm might get stuck trying to improve non-optimal partial solutions due to its greedy nature. Therefore, we need further investigations before we move on with the development of our algorithm to noisy real-world problems.

Another limitation is the poor scalability of grammar enumeration when increasing the number of features or the size of the search space. When increasing these parameters we can not expect to explore a significant part of the complete search space and must increasingly rely on the power of heuristics to hone in on relevant subspaces. Currently, we have only integrated a single heuristic which evaluates terms in partial solutions and prioritizes phrase which include well-fitting terms. However, the algorithm has no way to prioritize incomplete terms and is inefficient when trying to find complex terms.

5.7 Outlook

Even when considering the above mentioned limitations of the currently implemented algorithm we still see significant potential in the approach of more systematic and deterministic search for symbolic regression and we already have several ideas to improve the algorithm and overcome some of the limitations.

The integration of improved heuristics for guided search is our top-priority. An advantage of the concept is that it is extremely general and allows to experiment with many different heuristics. Heuristics can be as simple as prioritizing shorter expressions or less complex expressions. More elaborate schemes which guide the search based on prior knowledge about the data-generating process are easy to imagine. Heuristics could incorporate syntactical information (e.g. which variables already occur within the expression) as well as information from partial evaluation of expressions. We also consider dynamic heuristics which are adjusted while the algorithm is running and learning about the problem domain. Potentially, we could even identify and learn heuristics which are transferable to other problem instances and would improve efficiency in a transfer learning setting.

Getting trapped in local optima is less of a concern when we apply global search algorithms for coefficient values such as evolution strategies, differential evolution, or particle swarm optimization (cf. [11]). Another approach would be to reduce the ruggedness of the objective function through regularization of the coefficient optimization step. This could be helpful to reduce the potential of overfitting and getting stuck in sub-optimal subspaces of the search space.

Generally, we consider grammar enumeration to be effective only when we limit the search space to relatively short expressions—which is often the case in our industrial applications. Therein lies the main potential compared to the more general approach of genetic programming. In this context we continue to explore potential for segmentation of the search space [19] in combination with grammar enumeration in an offline phase for improving later search runs. Grammar enumeration with deduplication of structures could also be helpful to build large offline libraries of sub-expressions that could be used by GP [2, 8, 17, 18].

Acknowledgements The authors gratefully acknowledge support by the Christian Doppler Research Association and the Federal Ministry for Digital and Economic Affairs within the *Josef Ressel Center for Symbolic Regression.*

References

1. Affenzeller, M., Winkler, S., Wagner, S., Beham, A.: Genetic Algorithms and Genetic Programming - Modern Concepts and Practical Applications, *Numerical Insights*, vol. 6. CRC Press, Chapman & Hall (2009)
2. Angeline, P.J., Pollack, J.: Evolutionary module acquisition. In: Proceedings of the Second Annual Conference on Evolutionary Programming, pp. 154–163. La Jolla, CA, USA (1993)
3. Burlacu, B., Kammerer, L., Affenzeller, M., Kronberger, G.: Hash-based Tree Similarity and Simplification in Genetic Programming for Symbolic Regression. In: Computer Aided Systems Theory, EUROCAST 2019 (2019)
4. Chen, C., Luo, C., Jiang, Z.: A multilevel block building algorithm for fast modeling generalized separable systems. Expert Systems with Applications **109**, 25–34 (2018)
5. Hart, P.E., Nilsson, N.J., Raphael, B.: A formal basis for the heuristic determination of minimum cost paths. IEEE Transactions on Systems Science and Cybernetics **4**(2), 100–107 (1968)
6. Keijzer, M.: Improving symbolic regression with interval arithmetic and linear scaling. In: Genetic Programming, Proceedings of EuroGP'2003, *LNCS*, vol. 2610, pp. 70–82. Springer-Verlag, Essex (2003)
7. Keijzer, M., Babovic, V.: Genetic programming, ensemble methods and the bias/variance tradeoff - introductory investigations. In: Genetic Programming, Proceedings of EuroGP'2000, *LNCS*, vol. 1802, pp. 76–90. Springer-Verlag, Edinburgh (2000)
8. Keijzer, M., Ryan, C., Murphy, G., Cattolico, M.: Undirected training of run transferable libraries. In: Proceedings of the 8th European Conference on Genetic Programming, *Lecture Notes in Computer Science*, vol. 3447, pp. 361–370. Springer, Lausanne, Switzerland (2005)

9. Kommenda, M., Kronberger, G., Winkler, S., Affenzeller, M., Wagner, S.: Effects of constant optimization by nonlinear least squares minimization in symbolic regression. In: Proceedings of the 15th Annual Conference Companion on Genetic and Evolutionary Computation, GECCO '13 Companion, pp. 1121–1128. ACM (2013)
10. Korns, M.F.: Symbolic regression using abstract expression grammars. In: GEC '09: Proceedings of the first ACM/SIGEVO Summit on Genetic and Evolutionary Computation, pp. 859–862. ACM, Shanghai, China (2009)
11. Korns, M.F.: Abstract expression grammar symbolic regression. In: Genetic Programming Theory and Practice VIII, *Genetic and Evolutionary Computation*, vol. 8, chap. 7, pp. 109–128. Springer, Ann Arbor, USA (2010)
12. Korns, M.F.: Extreme accuracy in symbolic regression. In: Genetic Programming Theory and Practice XI, Genetic and Evolutionary Computation, chap. 1, pp. 1–30. Springer, Ann Arbor, USA (2013)
13. Korns, M.F.: Extremely accurate symbolic regression for large feature problems. In: Genetic Programming Theory and Practice XII, Genetic and Evolutionary Computation, pp. 109–131. Springer, Ann Arbor, USA (2014)
14. Korns, M.F.: Highly accurate symbolic regression with noisy training data. In: Genetic Programming Theory and Practice XIII, Genetic and Evolutionary Computation, pp. 91–115. Springer, Ann Arbor, USA (2015)
15. Kotanchek, M., Smits, G., Vladislavleva, E.: Trustable symbolic regression models: using ensembles, interval arithmetic and pareto fronts to develop robust and trust-aware models. In: Genetic Programming Theory and Practice V, Genetic and Evolutionary Computation, chap. 12, pp. 201–220. Springer, Ann Arbor (2007)
16. Kotanchek, M.E., Vladislavleva, E., Smits, G.: Symbolic Regression Is Not Enough: It Takes a Village to Raise a Model, pp. 187–203. Springer New York, New York, NY (2013)
17. Krawiec, K., Pawlak, T.: Locally geometric semantic crossover. In: GECCO Companion '12: Proceedings of the fourteenth international conference on Genetic and evolutionary computation conference companion, pp. 1487–1488. ACM, Philadelphia, Pennsylvania, USA (2012)
18. Krawiec, K., Swan, J., O'Reilly, U.M.: Behavioral program synthesis: Insights and prospects. In: Genetic Programming Theory and Practice XIII, Genetic and Evolutionary Computation, pp. 169–183. Springer, Ann Arbor, USA (2015)
19. Kronberger, G., Kammerer, L., Burlacu, B., Winkler, S.M., Kommenda, M., Affenzeller, M.: Cluster analysis of a symbolic regression search space. In: Genetic Programming Theory and Practice XVI. Springer, Ann Arbor, USA (2018)
20. Levenberg, K.: A method for the solution of certain non-linear problems in least squares. Quarterly of Applied Mathematics **2**(2), 164–168 (1944)
21. Marquardt, D.W.: An algorithm for least-squares estimation of nonlinear parameters. Journal of the Society for Industrial and Applied Mathematics **11**(2), 431–441 (1963)
22. McConaghy, T.: FFX: Fast, scalable, deterministic symbolic regression technology. In: Genetic Programming Theory and Practice IX, Genetic and Evolutionary Computation, chap. 13, pp. 235–260. Springer, Ann Arbor, USA (2011)
23. Merkle, R.C.: A digital signature based on a conventional encryption function. In: Advances in Cryptology — CRYPTO '87, pp. 369–378. Springer Berlin Heidelberg, Berlin, Heidelberg (1988)
24. Pagie, L., Hogeweg, P.: Evolutionary consequences of coevolving targets. Evolutionary Computation **5**(4), 401–418 (1997)
25. Poli, R.: A simple but theoretically-motivated method to control bloat in genetic programming. In: Genetic Programming, Proceedings of EuroGP'2003, *LNCS*, vol. 2610, pp. 204–217. Springer-Verlag, Essex (2003)
26. Salustowicz, R.P., Schmidhuber, J.: Probabilistic incremental program evolution. Evolutionary Computation **5**(2), 123–141 (1997)

27. Schmidt, M., Lipson, H.: Co-evolving fitness predictors for accelerating and reducing evaluations. In: Genetic Programming Theory and Practice IV, *Genetic and Evolutionary Computation*, vol. 5, pp. 113–130. Springer, Ann Arbor (2006)
28. Schmidt, M., Lipson, H.: Symbolic regression of implicit equations. In: Genetic Programming Theory and Practice VII, Genetic and Evolutionary Computation, chap. 5, pp. 73–85. Springer, Ann Arbor (2009)
29. Schmidt, M., Lipson, H.: Age-fitness pareto optimization. In: Genetic Programming Theory and Practice VIII, *Genetic and Evolutionary Computation*, vol. 8, chap. 8, pp. 129–146. Springer, Ann Arbor, USA (2010)
30. Smits, G., Kotanchek, M.: Pareto-front exploitation in symbolic regression. In: Genetic Programming Theory and Practice II, chap. 17, pp. 283–299. Springer, Ann Arbor (2004)
31. Stijven, S., Vladislavleva, E., Kordon, A., Kotanchek, M.: Prime-time: Symbolic regression takes its place in industrial analysis. In: Genetic Programming Theory and Practice XIII, Genetic and Evolutionary Computation, pp. 241–260. Springer, Ann Arbor, USA (2015)
32. Streeter, M.J.: Automated discovery of numerical approximation formulae via genetic programming. Master's thesis, Computer Science, Worcester Polytechnic Institute, MA, USA (2001)
33. Topchy, A., Punch, W.F.: Faster genetic programming based on local gradient search of numeric leaf values. In: Proceedings of the Genetic and Evolutionary Computation Conference (GECCO-2001), pp. 155–162. Morgan Kaufmann, San Francisco, California, USA (2001)
34. Uy, N.Q., Hoai, N.X., O'Neill, M., McKay, R.I., Galvan-Lopez, E.: Semantically-based crossover in genetic programming: application to real-valued symbolic regression. Genetic Programming and Evolvable Machines **12**(2), 91–119 (2011)
35. Vladislavleva, E.J., Smits, G.F., den Hertog, D.: Order of nonlinearity as a complexity measure for models generated by symbolic regression via Pareto genetic programming. IEEE Transactions on Evolutionary Computation **13**(2), 333–349 (2009)
36. Wagner, S., Affenzeller, M.: HeuristicLab: A generic and extensible optimization environment. In: Adaptive and Natural Computing Algorithms, pp. 538–541. Springer (2005)
37. White, D.R., McDermott, J., Castelli, M., Manzoni, L., Goldman, B.W., Kronberger, G., Jaśkowski, W., O'Reilly, U.M., Luke, S.: Better GP benchmarks: community survey results and proposals. Genetic Programming and Evolvable Machines **14**(1), 3–29 (2013)
38. Worm, T., Chiu, K.: Prioritized grammar enumeration: symbolic regression by dynamic programming. In: GECCO '13: Proceeding of the fifteenth annual conference on Genetic and evolutionary computation conference, pp. 1021–1028. ACM, Amsterdam, The Netherlands (2013)
39. Zou, H., Hastie, T.: Regularization and variable selection via the elastic net. Journal of the royal statistical society: series B (statistical methodology) **67**(2), 301–320 (2005)

Chapter 6
Temporal Memory Sharing in Visual Reinforcement Learning

Stephen Kelly and Wolfgang Banzhaf

6.1 Introduction

Reinforcement learning (RL) is an area of machine learning that models the way living organisms adapt through interaction with their environment. RL can be characterized as learning how to map situations to actions in the pursuit of a pre-defined objective [34]. A solution, or *policy* in RL is represented by an agent that learns through episodic interaction with the problem environment. Each episode begins in an initial state defined by the environment. Over a series of discrete timesteps, the agent observes the environment (via sensory inputs), takes an action based on the observation, and receives feedback in the form of a reward signal. The agent's actions potentially change the state of the environment and impact the reward received. The agent's goal is to select actions that maximize the long-term cumulative reward. Most real-world decision-making and prediction problems can be characterized as this type of environmental interaction.

Animal and human intelligence is partially a consequence of the physical richness of our environment, and thus scaling RL to complex, real-world environments is a critical step toward sophisticated artificial intelligence. In real-world applications of RL, the agent is likely to observe the environment through a high-dimensional, visual sensory interface (e.g. a video camera). However, scaling to high-dimensional input presents a significant challenge for machine learning, and RL in particular. As the complexity of the agent's sensory interface increases, there

S. Kelly (✉)
Department of Computer Science and Engineering & Beacon Center, Michigan State University, East Lansing, MI, USA
e-mail: kellys27@msu.edu

W. Banzhaf
Michigan State University, East Lansing, MI, USA
e-mail: banzhafw@msu.edu

© Springer Nature Switzerland AG 2020
W. Banzhaf et al. (eds.), *Genetic Programming Theory and Practice XVII*, Genetic and Evolutionary Computation, https://doi.org/10.1007/978-3-030-39958-0_6

is a significant increase in the number of environmental observations required for the agent to gain the breadth of experience necessary to build a strong decision-making policy. This is known as the curse of dimensionality (Section 1.4 of [6]). The temporal nature of RL introduces additional challenges. In particular, complete information about the environment is not always available from a single observation (i.e. the environment is partially observable) and delayed rewards are common, so the agent must make thousands of decisions before receiving enough feedback to assess the quality of its behaviour [22]. Finally, real-world environments are dynamic and non-stationary [9, 26]. Agents are therefore required to adapt to changing environments without 'forgetting' useful modes of behaviour that are intermittently important over time.

Video games provide a well-defined test domain for scalable RL. They cover a diverse range of environments that are designed to be challenging for humans, all through a common high-dimensional visual interface, namely the game screen [5]. Furthermore, video games are subject to partial observability and explicitly non-stationary. For example, many games require the player to predict the trajectory of a moving object. These calculations cannot be made from observing a single screen capture. To make such predictions, players must identify, store, and reuse important parts of past experience. As the player improves, new levels of play are unlocked which may contain completely new visual and physical dynamics. As such, video games represent a rich combination of challenges for RL, where the objective for artificial agents is to play the game with a degree of sophistication comparable to that of a human player [4, 30]. The potential real-world applications for artificial agents with these capabilities is enormous.

6.2 Background

Tangled Program Graphs (TPG) are a representation for Genetic Programming (GP) with particular emphasis on *emergent* modularity through compositional evolution: the evolution of hierarchical organisms that combine multiple agents which were previously adapted independently [38], Fig. 6.1.

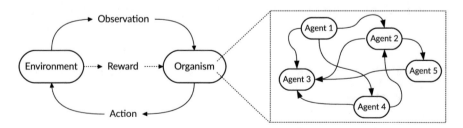

Fig. 6.1 A multi-agent organism developed through compositional evolution

This approach leads to automatic division of labour within the organism and, over time, a collective decision-making policy emerges that is greater than the sum of its parts. The system has three critical attributes:

1. **Adaptive Complexity.** Solutions begin as single-agent organisms and then develop into multi-agent organisms through interaction with their environment. That is, the complexity of a solution is an adapted property.
2. **Input Selectivity.** Multi-agent organisms are capable of decomposing the input space such that they can ignore sensory inputs that are not important at the current point in time. This is more efficient than assuming that the complete sensory system is necessary for every decision.
3. **Modular Task Decomposition.** As multi-agent organisms develop they may subsume a variable number of stand-alone agents into a hierarchical decision-making policy. Importantly, hierarchies emerge incrementally over time, slowly growing and breaking apart through interaction with the environment. This property allows a TPG organism to adapt in non-stationary environments and avoid unlearning behaviours that were important in the past but are not currently relevant.

In the Atari video game environment, TPG matches the quality of solutions from a variety of state-of-the-art deep learning methods. More importantly, TPG is less computationally demanding, requiring far fewer calculations per decision than any of the other methods [20]. However, these TPG policies were purely reactive. They represent a direct mapping from observation to action with no mechanism to integrate past experience (prior state observations) into the decision-making process. This might be a limitation in environments with temporal properties. For example, there are Atari games that explicitly involve predicting the trajectory of a moving object (e.g. Pong) for which TPG performed poorly. A temporal memory mechanism would allow agents to make these calculations by integrating past and present environmental observations.

6.2.1 Temporal Memory

Temporal memory in sequential decision-making problems implies that behavioural agent has the ability to identify, store, and reuse important aspects of past experience when predicting the best action to take in the present. More generally, temporal memory is essential to any time series prediction problem, and has thus been investigated extensively in GP (see Agapitos et. al [2] for a broad review). In particular, an important distinction is made between *static* memory in which the same memory variables are accessed regardless of the state of the environment, and *dynamic* memory in which different environmental observations trigger access to different variables.

Dynamic memory in GP is associated with indexed memory, which requires the function set to include parameterized operations for reading and writing to specific memory addresses. Teller [35] showed that GP with memory indexing is Turing complete, i.e. theoretically capable of evolving any algorithm. Brave [8] emphasized the utility of dynamic memory access in GP applied to an agent planning problem, while Haynes [15] discuses the value of GP with dynamic memory in sequential decision-making environments that are themselves explicitly dynamic. Koza [23] proposed Automatically Defined Stores, a modular form of indexable memory for GP trees, and demonstrated its utility for solving the cart centering problem *without* velocity sensors, a classic control task that requires temporal memory.

Static memory access is naturally supported by the register machine representation in linear genetic programming [7]. For example, a register machine typically consists of a sequence of instructions that read and write from/to memory, e.g. $Register[x] = Register[x] + Register[y]$. In this case, the values contained in registers x and y may change depending on environmental input, but reference to the specific registers x and y is determined by the instruction's encoding and is not affected by input to the program. If the register content is cleared prior to each program execution, the program is said to be *stateless*. A simple form of temporal memory can be implemented by not clearing the register content prior to each execution of the program. In the context of sequential decision-making, the program retains/accumulates state information over multiple timesteps, i.e., the program is *stateful*. Alternatively, register content from timestep t may be fed back into the program's input at time $t + 1$, enabling temporal memory through recurrent connections, e.g. [10, 16].

Smith and Heywood [32] introduced the first memory model for TPG in particular. Their method involved a single global memory bank. TPG's underlying linear GP representation was augmented with a probabilistic write operation, enabling long and short-term memory stores. They also included a parameterized read operation for indexed (i.e. dynamic) reading of external memory. Furthermore, the external memory bank is ecologically global. That is, sharing is supported among the entire population such that organisms may integrate their own past experience *and* experience gained by other (independent) organisms. The utility of this memory model was demonstrated for navigation in a partially observable, visual RL environment.

In this work we propose that multi-agent TPG organisms can be extended to support dynamic temporal memory *without* the addition of specialized read/write operations. This is possible because TPG organisms naturally decompose the task both spatially *and* temporally. Specifically, each decision requires traversing one path through the graph of agents, in which only agents along the path are 'executed' (i.e. a subset of the agents in the organism). Each agent will have a unique complement of static environmental input and memory references. Since the decision path at time t is entirely dependent on the state of the environment, both state and memory access are naturally dynamic.

The intuition behind this work is that dynamic/temporal problem decomposition w.r.t input and memory access is particularly important in visual RL tasks because: (1) High-dimensional visual input potentially contains a large amount of information that is intermittently relevant to decision making over time. As such, it is advantageous if the model can parse out the most salient observational data for the current timestep and ignore the rest; and (2) RL environments with real-world complexity are likely to exhibit partial observability at multiple times scales. For example, predicting the trajectory of a moving object may require an agent to integrate a memory of the object's location at $t - 1$ with the current observation at time t, or short-term memory. Conversely, there are many tasks that require integration over much longer periods of time, e.g. maze navigation [12]. These two points clearly illustrate the drawback of *autoregressive* models [2, 30], in which the issue of temporal memory is side-stepped by stacking a fixed number of the most recent environmental observations into a single 'sliding window' world view for the agent. This approach potentially increases the amount of redundant input information and limits temporal state integration to a window size fixed a priori.

6.2.2   Heterogeneous Policies and Modularity

Heterogeneous policies provide a mechanism through which multiple types of active device, entities which accept input, perform some computation, and produce output, may be combined within a stand-alone decision-making agent, e.g. [17, 25]. In this work, we investigate how compositional evolution can be used to adaptively combine general-purpose and task-specific devices. Specifically, GP applied to visual problems can benefit from the inclusion of specialized image-progressing operators, e.g. [3, 24, 39]. Rather than augmenting the instruction set of *all* programs with additional operators, compositional evolution of heterogeneous policies provides an opportunity to integrate task-specific functionality in a modular fashion, where modularity is likely to improve the evolvability of the system [1, 36, 38].

6.3   Evolving Heterogeneous Tangled Program Graphs

The algorithm investigated in this work is an extension of Tangled Program Graphs [20] with additional support for temporal memory and heterogeneous policies. This section details the extended system with respect to three system components, each playing a distinct role within the emergent hierarchical structure of TPG organisms:

- A **Program** is the simplest active device, capable of processing input data and producing output, but *not* representing a stand-alone RL agent.
- A **Team of Programs** is the smallest independent decision-making organism, or agent, capable of observing the environment and taking *actions*.
- A **Policy Graph** adaptively combines multiple teams (agents) into a single hierarchical organism through open-ended compositional evolution. In this context, *open-ended* refers to the fact that hierarchical transitions are not planned a priori. Instead, the hierarchical complexity of each policy graph is a property that emerges through interaction with the problem environment.

6.3.1 Programs and Shared Temporal Memory

In this work, all programs are linear register machines [7]. Two types of program are supported: Action-value programs and Image processors.

Action-value programs have a pointer to one action (e.g. a joystick position from the video game domain) and produce one scalar *bidding* output, which is interpreted as the program's confidence that its action is appropriate given the current state of the environment. As such, the role of an action-value program is to define environmental *context* for one action, Algorithm 1. In order to support shared temporal memory, action-value programs have two register banks; one *stateless* bank that is reset prior to each program execution, and a pointer to one *stateful* bank that is only reset at the start of each episode (i.e. game start). Stateless register banks are private to each program, while stateful banks are stored in a dedicated *memory* population and may be shared among multiple programs (This relationship is illustrated in the lower-left of Fig. 6.2). Shared memory allows multiple programs to communicate within a single timestep or integrate information across multiple timesteps. In effect, shared memory implies that each program has a parameterized number of *context* outputs (see $Registers_{shared}$ in Table 6.2).

Image processing programs have *context* outputs only. They perform matrix manipulations on the raw pixel input and store the result in shared memory accessible to all other programs, Algorithm 2. In addition to private and shared register memory, image processors have access to a task-specific type of shared memory

---

Algorithm 1 Example **action-value program**. Each program contains one *private stateless* register bank, R_p, and a pointer to one *shareable stateful* register bank, R_s. R_p is reset prior to each execution, while R_s is reset (by an external process) at the start of each episode. Programs may include two-argument instructions of the form $R[i] \leftarrow R[j] \circ R[k]$ in which $\circ \in \{+, -, x, \div\}$; single-argument instructions of the form $R[i] \leftarrow \circ(R[k])$ in which $\circ \in \{cos, ln, exp\}$; and a conditional statement of the form IF $(R[i] < R[k])$ THEN $R[i] \leftarrow -R[i]$. The second source variable, $R[k]$, may reference either memory bank or a state variable (pixel), while the target and first source variables $(R[i]$ and $R[j])$ may reference either the stateless or stateful memory bank only. Action-value programs always return the value stored in $R_p[0]$ at the end of execution

---

1: $R_p \leftarrow 0$ # reset private memory bank R_p

2: $R_p[0] \leftarrow R_s[0] - Input[3]$
3: $R_s[1] \leftarrow R_p[0] \div R_s[7]$
4: $R_s[2] \leftarrow Log(R_s[1])$
5: **if then**$(R_p[0] < R_s[2])$
6:     $R_p[0] \leftarrow -R_p[0]$
7: **end if**
8: **return** $R_p[0]$

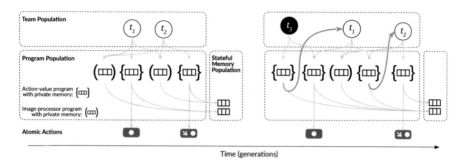

Fig. 6.2 Illustration of the relationship between teams, programs, and shared memory in heterogeneous TPG

Algorithm 2 Example **image-processor program**. As with action-value programs, image-processor programs contain one *private stateless* register bank, R_p, and a pointer to one *shareable stateful* register bank, R_s. In addition, all image processors have access to a global buffer matrix, S, which is reset (by an external process) at the start of each episode. Image-processor programs accept either the raw pixel screen or the shared buffer as input. Depending on the operation, the result is stored in R_p, R_s, or back into the shared buffer S. Unlike the operations available to action-value programs, some image processing operations are parameterized by values stored in register memory or sampled directly form input. This opens a wide range of possibilities for image processing instructions, a few of which are illustrated in this algorithm. Table 6.1 provides a complete list of image processing operations used in this work

1: $R_p \leftarrow 0$	# reset private memory bank R_p
2: $S \leftarrow AddC(Screen, R_s[0])$	# Add an amount to each image pixel
3: $S \leftarrow Div(Screen, S)$	# Divide the pixel values of two images
4: $S \leftarrow Sqrt(S)$	# Take the square root of each pixel
5: $R_p[2] \leftarrow MaxW(S, R_s[0], R_s[5], R_s[7])$	# Store max of parameterized window
6: $R_s[2] \leftarrow MeanW(S, R_s[3], R_s[1], R_p[2])$	# Store mean of parameterized window

---

in the form of a single, global image buffer matrix with the same dimensionality as the environment's visual interface. The buffer matrix is stateful, reset only at the start of each episode. This allows image processors to accumulate full-screen manipulations over the entire episode. Note that image-processor programs have no bidding output because they do not contribute directly to action selection. Their role is to preprocess input data for action-value programs, and this contribution is communicated to action-value programs through shared register memory (This relationship is illustrated in the lower-left of Fig. 6.2).

Memory sharing implies that a much larger proportion of program code is now *effective*, since an effective instruction is one that effects the final value in the bidding output ($R_p[0]$) or any of the shared registers, R_s. As a result, sharing memory incurs a significant computational cost relative to programs without shared

Table 6.1 Operations available to image-processor programs

Operation	Parameters	Description
Add	Image, Image	Add pixel values of two images
Sub	Image, Image	Subtract pixel values of two images
Div	Image, Image	Divide pixel values of two images
Mul	Image, Image	Multiply pixel values of two images
Max2	Image, Image	Pixel-by-pixel max of two images
Min2	Image, Image	Pixel-by-pixel min of two images
AddC	Image, x	Add integer x to each pixel
SubC	Image, x	Subtract integer x from each pixel
DivC	Image, x	Divide each pixel by integer x
MulC	Image, x	Multiply each pixel by integer x
Sqrt	Image	Take the square root of each pixel
Ln	Image	Take the natural log of each pixel
Mean	Image, x	Uses a sliding window of size x and replaces the centre pixel of the window with the mean of the window
Max	Image, x	As Mean, but takes the maximum value
Min	Image, x	As Mean, but takes the minimum value
Med	Image, x	As Mean, but takes the median value
MeanA	Image, 3 Int (x, y, size)	Returns the mean value of the pixels contained in a window of $size$, centred at x, y in the image
StDevA	Image, 3 Int (x, y, size)	Returns standard deviation
MaxA	Image, 3 Int (x, y, size)	Returns maximum value
MinA	Image, 3 Int (x, y, size)	Returns minimum value

These operations were selected based on their previous application in GP applied to image classification tasks [3]. See Algorithm 2 for an example program using a subset of these operations

memory, since fewer ineffective instructions, or *introns*, can be removed prior to program execution. In addition, shared memory implies that the order of program execution within a team now potentially impacts bidding outputs. In this work, the order of program execution within a team remains fixed, but future work could investigate mutation operators that modify execution order. Program variation operators are listed in Table 6.2, providing an overview of how evolutionary search is focused on particular aspects of program structure. In short, program length and content, as well as the degree of memory sharing, are all adapted properties.

6.3.2 Cooperative Decision-Making with Teams of Programs

Individual action-value programs have only one action pointer, and can therefore never represent a complete solution independently. A team of programs represents a complete solutions by grouping together programs that collectively map

Table 6.2 Parameterization of team and program populations

Parameter	Value	Parameter	Value
Team population			
R_{size}	1000	R_{gap}	50% of *Root* Teams
p_{md}, p_{ma}	0.7	ω	60
p_{mm}	0.2	p_{mn}, p_{ms}	0.1
Program population			
$Registers_{private}$	8	$maxProgSize$	100
$Registers_{shared}$	8	p_{atomic}	0.99
p_{delete}, p_{add}	0.5	p_{mutate}, p_{swap}	1.0

For the team population, p_{mx} denotes a mutation operator in which: $x \in \{d, a\}$ are the prob. of deleting or adding a program respectively; $x \in \{m, n\}$ are the prob. of creating a new program, changing the program action pointer, and changing the program shared memory pointer respectively. ω is the max initial team size. For the program population, p_x denotes a mutation operator in which $x \in \{delete, add, mutate, swap\}$ are the prob. for deleting, adding, mutating, or reordering instructions within a program. p_{atomic} is the probability of a modified action pointer referencing an atomic action

environmental observations (e.g. the game screen) to *atomic* actions (e.g. joystick positions). This is achieved through a bidding mechanism. In each timestep, every program in the team will execute, and the team then takes the action pointed to by the action-value program with the highest output. This process repeats at each timestep from the initial episode condition to the end of an episode. When the episode ends, due to a *GameOver* signal from the environment or an episode time constraint is reached, the team as a whole is assigned a fitness score from the environment (i.e. the final game score). Since decision-making in TPG is a strictly collective process, programs have no individual concept of fitness. Team variation operators may add, remove, or modify programs in the team, with parameters listed in Table 6.2.

In this work, new algorithmic extensions at the team level are twofold: (1) Teams are heterogeneous, containing action-value programs *and* image processing programs; and (2) All programs have access to shareable stateful memory. Figure 6.2 illustrates these extensions.

6.3.3 Compositional Evolution of Tangled Program Graphs

This section details how teams and programs are coevolved, paying particular attention to emergent hierarchical transitions. Parameters are listed in Table 6.2.

Evolution begins with a population of R_{size} teams, each containing at least one program of each type, and a max of ω programs in total. Programs are created in pairs with one shared memory bank between them (See left-hand-side of Fig. 6.2). Program actions are initially limited to task-specific (atomic) actions, Fig. 6.2. Throughout evolution, program variation operators are allowed to introduce actions that index other teams within the team population. To do so, when a program's action is modified, it may reference either a different atomic action or any team created in a previous generation. Specifically, the action set from which new program actions are sampled will correspond to the set of atomic actions, A, with probability p_{atomic}, and will otherwise correspond to the set of teams present from any previous generation. In effect, action pointer mutations are the primary mechanism by which TPG supports compositional evolution, adaptively recombining multiple (previously independent) teams into variably deep/wide directed graph structures, or *policy graphs*, right-hand-side of Fig. 6.2. The hierarchical interdependency between teams is established entirely through interaction with the task environment. Thus, more complex structures can emerge as soon as they perform better than simpler solutions. The utility of compositional evolution is empirically demonstrated in Sect. 6.4.2.

Decision-making in a policy graph begins at the root team (e.g. t_3 in Fig. 6.2), where each program in the team will produce one bid relative to the current state observation, $\overrightarrow{s}(t)$. Graph traversal then follows the program with the largest bid, repeating the bidding process for the same state, $\overrightarrow{s}(t)$, at every team along the path until an atomic action is reached. Thus, in sequential decision-making tasks, the policy graph computes one path from root to atomic action at every time step, where only a subset of programs in the graph (i.e. those in teams along the path) require execution.

As hierarchical structures emerge, only root teams (i.e. teams that are not referenced as any program's action) are subject to modification by the variation operators. As such, rather than pre-specify the desired team population size, only the number of root teams to maintain in the population, or R_{size}, requires prior specification. Evolution is driven by a generational GA such that the worst performing root teams (50% of the root population, or R_{gap}) are deleted in each generation and replaced by offspring of the surviving roots. After team deletion, programs that are not part of any team are also deleted. As such, selection is driven by a symbiotic relationship between programs and teams: teams will survive as long as they define a complementary group of programs, while individual programs will survive as long as they collaborate successfully within a team. The process for generating team offspring uniformly samples and clones a root team, then applies mutation-based variation operators to the cloned team, as listed in Table 6.2. Complete details on TPG are available in [20] and [19].

6.4 Empirical Study

The objective of this study is to evaluate heterogeneous TPG with shared temporal memory for object tracking in visual RL. This problem explicitly requires an agent to develop short-term memory capabilities. For an evaluation of TPG in visual RL with longer-term memory requirements see [32].

6.4.1 Problem Environments

In order to compare with previous results, we consider the Atari video game Break-out, for which the initial version of TPG failed to learn a successful policy [20]. Breakout is a vertical tennis-inspired game in which a single ball is released near the top of the screen and descends diagonally in either left or right direction, Fig. 6.3a. The agent observes this environment through direct screen capture, an 84×64 pixel matrix[1] in which each pixel has a colour value between 0 and 128. The player controls the horizontal movement of a paddle at the bottom of the screen. Selecting form 4 *atomic* actions in each timestep, $A \in \{Serve, Left, Right, NoAction\}$, the goal is to maneuver the paddle such that it makes contact with the falling ball, causing it to ricochet up towards the brick ceiling and clear bricks one at a time. If the paddle misses the falling ball, the player looses a turn. The player has three turns to clear two layers of brick ceiling. At the end of each episode, the game returns a reward signal which increases relative to the number of bricks eliminated. The primary skill in breakout is simple: the agent must integrate the location of the ball over multiple timesteps in order to predict its trajectory and move the paddle to the correct horizontal position. However, the task is dynamic and non-trivial because, as the game progresses, the ball's speed increases, its angle varies more widely, and the width of the paddle shrinks. Furthermore, *sticky actions* [27] are utilized such that agents stochastically skip screen frames with probability $p = 0.25$, with the previous action being repeated on skipped frames. Sticky actions have a dual purpose in the ALE: (1) artificial agents are limited to roughly the same reaction time as a human player; and (2) stochasticity is present throughout the entire episode of gameplay.

The Atari simulator used in this work, The Arcade Learning Environment (ALE) [5], is computational demanding. As such, we conduct an initial study in a custom environment that models only the ball tracking task in breakout. This "ball catching" task is played on a 64×32 grid (i.e. representing roughly the bottom 3/4 of the Breakout game screen) in which each tile, or pixel, can be one of two colours represented by the values 0 (no entity present) and 255 (indicating either the ball or paddle is present at this pixel location), Fig. 6.3b. The ball is one pixel large and is stochastically initialized in one of the 64 top-row positions at the start of each episode. The paddle is 3 pixels wide and is initialized in the centre of the bottom row. The ball will either fall straight down (probability $= 0.33$) or diagonally, moving one pixel down and one pixel to the left or right (chosen with equal probability at time $t = 1$) in each timestep. If a diagonally-falling ball hits either wall, its horizontal direction is reversed. The agent's objective is to select one of 3 paddle movements in each timestep, $A \in \{Left, Right, NoAction\}$, such that the paddle makes contact with the falling ball. The paddle moves twice as fast as the ball, i.e. 2

---

[1]This screen resolution corresponds to 40% of the raw Atari screen resolution. TPG has previously been shown to operate under the full Atari screen resolution [21]. The focus of this study is temporal memory, and the down sampling is used here to speed up empirical evaluations.

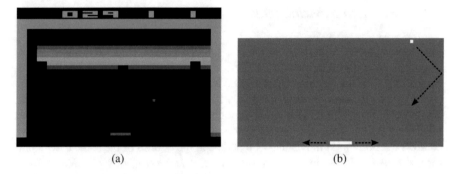

Fig. 6.3 Screenshots of the two video game environments utilized in this work. (**a**) Breakout. (**b**) Ball catching

pixels at a time in either direction. An episode ends when the ball reaches the bottom row, at which point the game returns a reward signal of 1.0 if the ball and paddle overlap, and 0 otherwise. As in Breakout, success in this task requires the agent to predict the trajectory of the falling ball *and* correlate this trajectory with the current position of the paddle in order to select appropriate actions.

6.4.2 Ball Catching: Training Performance

Four empirical comparisons are considered in the ball catching environment, with 10 independent runs performed for each experimental case. As discussed in Sect. 6.1, visual RL policies require a breadth of experience interacting with the problem environment before their fitness can be estimated with a sufficient degree of generality. As such, in each generation we evaluate every TPG policy in 40 episodes and let the mean episode score represent their fitness. Curves in Fig. 6.4 represent the fitness of the champion individual over 2000 generations, which is equivalent to roughly 12 h of wall-clock time. Each line is the median over 10 independent runs.

Figure 6.4a is the training curve for heterogeneous TPG with the capacity for shared stateful memory as described in Sect. 6.3. For reference, a mean game score of 0.65 indicates that the champion policy successfully maneuvered the paddle to meet the ball 65% of the time over 40 episodes.

Figure 6.4b is the training curve for TPG with shared memory but *without* image-processor programs. While the results are not significantly different than case (a), heterogeneous TPG did not hinder progress in any way. Furthermore, the single best policy in either (a) or (b) was heterogeneous and *did* make use of image-processor programs, indicating that the method has potential.

Figure 6.4c is the training curve for heterogeneous TPG without the capacity for *shared* memory. In this case, each program is initialized with a pointer to one *private* stateful register bank. Mutation operators are not permitted to modify memory

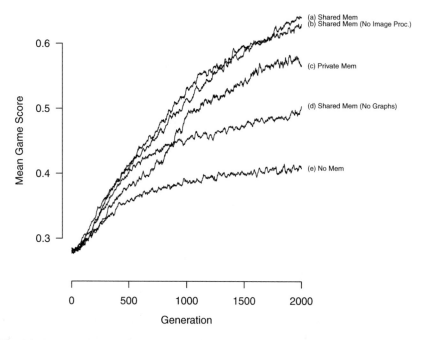

Fig. 6.4 Fraction of successful outcomes (Mean Game Score) in 40 episodes for the champion individual at each generation. Each line is the median over 10 independent runs for experimental cases (**a**)–(**e**). See Sect. 6.4.2 text for comparative details

pointers ($p_{ms} = 0$). The case *with* memory sharing exhibits significantly better median performance after very few generations ($\approx$100).

Figure 6.4d shows the training curve for heterogeneous TPG without the capacity to build policy graphs. In this case, team hierarchies can never emerge ($p_{atomic} = 1.0$). Instead, adaptive complexity is supported by allowing root teams to acquire an unbounded number of programs ($\omega = \infty$). The weak result clearly illustrates the advantage of emergent hierarchical transitions for this task. As discussed in Sect. 6.2.1, one possible explanation for this is the ability of TPG policy graphs to decompose the task spatially by defining an appropriate subset of inputs (pixels) to consider in each timestep, and to decompose the task temporally by identifying, storing, and reusing subsets of past experience through dynamic memory access. Without the ability to build policy graphs, input and memory indexing would be static, i.e. the same set of inputs and memory registers would be accessed in every timestep regardless of environmental state.

Figure 6.4e shows the training curve for heterogeneous TPG without the capacity for stateful memory. In this case, both private and shared memory registers are cleared prior to each program execution. This is equivalent to equipping programs with 16 *stateless* registers. The case with shared temporal memory achieves significantly better policies after generation $\approx$500, and continues to gradually discover increasingly high scoring policies up until the runs terminate at generation 2000.

This comparison clearly illustrates the advantage that temporal memory provides for TPG organisms in this domain. Without the ability to integrate observations over multiple timesteps, even the champion policies are only slightly better than random play.

6.4.3   Ball Catching: Solution Analysis

Section 6.4.2 established the effectiveness of heterogeneous TPG with shared temporal memory in a visual object tracking task (i.e. ball-catching). The following analysis confirms that champion policies rely on shared temporal memory to succeed at this task under test conditions. Box plots in Fig. 6.5 summarize the mean game score (over 30 episodes) for the single champion policy from 20 runs.[2] The distribution labeled 'Shared Mem' indicates that the median success rate for these 20 champions is ≈76%. The box labeled 'Mem (No Sharing)' summarizes scores for the *same* 20 policies when their ability to share memory is suppressed. The decrease in performance indicates that policies are indeed exploiting shared temporal memory in solving this task. The box labeled 'No Mem' provides test scores for these policies when all memory registers are stateless (i.e. reset prior to each program execution), again confirming that temporal memory plays a crucial role in the behaviour of these policies. Without temporal memory, the champion policies are often no better than a policy that simply selects actions at random, or 'Rand' in Fig. 6.5.

6.4.4   Atari Breakout

In this section, the most promising TPG configuration identified under the ball-catching task, or heterogeneous TPG with shared temporal memory, is evaluated in the Atari game Breakout (Sect. 6.4.1). The computational cost of game simulation in the ALE precludes evaluating each individual policy in 40 episodes per generation during training. In Breakout, each policy is evaluated in only 5 episodes per generation. This limits the generality of fitness estimation but is sufficient for a proof-of-concept test of our methodology in a challenging and popular visual RL benchmark. Figure 6.6 provides the training curves for 10 independent Breakout runs. In order to score any points, policies must learn to serve the ball (i.e. select the *Serve* action) whenever the ball does not appear on screen. This skill appears relatively quickly in most of the runs in Fig. 6.6. Next, static paddle locations (e.g. moving the paddle to the far right after serving the ball and leaving it there) can

---

[2]An additional 10 runs were conducted for this analysis relative to the 10 runs summarized in Fig. 6.4a.

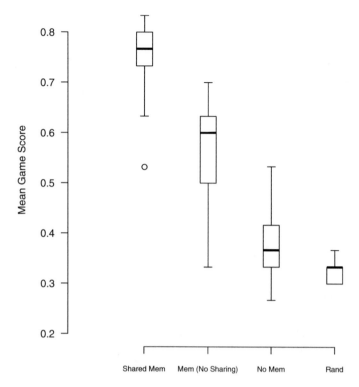

Fig. 6.5 Fraction of successful outcomes (Mean Game Score) in 30 test episodes for the champion policies from the case of heterogeneous TPG with shared temporal memory, Fig. 6.4a. Box plots summarize the results from 20 independent runs. See Sect. 6.4.3 text for details on each distribution

occasionally lead to ≈11–15 points. In order to score ≈20–50 points, policies must surpass this somewhat degenerate local optima by discovering a truly responsive strategy in which the paddle and ball make contact several times at multiple horizontal positions. Finally, policies that learn to consistently connect the ball and paddle will create a hole in the brick wall. When this is achieved, the ball can pass through all layers of brick and become trapped in the upper region of the world where it will bounce around clearing bricks from the top down and accumulating scores above 100.

Table 6.3 lists Breakout test scores for several recent visual RL algorithms. Previous methods either employed stateless models that failed to achieve a high score (TPG, CGP, HyperNeat) or side-stepped the requirement for temporal memory by using autoregressive, sliding window state representations. Heterogeneous TPG with shared memory (HTPG-M) is the highest scoring algorithm that operates directly from screen capture *without* an autoregressive state, and roughly matches the test scores from 3 of the 4 deep learning approaches that *do* rely on autoregressive state.

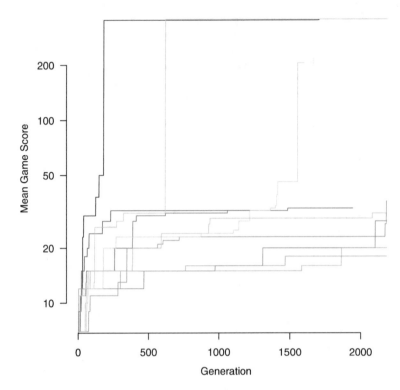

Fig. 6.6 Fitness (Mean Game Score) for the single champion policy in each of 10 independent Breakout runs. Scores are averaged over 5 episodes. For clarity, line plots show the fitness of the best policy discovered up to each generation

Table 6.3 Comparison of Breakout test scores (mean game score over 30 episodes) for state-of-the art methods that operate directly from screen capture

Human	Double	Dueling	Prioritized	A3C LSTM	TPG	HTPG-M	HyperNeat	CGP
31.8	368.9	411.6	371.6	766.8	12.8	374.2	2.8	13.2

Heterogeneous TPG with shared memory (HTPG-M) is the highest scoring algorithm that operates *without* an autoregressive, sliding window state representation. 'Human' is the score achieved by a "professional" video game tester reported in [30]. Scores for comparator algorithms are from the literature: Double [13], Dueling [37], Prioritized [31], A3C FF [28], A3C LSTM [28], TPG [20], HyperNeat [14], CGP [39]

6.5   Conclusions and Future Work

We have proposed a framework for shared temporal memory in TPG which significantly improves the performance of agents in a partially observable, visual RL problem with short-term memory requirements. This study confirms the significance of private temporal memory for individual programs as well as the added benefit of memory sharing among multiple programs in a single organism, or policy graph. No specialized program instructions are required to support dynamic memory access. The nature of temporal memory access and the degree of memory sharing among programs are both emergent properties of an open-ended evolutionary process. Future work will investigate this framework in environments with long-term partial observability. Multi-task RL is a specific example of this [20], where the agent must build short-term memory mechanisms *and* integrate experiences from multiple ongoing tasks. Essentially, the goal will be to construct shared temporal memory at multiple time scales, e.g. [18]. This will most likely require a mechanism to trigger the erosion of non-salient memories based on environmental stimulus, or active forgetting [11].

Supporting multiple program representations within a single heterogeneous organism is proposed here as an efficient way to incorporate domain knowledge in TPG. In this study, the inclusion of domain-specific image processing operators was not crucial to building strong policies, but it did not hinder performance in any way. Given the success of shared memory as a means of communication within TPG organisms, future work will continue to investigate how heterogeneous policies might leverage specialized capabilities from a wider of variety bio-inspired virtual machines. Image processing devices that model visual attention are of particular interest, e.g. [29, 33].

Acknowledgements Stephen Kelly gratefully acknowledges support from the NSERC Postdoctoral Fellowship program. Computational resources for this research were provided by Michigan State University through the Institute for Cyber-Enabled Research (icer.msu.edu) and Compute Canada (computecanada.ca).

References

1. A. Simon, H.: The architecture of complexity. Proceedings of the American Philosophical Society **106**, 467–482 (1962)
2. Agapitos, A., Brabazon, A., O'Neill, M.: Genetic programming with memory for financial trading. In: G. Squillero, P. Burelli (eds.) Applications of Evolutionary Computation, pp. 19–34. Springer International Publishing (2016)
3. Atkins, D., Neshatian, K., Zhang, M.: A domain independent genetic programming approach to automatic feature extraction for image classification. In: 2011 IEEE Congress of Evolutionary Computation (CEC), pp. 238–245 (2011)
4. Beattie, C., Leibo, J.Z., Teplyashin, D., Ward, T., Wainwright, M., Küttler, H., Lefrancq, A., Green, S., Valdés, V., Sadik, A., Schrittwieser, J., Anderson, K., York, S., Cant, M., Cain, A., Bolton, A., Gaffney, S., King, H., Hassabis, D., Legg, S., Petersen, S.: Deepmind lab. arXiv preprint arXiv:1612.03801 (2016)
5. Bellemare, M.G., Naddaf, Y., Veness, J., Bowling, M.: The arcade learning environment: An evaluation platform for general agents. Journal of Artificial Intelligence Research **47**, 253–279 (2013)
6. Bishop, C.M.: Pattern Recognition and Machine Learning (Information Science and Statistics). Springer-Verlag (2006)
7. Brameier, M., Banzhaf, W.: Linear Genetic Programming, 1st edn. Springer (2007)

8. Brave, S.: The evolution of memory and mental models using genetic programming. In: Proceedings of the 1st Annual Conference on Genetic Programming, pp. 261–266. MIT Press (1996)

9. Choi, S.P.M., Yeung, D.Y., Zhang, N.L.: An environment model for nonstationary reinforcement learning. In: S.A. Solla, T.K. Leen, K. Müller (eds.) Advances in Neural Information Processing Systems 12, pp. 987–993. MIT Press (2000)

10. Conrads, M., Nordin, P., Banzhaf, W.: Speech sound discrimination with genetic programming. In: W. Banzhaf, R. Poli, M. Schoenauer, T.C. Fogarty (eds.) Genetic Programming, pp. 113–129. Springer Berlin Heidelberg (1998)

11. Davis, R.L., Zhong, Y.: The Biology of Forgetting – A Perspective. Neuron **95**(3), 490–503 (2017)

12. Greve, R.B., Jacobsen, E.J., Risi, S.: Evolving neural turing machines for reward-based learning. In: Proceedings of the Genetic and Evolutionary Computation Conference 2016, GECCO '16, pp. 117–124. ACM (2016)

13. Hasselt, H.v., Guez, A., Silver, D.: Deep reinforcement learning with double q-learning. In: Proceedings of the Thirtieth AAAI Conference on Artificial Intelligence, AAAI' 16, pp. 2094–2100. AAAI Press (2016)

14. Hausknecht, M., Lehman, J., Miikkulainen, R., Stone, P.: A neuroevolution approach to general Atari game playing. IEEE Transactions on Computational Intelligence and AI in Games **6**(4), 355–366 (2014)

15. Haynes, T.D., Wainwright, R.L.: A simulation of adaptive agents in a hostile environment. In: Proceedings of the 1995 ACM Symposium on Applied Computing, SAC '95, pp. 318–323. ACM (1995)

16. Hintze, A., Edlund, J.A., Olson, R.S., Knoester, D.B., Schossau, J., Albantakis, L., Tehrani-Saleh, A., Kvam, P.D., Sheneman, L., Goldsby, H., Bohm, C., Adami, C.: Markov brains: A technical introduction. arXiv preprint 1709.05601 (2017)

17. Hintze, A., Schossau, J., Bohm, C.: The evolutionary buffet method. In: W. Banzhaf, L. Spector, L. Sheneman (eds.) Genetic Programming Theory and Practice XVI, Genetic and Evolutionary Computation Series, pp. 17–36. Springer (2018)

18. Jaderberg, M., Czarnecki, W.M., Dunning, I., Marris, L., Lever, G., Castañeda, A.G., Beattie, C., Rabinowitz, N.C., Morcos, A.S., Ruderman, A., Sonnerat, N., Green, T., Deason, L., Leibo, J.Z., Silver, D., Hassabis, D., Kavukcuoglu, K., Graepel, T.: Human-level performance in 3d multiplayer games with population-based reinforcement learning. Science **364**(6443), 859–865 (2019)

19. Kelly, S.: Scaling genetic programming to challenging reinforcement tasks through emergent modularity. Ph.D. thesis, Faculty of Computer Science, Dalhousie University (2018)

20. Kelly, S., Heywood, M.I.: Emergent solutions to high-dimensional multitask reinforcement learning. Evolutionary Computation **26**(3), 347–380 (2018)

21. Kelly, S., Smith, R.J., Heywood, M.I.: Emergent Policy Discovery for Visual Reinforcement Learning Through Tangled Program Graphs: A Tutorial, pp. 37–57. Springer International Publishing (2019)

22. Kober, J., Peters, J.: Reinforcement learning in robotics: A survey. In: M. Wiering, M. van Otterio (eds.) Reinforcement Learning, pp. 579–610. Springer (2012)

23. Koza, J.R., Andre, D., Bennett, F.H., Keane, M.A.: Genetic Programming III: Darwinian Invention & Problem Solving, 1st edn. Morgan Kaufmann Publishers Inc. (1999)

24. Krawiec, K., Bhanu, B.: Visual learning by coevolutionary feature synthesis. IEEE Transactions on Systems, Man, and Cybernetics, Part B (Cybernetics) **35**(3), 409–425 (2005)

25. Lalejini, A., Ofria, C.: What Else Is in an Evolved Name? Exploring Evolvable Specificity with SignalGP. In: W. Banzhaf, L. Spector, L. Sheneman (eds.) Genetic Programming Theory and Practice XVI, pp. 103–121. Springer International Publishing (2019)

26. Lughofer, E., Sayed-Mouchaweh, M.: Adaptive and on-line learning in non-stationary environments. Evolving Systems **6**(2), 75–77 (2015)

27. Machado, M.C., Bellemare, M.G., Talvitie, E., Veness, J., Hausknecht, M., Bowling, M.: Revisiting the arcade learning environment: Evaluation protocols and open problems for general agents. J. Artif. Int. Res. **61**(1), 523–562 (2018)
28. Mnih, V., Badia, A.P., Mirza, M., Graves, A., Lillicrap, T., Harley, T., Silver, D., Kavukcuoglu, K.: Asynchronous methods for deep reinforcement learning. In: M.F. Balcan, K.Q. Weinberger (eds.) Proceedings of The 33rd International Conference on Machine Learning, *Proceedings of Machine Learning Research*, vol. 48, pp. 1928–1937. PMLR (2016)
29. Mnih, V., Heess, N., Graves, A., Kavukcuoglu, K.: Recurrent models of visual attention. In: Proceedings of the 27th International Conference on Neural Information Processing Systems - Volume 2, NIPS'14, pp. 2204–2212. MIT Press (2014)
30. Mnih, V., Kavukcuoglu, K., Silver, D., Rusu, A.A., Veness, J., Bellemare, M.G., Graves, A., Riedmiller, M., Fidjeland, A.K., Ostrovski, G., Petersen, S., Beattie, C., Sadik, A., Antonoglou, I., King, H., Kumaran, D., Wierstra, D., Legg, S., Hassabis, D.: Human-level control through deep reinforcement learning. Nature **518**(7540), 529–533 (2015)
31. Schaul, T., Quan, J., Antonoglou, I., Silver, D.: Prioritized experience replay. In: International Conference on Learning Representations (2016)
32. Smith, R.J., Heywood, M.I.: A model of external memory for navigation in partially observable visual reinforcement learning tasks. In: L. Sekanina, T. Hu, N. Lourenço, H. Richter, P. García-Sánchez (eds.) Genetic Programming, pp. 162–177. Springer International Publishing (2019)
33. Stanley, K.O., Miikkulainen, R.: Evolving a Roving Eye for Go. In: T. Kanade, J. Kittler, J.M. Kleinberg, F. Mattern, J.C. Mitchell, M. Naor, O. Nierstrasz, C. Pandu Rangan, B. Steffen, M. Sudan, D. Terzopoulos, D. Tygar, M.Y. Vardi, G. Weikum, K. Deb (eds.) Genetic and Evolutionary Computation — GECCO 2004, vol. 3103, pp. 1226–1238. Springer Berlin Heidelberg, Berlin, Heidelberg (2004)
34. Sutton, R.R., Barto, A.G.: Reinforcement Learning: An introduction. MIT Press (1998)
35. Teller, A.: Turing completeness in the language of genetic programming with indexed memory. In: Proceedings of the First IEEE Conference on Evolutionary Computation. IEEE World Congress on Computational Intelligence, vol. 1, pp. 136–141 (1994)
36. Wagner, G.P., Altenberg, L.: Perspective: Complex adaptations and the evolution of evolvability. Evolution **50**(3), 967–976 (1996)
37. Wang, Z., Schaul, T., Hessel, M., Van Hasselt, H., Lanctot, M., De Freitas, N.: Dueling network architectures for deep reinforcement learning. In: Proceedings of the 33rd International Conference on International Conference on Machine Learning - Volume 48, ICML'16, pp. 1995–2003. JMLR.org (2016)
38. Watson, R.A., Pollack, J.B.: Modular interdependency in complex dynamical systems. Artificial Life **11**(4), 445–457 (2005)
39. Wilson, D.G., Cussat-Blanc, S., Luga, H., Miller, J.F.: Evolving simple programs for playing atari games. In: Proceedings of the Genetic and Evolutionary Computation Conference, GECCO '18, pp. 229–236. ACM (2018)

Chapter 7
The Evolution of Representations in Genetic Programming Trees

Douglas Kirkpatrick and Arend Hintze

7.1 Introduction

One of the most promising approaches in the quest for general purpose artificial intelligence (AI) is neuroevolution [9, 35, 39]. The idea is to use a genetic algorithm to optimize the AI that controls an embodied agent to solve one or many cognitive tasks. Many different AI systems have been proposed for this task, such as Artificial Neural Networks (ANNs) and their recurrent versions (RNNs) [31], Genetic Programs (GPs) [16], Cartesian Genetic Programming (CGP) [22], Neuro Evolution of Augmenting Topologies (NEAT) [36], and Markov Brains (MBs) [11], among many others. One of the most pressing problems is that AIs have difficulty generating internal models about the environment they act in. Humans seem to be particularly good at creating these models for themselves, but we struggle greatly when it comes to imbuing artificial systems with such models, also called representations.[1] It even has been argued that one should not even try to do so and

---

[1]Observe that the term representations in computer science sometimes also refers to the structure of data, or how an algorithm, for example, is encoded. We mean neither, but instead use the term representation to be about the information a cognitive system has about its environment as

D. Kirkpatrick (✉)
Department of Computer Science and Engineering, BEACON Center for the Study of Evolution in Action, Michigan State University, East Lansing, MI, USA
e-mail: kirkpa48@msu.edu

A. Hintze
Department of Integrative Biology, Michigan State University, East Lansing, MI, USA

Department of Computer Science and Engineering, BEACON Center for the Study of Evolution in Action, Michigan State University, East Lansing, MI, USA
e-mail: hintze@msu.edu

© Springer Nature Switzerland AG 2020
W. Banzhaf et al. (eds.), *Genetic Programming Theory and Practice XVII*, Genetic and Evolutionary Computation, https://doi.org/10.1007/978-3-030-39958-0_7

instead create "intelligence without representation" [6]. However, we found earlier that neuroevolution is very capable of solving this problem of creating structures to store representations.

To prove that a system indeed has representations, we previously introduced an information theoretic measure R that quantifies exactly how much information the internal states of a system store about its environment independent of the information coming directly from its sensors [19]. Over the course of evolution Markov Brains stored more information about the environment necessary to solve complex tasks. They did so by associating specific hidden nodes with different environmental concepts. These sparse and distributed representations are believed to be similar to how humans for example store information: not all neurons fire, but certain areas perform computations based on specific firing patterns of other neurons. We also found that RNNs, when evolved to perform a task that requires memory, form less sparse representations and smear the information across all their hidden states in contrast to MBs.

However, it remains to be shown that genetic programming (GP) or systems similar to GP (such as CGPs) also evolve to solve tasks that require representations, to what extent they evolve the ability to form representations, and to what degree they smear their representations.

While it is academically interesting to find how much a system knows about its environment, quantifying representations has an application for genetic algorithms. Similar to multiple objective optimization [40] which improves at least two different qualities of a system at the same time, one can not only use the performance of an agent but also use the degree to which the system has developed representations to define the mean number of offspring in a genetic algorithm (GA) for augmentation of the GA [32]. While it has been shown how this helps to optimize Markov Brains and RNNs during evolution, again we do not know if the same is true for genetic programming systems. Lastly, representations can not only be measured for the whole system but also for individual sub-components. By testing all sub-components versus all concepts in the environment, we can pinpoint where these representations are exactly stored in the system. This results in a matrix defining the amount or information each element has about each concept. We found earlier [12] that substrates differ in how distributed or condensed these representations are. Markov Brains create sparsely distributed representations, while RNNs smear the knowledge about the environment across all nodes. In addition, we found that for Markov Brains, RNNs, and LSTMs [14] that smearedness and robustness correlate negatively. Perhaps the difference in representation structure between Markov Brains and RNNs is not surprising when one considers the topological features of these systems. RNNs receive their inputs and sequentially propagate this information through one layer after the other by connecting each node from one layer to all nodes to the next. While it obviously ensures that the information can reach all components it presumably also prevents representations from becoming

---

defined in Marstaller et al. [19]. The term representation, as we use it, is adapted from the fields of psychology and philosophy.

local and sparse. Markov Brains, on the other hand, start with sparse connections and only retain additional ones if there is an evolutionary benefit at the time of addition. This results in sparsely interconnected graphs, which might be the reason why these kinds of networks evolve to have sparse and condensed representations.

Genetic programming trees present a great tool to evolve and fit almost arbitrary mathematical functions to data. They can take a lot of different inputs, perform computations on them, and due to their tree structure, condense the result of all those computations into a single result at the root of the tree. It stands to reason that this topology provides an advantage to the question of condensing representations as well. These trees can be as wide as the input layers of an RNN, but do not have the property of further distributing the information at every step (see Fig. 7.1 for a comparison of topologies and representational smearedness across different kinds of computational substrates).

The genetic programming trees, however, have the disadvantage that they do not have an intuitive way to implement recurrence. One way of dealing with the exact problem of forming memory has been addressed by adding *indexed memory* and additional computational nodes into the tree that can read from and write to said memory [37]. Similarly, the nodes that perform the read and write operations, or push and pop in the case of a memory stack, can themselves be evolved by genetic

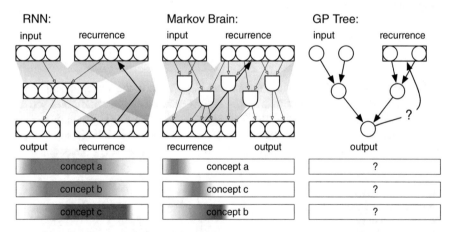

Fig. 7.1 Three different cognitive architectures—RNN, Markov Brain, and GP tree—and how they distribute or condense representations. RNNs have a joint input and recurrence layer, feed into arbitrary many hidden layers of arbitrary size, compute outputs which are also recurred back to the input layer. These networks tend to smear representations about different context over all recurrent nodes. The red, blue, and green color bars illustrate that. Markov Brains use inputs and hidden states as inputs to the logic gates or to other kinds of computational units. The result of the computations creates recurrent states and outputs. These networks tend to form sparse condensed representations, illustrated by narrower red, green, and blue bands. GP trees use inputs and potentially recurring information and combine them over several computational nodes into a result. For these structures, it is not clear to which degree they are smearing or condensing representations. It is also not obvious how to recur them, illustrated by the "?"

programming trees. These trees implement commands that can then be included into linear forms of genetic programming [18]. Finally, the genetically evolvable push programming language [33] provides a more complex solution for memory stored in stacks, specifically for linear genetic programs. While these methods all solve the problem of adding memory to the system, they essentially deviate from the idea of using the property of the tree to potentially condense information. Therefore, we are testing two alternative ways to create recurrence in genetic programming trees. We test if, as a consequence, the tree topology provides any advantage with respect to the ability to evolve sparsely condensed or smeared representations.

7.2 Material and Methods

7.2.1 Representations and the Neuro-Correlate R

The neuro-correlate R seeks to quantify the amount of information an agent has about its environment that is not currently provided by its sensors. The information-theoretic Venn diagram (see Fig. 7.2) illustrates the relation between environment or world states (W), memory or brain states (B), and sensor states (S) and defines R to be quantified as:

$$R = H(W : B|S) = H(W : B) - I(W : B : S) . \tag{7.1}$$

In order to actually measure R one needs to record the states of all three random variables. In the case of Markov Brains, this means the brain and sensor states are just the state of the hidden and input nodes. The world states, however, are not that obvious. In fact, they need to be defined and well-chosen by the experimenter and depend highly on the environment. Here we use a block catching

Fig. 7.2 Venn diagram of entropies and information for the three random variables W, S, and B, describing the world, sensor, and agent internal (brain) states. The representation $R = H(W : B|S)$ is shaded

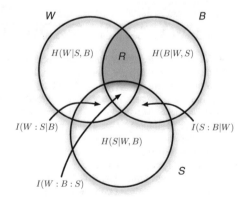

task (active categorical perception, see Sect. 7.2.3) and number comparison task (number discrimination, see Sect. 7.2.4), and thus use the relevant concepts of these environments to define the world states. Here, for the block catching task, we chose the size of the block, the direction of the block falling, and whether the block is to the left or right of the agent to be the world states. This is different from Marstaller [19] where they also measured if the block is currently above or next to the agent. It was observed that agents typically do not evolve representations about hitting or missing the block, and thus this state was omitted. For the number comparison task, we chose the actual numbers to be the world states. As such, we ask to what degree the agent evolves representations about each number explicitly. In addition, we also defined a state to represent whether the first or second number was larger, to see if the agent is able to represent the difference between the two numbers.

With regards to the actual computation of R, a reduction was found that decreases the number of steps needed to compute R. Where the original Eq. (7.1) expanded in practice to

$$R = H(W) + H(B) + H(S) - H(W, B, S) - I(S : W) - I(S : B) , \qquad (7.2)$$

we found a reduction to only four entropy calculations:

$$R = H(S, W) + H(S, B) - H(S) - H(W, B, S) . \qquad (7.3)$$

The proof for this reduction is as follows:

Proof We start with the original formula for R from [19]:

$$R = H_{Corr} - I(S : W) - I(S : B),$$

where H_{Corr} is defined as

$$H_{Corr} = H(W) + H(B) + H(S) - H(W, B, S).$$

This expands to

$$R = H(W) + H(B) + H(S) - H(W, B, S) - I(S : W) - I(S : B).$$

We then use the formula for conditional information from [19]

$$I(X : Y) = H(X) + H(Y) - H(X, Y)$$

to expand

$$I(S : W) = H(S) + H(W) - H(S, W)$$

and

$$I(S:B) = H(S) + H(B) - H(S, B).$$

Substituting these leads to

$$R = H(W) + H(B) + H(S) - H(W, B, S) - (H(S) + H(W) - H(S, W))$$
$$- (H(S) + H(B) - H(S, B)).$$

By distributing the negative signs we get

$$R = H(W) + H(B) + H(S) - H(W, B, S) - H(S)$$
$$- H(W) + H(S, W) - H(S) - H(B) + H(S, B).$$

The positive and negative entropies cancel, leaving

$$R = -H(S) - H(W, B, S) + H(S, W) + H(S, B).$$

A slight reordering gives the final equation,

$$R = H(S, B) + H(S, W) - H(S) - H(W, B, S).$$

As the reduced Eq. (7.3) requires fewer mathematical operations than the original equation (7.2), there is less chance for computational error and round-off in the reduced equation, which makes it more accurate in practice than the original. In addition, the reduced equation is more computationally efficient to calculate.

7.2.2   Smearedness of Representations

While the neuro-correlate R calculates the amount of mental representations as a whole, it does little to examine the structure or layout of said representations. Recent work [12] proposes a new measure, referred to as smearedness, to work around this deficiency. Smearedness quantifies the amount of representation that each memory state in the brain has about each individual concept in the environment, and takes a pairwise minimum to see how much the representations are spread across the different nodes. The equation for smearedness is seen in Eq. (7.4), where M is the measure of the representation in a specific node and concept, taken over all nodes i and for all combinations of concepts j and k.

$$S_N = \sum_i \sum_{j>k} min(M_{ji}, M_{ki}) \tag{7.4}$$

7.2.3   Active Categorical Perception Task

In order to augment the performance of a genetic algorithm using the neuro-correlate R the environmental states need to be described well. Since the neuro-correlate R has already been shown to correlate positively with performance on the active categorical perception task [19], and been shown to boost performance in learning the task when the GA was augmented with R [32], the same task was used here as well to allow for a direct comparison. Augmentation is done by multiplying the attained score for the task by the amount of representations the agent has about the environment (see Sect. 7.2.11).

In this task [2, 3] an agent who can move left and right has to either catch or avoid a block that is falling towards it. The agent has 4 sensors, split into two groups of 2 sensors, separated by a 2 or 3 unit space between them. Blocks of various size are dropped one at a time, with the agent having 34 time steps to catch or avoid the block. The agents in the original experiments were evolved to catch blocks of width 2, and avoid blocks of width 4. We extended the task to try different combinations of block sizes. The blocks move 1 unit to the right or left on any given time step. On each time step, the agent receives the sensory input and can move 1 unit to the right or to the left. The agent is expected to determine the size of the block, and its direction of movement, so that it can decide whether to avoid or catch the specific block being dropped. The agents were tested over both block types and all possible permutations of starting positions relative to the sensors and block movement patterns.

Due to the configuration of the sensors and the relative location of the agent to the block, agents can only very rarely decide if a block is large or small without movement and thus need to navigate to be directly under the block—hence the "active" in active categorical perception. This can only be done if the block was observed before for at least two updates in a row, since otherwise the direction of the fall could not be determined. All this information needs to be integrated into a decision about catching or avoiding the block. It is this integration of sensor information over multiple time steps that requires agents to first evolve memory and with it representations in order to make proper decisions.

7.2.4   Number Discrimination Task

In addition to the active categorical perception task, recent work [15] has identified a different task in which we can also investigate the role of representations and smearedness in the neuroevolutionary process. This task, referred to as the Number Discrimination Task, has agents receive two values in sequence, and require the agent to identify the larger one. This task is inspired by previous biological [21, 24] and psychological [20] studies. When adapted for use *in silico*, the agents are presented with every possible pair of numbers between 0 and 5, with the values being shown as bit strings where the number of ones is the value the agent must comprehend (e.g., 00000 is 0, 01000 is 1, 11000 is 2, etc.). Agents are additionally presented with all possible permutations of each pair (e.g. 10001 is equivalent to 01100 and both are used to represent 2). This task requires agents to store the first number and then perform a comparison to the second. The agents are first shown one value, then they are given 3 updates to process the information, followed by a second value and an additional 3 updates to process the information, before being required to make a determination of which number is larger. The world states used in calculating R are defined by the individual numbers used, and the information of whether the first or second value presented is larger.

7.2.5   The Perception-Action Loop for Stateful Machines

Genetic programming has been used before to control robots in virtual [10, 29, 30] as well as in actual environments [26–28]. In order to create a meaningful controller, the evolved machine or program needs to map the sensor inputs it receives to motor outputs which control the actions of the robot (see Fig. 7.3). A purely reactive machine would have a direct mapping of inputs to outputs, while a stateful machine would use hidden states (memory) to store information from the sensors to use it for later decisions.

Previous approaches used genetic programming methods to evolve computer-like code running on register machines [25]. Some other approaches encoded the program controlling the robot as a tree of binary operations executed on only the sensor inputs [17]. While it is these kinds of tree-like encodings that we seek to investigate here, this specific approach did not use hidden states and

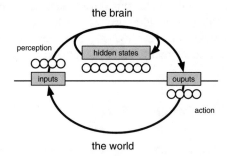

Fig. 7.3 Illustration of the perception-action loop. The world is in a particular state and can be perceived by the agent. This perception defines specific input states. These states now allow the brain (computational substrate) to perform a computation, which leads to new output states. These states, in turn, cause the agent to act in the environment and thus can change the world state. In case the agent needs to remember information about the past, it has to have the ability to store these memories somewhere. Thus, any machine that needs to be stateful has to have internal (here hidden states) that can be set depending on the current input and the current hidden states. This scheme of a brain resembles the structure of a recurrent neural network, but can be generally applied to all sort of computational machines that have to be stateful and act in an environment

thus created purely reactive machines. Adding memory has been done differently before [18, 33, 37]. However, we think that the tree structure is of particular importance. This structure allows the information from all sensor and hidden states to be funneled together into a coherent representation, like all the leaves of a tree feed into the root (see Fig. 7.1 for an illustration). Although intuition may indicate that this structure is correct, we must empirically test the viability of these structures. This does not imply that other substrates, for example linear programming structures or other memory models, cannot perform such functions. As such, the task here is to use genetic programming to encode tree like computational structures that can take the inputs of the machine and the hidden states into account to create new outputs while also setting hidden states in the process. A typical recurrent neural network [31] does this by extending the input and output layer of an artificial neural network and then recurring the additionally computed output states back to the inputs (see Fig. 7.1 RNN). This is exactly the way how Cartesian genetic programming [23] would be used to create systems that can take advantage of hidden states. Similarly, Markov Brains [11] use the same concept of recurrence to store hidden states (see Fig. 7.1 Markov Brain). In fact, when Markov Brains use the same nodes that CGPs use [13], they become very similar to each other. This kind of recurrence is very capable of creating systems that are stateful and have representations that are condensed, but again are not necessarily tree-like, and rather are more arbitrarily connected networks. Beyond using Markov Brains equipped with the computational nodes found in CGPs we use two other methods to create genetic programming trees that are stateful. GP-forests use one tree per hidden and output state and can use default values as well as inputs and hidden states from the last update.[2] GP-vector-trees are generic trees but instead of performing computations on single continuous variables, they execute vector operations. The vector supplied contains the hidden states of the last update as well as the current inputs, and part of the vector defines the output computed. We assume that this obvious extension of normal variables to vectors has certainly been done previously, but we can not find an explicit reference for this approach. Generally speaking, here we introduce one form of recurrence into genetic programming tree structures. The programming language push [33] should in principle allow for the same structures to emerge. However, these structures would need to arise by chance, in comparison to the models that we introduce here which achieve the desired structural goals by their definition.

7.2.6 Markov GP Brains Using CGP Nodes

We used Markov Brains extensively on similar tasks and also to introduce and confirm our definitions of representations [19]. In a nutshell, a genome is used to

---

[2]Inspired by the multiple trees used to encode memory in Langdon [18].

encode the computational units of a Markov Brain. The genome defines the function and connectivity of these units, which can connect to input, output, and hidden states. The connections allow the computational units to read from these states and compute new outputs and new hidden states. As such a Markov Brain defines the update function of a state vector from time point t to $t + 1$. Some values of this vector are inputs, some outputs, the rest hidden states. This is very similar to how an RNN is constructed. The number of hidden states in a Markov Brain or RNN can be chosen by the experimenter arbitrarily. Generally speaking, too many states slows evolution down, too few states makes evolving the task impossible. The number of hidden states used here (8) was found to be relatively optimal for the task at hand [19]. The computational units a Markov Brain uses were originally probabilistic and deterministic logic gates. As such, a Markov Brain defines a hidden Markov Model conditional on an input sequence [4]—hence the name Markov Brain. For a more detailed description of Markov Brains see Hintze et al. [11]. In this work, we use Markov Brains with CGP gates to approximate a CGP network. This creates a dynamic, GP-like structure that can be contrasted with the more rigid tree structures.

7.2.7 Genetic Encoding of GP Brains in a Tree-Like Fashion

Genetic programming trees and also Cartesian Genetic Programming define the computations they perform by identifying computational nodes, their connections, and how they receive inputs. When evaluated, they take those inputs to perform all computations defined by their structure and return a solution. When mutated, the nodes and inputs can change, as well as rewiring between the components can occur. In addition, when for example genetic programming trees are recombined, sub-branches of the trees are identified and exchanged between the trees [1]. Since mutations directly identify the components to be changed, this could be called a *direct encoding*. An alternative to this approach are *indirect encoding* schemes which can sometimes provide an advantage during evolutionary adaptation, such as the hyper-geometric encoding for NEAT [7]. Markov Brains, for example, use a type of indirect encoding where a genome specifies genes, and these genes then define the computational components and their connectivity. Mutations happen to the genome and not on the components themselves. The genetic programming substrates used here are defined in a similar *genetic encoding* scheme. We use the term *genetic encoding* to imply that it is not entirely direct nor as indirect as other systems might be.

Here different kinds of computational structures (GP-Forest and GP-Vector Brains) are defined by such a genetic encoding scheme. Starting from a root node, new nodes have to be sequentially added. Therefore, a linear genome (vector of bytes) is scanned for start codons (a set of two numbers) which define the start of a gene. The numbers following such a start codon are then used to define the mathematical operation of a node, and how this node is inserted into the genetic programming tree (see Fig. 7.4). Observe that the order of genes in the genome

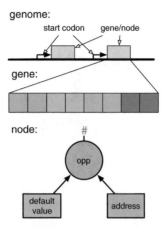

Fig. 7.4 Schematic overview of the genetic encoding for computational nodes. The genome is a sequence of numbers (bytes), and specific subsets of numbers (42, 213) define start codons. The sequence of numbers behind the start codon define a gene. As such, a gene stores all information necessary to define all properties of a computational node as it is needed to function within a genetic programming tree. Specifically: the computation the node has to perform (orange), the default for its left input (green), the default for its right input (blue), and the address where it needs to be inserted (purple). Observe that the defaults can be specific values (as seen for example for the left input of the node), or the address of a hidden node as seen for example in the right input of the node

matters: genes found earlier encode nodes that will be inserted into a tree earlier. Mutations affect the genome and can not only shuffle genetic material around but also insert and delete whole parts. As such, genes can disappear or be created *de novo*. This means, that at the time when a node is created by reading the gene defining it, it is not known if other nodes will be appended to it later or not. Since each node in our implementations has two possible inputs, each gene encodes default values or addresses for hidden states or input values. If another node becomes added later it replaces the default value with it's own output.

Since nodes are sequentially added, they always get appended to an already existing node. To know where, each gene also encodes a continuous number [0.0, 1.0] that specifies where in the tree the node has to be added (see Fig. 7.5 for an illustration of that process).

This scheme applies to both GP-Forest and GP-Vector. However, what each gene encodes differs with respect to the function of the node or the specific default values or addresses.

7.2.8   GP-Forest Brain

Any kind of computational system controlling an agent that needs to be stateful, cannot just map outputs to inputs but instead has to rely on hidden states which

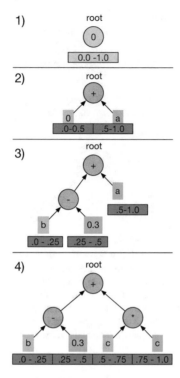

Fig. 7.5 Illustration of tree growth by adding nodes. (1) In the beginning the tree is defined as an empty root, that defaults to an output (here 0.0). (2) Once the genome is parsed and a new gene that encodes a node has been found, the node replaces the root node. The left and right inputs now define two ranges [0.0, 0.5] and]0.5, 1.0]. (3) Each new node to be added has a specific address (a value from [0.0, 1.0] as defined by the gene, see Fig. 7.4 the purple component of the gene). This number defines which default input to replace. The added node itself subdivides the range it connected to, in order to allow new nodes to again be added. (4) The concept of adding nodes and subdividing the connection ranges continues until the tree is built. In this way, a new node will always have a place in the tree. However, since nodes are added sequentially, deletions of genes/nodes will affect the topology greatly, but will not make it impossible to add nodes

are also sometimes called recurrent states. In the case of an RNN, for example, extra outputs that the network produces are just mapped back as additional inputs. Similarly, Markov Brains use inputs, outputs, hidden states upon which all computations are performed. While the inputs are generated by the environment the Markov Brain generates the outputs and the new hidden states to complete a perception-action loop. In order to allow for the same computational complexity in a genetic programming tree, we need to somehow make it recurrent. Conventional trees compute a single output based on a selection of inputs. However, this would allow us to only create systems without hidden states. Here we use, what we call a genetic programming forest (GP-Forest), where each output, as well as each hidden state, is computed by an individual genetic programming tree (hence the forest

Table 7.1 The possible
mathematical operations of
each node in a GP-Forest
brain

Operation	Return value
NOP	0.0
LEFT	L
RIGHT	R
ADD	L+R
SUB	L−R
MUL	L*R
DIV	L/R, 0.0 if R==0.0
EQU	1.0 if L==R, else 0.0
NEQU	1.0 if L!=R, else 0.0
LOW	1.0 if L<=R, else 0.0
HIGH	1.0 if L>=R, else 0.0
LOW	1.0 if L<R, else 0.0
+16	All 16 possible Boolean logic
LOGIC	operations performed on L and R

Each node has two possible inputs: L and R,
performs a computation on them, and returns the
value of said computation. The last category of
LOGIC consists of 16 independent binary logic
operations (e.g., AND, NAND, NOR, XOR, OR,
etc.) performed on the L and R input

analogy). Each tree has access to all inputs and hidden states and computes new
outputs and new hidden states at every update. The tree itself does not allow for
cycles. However, after each update of the entire tree, the new hidden states as well as
new inputs are again made available to the tree. This is what creates the recurrence,
which happens at the time step boundaries. For a list of computations that each node
can potentially compute see Table 7.1. The number of hidden states plus the number
of output nodes to be computed defines the number of trees needed to be specified.
To allow for a proper comparison of GP-forests with Markov Brains the number of
hidden states used was kept the same in all brains (Markov Brain, GP-forests, and
GP-Vector Brain; see below).

The genetic encoding specifies for each newly added node which of the trees it
should be appended to as well as the location in the tree (see Fig. 7.5 to see how
nodes are added, see Fig. 7.6 for a structural overview of a GP tree brain).

7.2.9   GP-Vector Brain

While genetic programming forests can deal with the hidden state problem by
adding many trees, one for each new state to be computed, GP-Vector brains try
to solve this problem differently. We assume that the input states, output states, and
hidden states form a continuous vector. At each update of the brain, a new vector

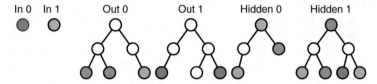

Fig. 7.6 Illustration of a GP-Forest. In this example, we assume the system to have two input nodes (red and orange), two output nodes (white), and two hidden nodes (green and blue). When decoding the genome, nodes can get added to hidden and output nodes sequentially. Each node can either use a default input (white) or use the value specified by an input or hidden node. For example, the tree added to the output node 1 reads from the orange input node, and twice from the blue hidden node. When a new brain state has to be computed, all trees (outputs, hidden) are evaluated simultaneously, and consequently a new set out outputs and hidden states become defined, based on previous inputs, and hidden states

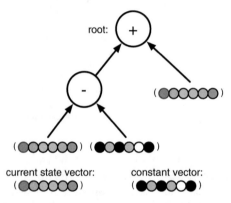

Fig. 7.7 Illustration of a GP-Vector Brain. This brain is structurally identical to conventional genetic programming trees, except that it can execute computations on vectors instead of single values. The inputs to nodes can be the state vector, including all inputs (red and orange) and all hidden states (blue to green). Alternatively, nodes can use default vectors as inputs. These defaults are defined by the genes encoding each node

has to be computed. Therefore, we extend the genetic programming paradigm that conventionally allows nodes to be mathematical operators for single variables to now compute vectors. Each node is now defining a vector operation, and the inputs of each node can be constant vectors or the current state vector. Regardless, such a tree of vector operations will result in a newly computed vector which itself contains the values of the current hidden state as well as the outputs of the agent's brain. Inputs from the environment are fed into this vector before each update of the brain. Each node is again encoded by a gene which provides the mathematical operation to be performed, where in the tree it gets added, and it defines if the default input is the state vector, or defines an individual constant vector. Remember that when a node is added, it replaces one of the defaults of the current node with its own output (see Fig. 7.7 for a structural overview of a GP-Vector brain).

Each node can now perform vector operations on its inputs and will return a vector as a result of that computation. However, such vector operations are often computationally expensive, and maybe not always necessary. Thus, nodes are also allowed to perform operations on the individual elements of the vectors. Each gate uses either the result of other nodes, a default vector of zeroes, or the state vector. These vectors are called L and R to capture the idea that each node in the tree can have two inputs, one from a left one from a right branch. If an addition on these vectors is performed the resulting output O is just $L + R$. Which means that every element O_i is the result of $L_i + R_i$ for all i. To allow for individual elements to be manipulated by each node, all operations are also defined as "point" operations. An $ADD_{i,j}$ would now compute an addition only on the i element of vector L and the j element of vector R. The result would be $O_i = L_i + R_j$ while all other elements of O would be those defined by L (For all operations allowed in the GP-Vector Brain, see Table 7.2).

The size of the vector is defined by the number of inputs, outputs, and hidden states. While the number of input and output states is defined by the task, the number of hidden states for this brain was set to be equal to the number of hidden states in the Markov Brain, to allow all brains to be directly comparable.

Table 7.2 The allowed mathematical operation for the nodes of a GP-vector-brain

Operation	Return value
ADD	$O = L + R$
SUB	$O = L - R$
MUL	$O = L * R$
DIV	$O = L/R$
SIN	$O = \sin L$
COS	$O = \cos L$
TANH	$O = \tanh L$
MIN	$O = \min L$
MAX	$O = \max L$
MEAN	$O = 0.5(L + R)$
ABS	$O = \mid L \mid$
NEG	$O = \mid -L \mid$
LOGIC	$O_i = \text{logic}(L_i, R_j)$

Observe that all operations are either performed on all elements of the L and R vector resulting in an output vector O, or on individual elements of L and R. Only the LOGIC operation is only defined as "point" operation and not on the entire vector

7.2.10 Evolutionary Process

Each of the agent populations with different brain types was evolved for 40,000 generations in populations of 100 organisms. The start population was initialized with random genomes of length 5000. The genome size was capped at a maximum of 20,000 sites and a minimum of 2000 sites. For each experimental condition, we ran 400 independent replicate experiments. At the end of the evolutionary process, the line of descent was reconstructed, and all analyses were performed on these agents. All experiments, using each of the three agent types and two worlds, were implemented using the MABE code framework [5] which among many other things includes an extensive version of Markov Brains as well as all other tools required to perform neuroevolution on arbitrary substrates and tasks.

In both worlds, we used the number of correct decisions C (e.g. blocks caught or avoided correctly, pairs of numbers correctly ordered) and incorrect decisions I (e.g. blocks caught that should have been avoided, pairs of numbers incorrectly ordered) to define the fitness function as:

$$W = 1.1^{(C-I)} \tag{7.5}$$

This function is an exponential fitness function where an additional correct decision of the agent does not linearly but exponentially improves fitness. This increases the chances of an agent to be selected for reproduction exponentially, which in our experience allows the GA to select for functional agents better. The exponential base value (1.1) was taken from previous work [19].

After the fitness has been computed for every organism in the population, we use roulette wheel selection to determine which organisms contribute offspring to the next generation [38]. Roulette wheel selection is a method that ensures selection to be directly proportional to fitness. Every time an agent made an offspring the genome experienced gene duplications (with probability $2 * 10^{-5}$ per nucleotide of the parent genome, duplicate the next 128–512 contiguous sites), deletions (with probability $2 * 10^{-5}$ per nucleotide of the parent genome, delete the next 128–512 contiguous sites), and point mutations (with probability $5 * 10^{-3}$ per nucleotide of the parent genome, randomize the site).

7.2.11 Augmenting with R

In order to augment the performance of the GA, not only the performance of each agent is measured but also the amount of representations each agent has about its environment. In the block catching task these environmental conditions are specifically the size of the block, the direction of it, and whether it is to the left or right. In the number comparison task, representations are about the individual values of the numbers to be compared, and the ordering of the pair (i.e. if the first or the second was smaller).

To assess performance the number of correct catches (C) and misses (I), or the number of correct (C) versus incorrect answers (I) are counted. R is quantified as described above, and R_{max} is the maximum hypothetical value for R, which is dependent on the environmental and memory states. The total fitness of the agent is determined by the following equation:

$$W = 1.1^{(C-I)} \left(1 + \frac{R}{R_{max}}\right) \tag{7.6}$$

Alternatively we could have used a multi-objective optimization method [8] that would allow some agents in the population to have a high score and low R, while others might perform poorly but have proper representations. This keeps the diversity high, but only a recombination step might take full advantage of such diversity. Recombination however, would introduce another level of complexity that might have confounded the results. As such, we did not pursue this option here.

7.3   Results

7.3.1   GP Trees Evolve to Have Representations

Each system responds differently to optimization by a genetic algorithm (GA), and as such we expect GP-Vector, GP-Forest, and Markov Brains using GP gates (further called MBwGP) to generally evolve to solve the task at hand but be optimized at different speeds. Similarly, systems might tend to get stuck in local optima differently, and thus, the resulting performance might differ between brains.

As expected, we find the different brains to respond differently to the optimization of the GA (see Fig. 7.8). We find all systems to evolve generally well, however, GP-Forests tend to perform better on the number task (see Fig. 7.8 right panel). MBwGP and GP-Forest are almost indistinguishable with respect to their performance in the active categorical perception (APC) task (see Fig. 7.8 left panel). GP-Vector brains, however, perform poorest on both tasks.

Interestingly, we find a clearer delineation of result with respect to the degree in which these systems evolve the ability to have representations (see Fig. 7.9). GP-Forests evolve the highest degree of representations, followed by MBwGP, trailed by GP-Vector Brains. This is particularly interesting since we found earlier [19] that RNNs have it evolve representations early, but struggle with improving their performance, even though they already have a high level of representations. Markov Brains on the other hand evolve representations slowly over time, and thus can only improve their performance later, after they have sufficient representations. GP-Forest Brains seem to evolve like Markov Brains but store more representations.

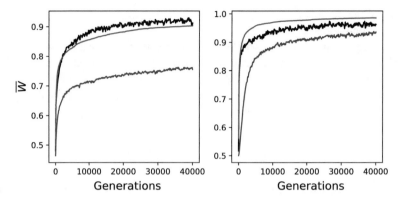

Fig. 7.8 Performance over time for all 3 brain types (GP-Forest = Black, MBwGP = Red, GP-Vector = Blue). Left plot is active categorical perception, right is number comparison

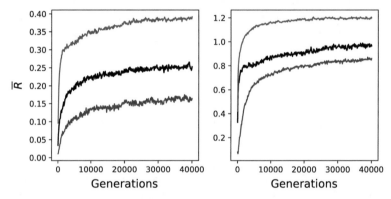

Fig. 7.9 R over time for all 3 brain types (GP-Forest = Black, MBwGP = Red, GP-Vector = Blue). Left plot is active categorical perception, right is number comparison

7.3.2 Does Augmentation Using R Improve the Performance of a GA?

We know that one can augment the optimization a GA performs by using R (see Augmenting with R). The idea is that mutations that improve an agents ability to store more representation would normally be neutral as they do not necessarily also immediately provide an advantage to performance. However, in order to improve performance, often representations are needed. The augmentation process, therefore, turns mutations that improve R but are neutral into beneficial ones. While this is true for Markov Brains and RNNs, it is not clear is the GP-forest or GP-vector brains will behave the same.

We tested this hypothesis by evolving all three types of brains again on both environments, but also measured the representations they have during evolution, and used this quantity to augment the GA. This is in contrast to the results above,

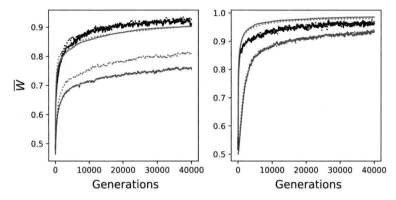

Fig. 7.10 Performance over time for all 3 brain types (GP-Forest = Black, MBwGP = Red, GP-Vector = Blue). Left plot is active categorical perception, right is number comparison. Solid line is control, dotted line is augmented with R

which merely examined the unaugmented GA. We find that this augmentation has almost no effect on GP-Forests and only little on MBwGP brains (see Fig. 7.10). However, it significantly improves the performance of GP-Vector Brains in the active categorical perception task. We also find that GP-Forests and MBwGP already evolve the ability to store representations well, but GP-Vector Brains struggle. This supports the hypothesis that augmenting a GA by using R works only for systems that struggle to form representations in the first place.

We further find that the augmentation with R might not improve performance, but still had an effect on the ability to store representations in each system. Each brain when evolved using R-augmentation evolved to have more representation over evolutionary time than the unaugmented control (see Fig. 7.11). This proves that the augmentation indeed has an effect, but again, only improves the performance of the GA when a system struggled to evolve representations.

The most notable finding is that GP-Forest Brains do evolve a large degree of representations, and respond to the augmentation with an increase in R but do not significantly benefit from the augmentation with respect to their performance.

7.3.3   Smeared Representations

We found earlier [12] that brains who evolve to have sparse and condensed representation are more robust to noise, and conjecture that overly smeared representations negatively affect performance and evolvability. When representations are smeared, every mutation changing the system would affect the computation of all concepts. A similar argument has been made explaining the benefit of the dropout method in deep learning [34]. By removing nodes while training an ANN, representations might become more condensed, improving the ANNs ability to generalize. Another

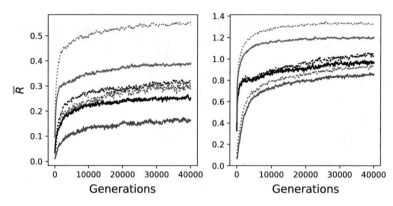

Fig. 7.11 R over time for all 3 brain types (GP-Forest = Black, MBwGP = Red, GP-Vector = Blue). Left plot is active categorical perception, right is number discrimination. Solid line is control, dotted line is augmented with R

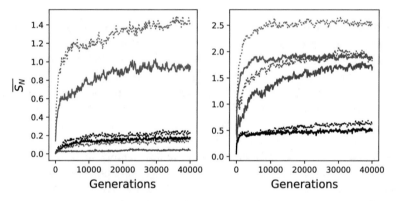

Fig. 7.12 Smearedness of Nodes over time for all 3 brain types (GP-Forest = Black, MBwGP = Red, GP-Vector = Blue). Left plot is active categorical perception, right is number comparison. Solid line is control, dotted line is augmented with R

way of thinking about this is that the more clearly separated the concepts in the cognitive system are, the better it can classify. We already showed that neuroevolution of RNNs and LSTMs results in systems that smear representations greatly, while Markov Brains evolve to have sparse condensed representations [12]. The question here is, if the structures of genetic programming trees helps to further condense representations, or if these systems smear representations nevertheless?

We find that the degree to which these systems evolve smeared representations again varies greatly between the systems. While MBwGP and GP-Vector Brains only smear their representations marginally when evolved to solve the active categorical perception task, GP-Forest Brains smear their representations much more (see Fig. 7.12 left panel). When evolve to solve the number task, both GP-Forest and GP-Vector smear more than the MBwGP (see Fig. 7.12 right panel).

7.4   Discussion

We used neuroevolution to optimize the performance of three different computational systems. Markov Brains that use GP nodes are similar to CGPs and evolve a network topology that we know is capable of forming representations which are not smeared. We confirmed that Markov Brains using GP nodes benefit from augmenting the GA using R, and asked if the same principles apply to other cognitive topologies. Of particular interest are the tree-like structures from genetic programming trees. Other than the topology of ANNs which connects every node with every other node and leads to extreme smearing of representations [12] we assumed that tree structures would behave differently in this regard. The tree structure can take advantage of all possible inputs and hidden states but converges them into a single output at the end. Our assumption was that these systems would condense representations better and thus smear less. The hope was that such a construction allows a system to take full advantage of having a high degree of representations while at the same time not smearing them.

While the GP-Vector implementation performed poorly compared to GP-Forests and MBwGP, we found that GP-Forests evolve equally well or even better when compared to MBwGP (see Fig. 7.8). GP-Forests also evolve to have more representations than the less structured MBwGP Brains (see Fig. 7.9), and these representations do smear (see Fig. 7.12). However, augmenting the GA with R has no effect on their performance (see Fig. 7.10) and only increases the amount of representations they evolve to have as expected (see Fig. 7.11). As such, GP-Forests have the ability to evolve representations and take advantage of them by construction, and are thus "immune" to further augmentation by using R.

Lastly, the smearing of representation in GP-Forests is much less than it has been observed in RNNs before [12]. This further supports the idea that the tree structure provides a benefit over either the networks that Markov Brains or CGPs form and over the fully connected regular structures that RNNs have when it comes to having representations and taking advantage of them over the course of evolution.

In other words, GP-Forest Brains evolve equally well or better than Markov Brains and form representations easily by their construction, and thus do not benefit from further augmentation with R. At the same time, they do smear their representation but not to the degree that it becomes a problem as it does in RNNs.

7.5   Conclusions

The two genetic programming systems (GP-Forest and GP-Vector) solve the quest to make genetic programming trees have recurrence, thus allowing them to become stateful machines. In the case of GP-Forests, their structure provides an advantage over other regular (ANN) or network structures (Markov Brains) with respect to evolving meaningful representations. The immunity of GP-Forests to the augmentation with R methods shows that this system automatically retains representations sufficiently well, without smearing them, by nature of its structure.

This insight suggests that one should take further advantage of tree-like structures in neuroevolution. The capabilities of the GP-Forests also suggests investigating tree structures and how they affect the concept of smearedness in the context of deep learning.

Acknowledgements We thank Stephan Winkler for insightful discussions on hidden states in genetic programming trees, and for implementing the prototype for GP-Forests in EvoSphere.

References

1. Banzhaf, W., Nordin, P., Keller, R.E., Francone, F.D.: Genetic Programming - An Introduction. Morgan Kaufmann, San Francisco CA (1998)
2. Beer, R.D.: The dynamics of active categorical perception in an evolved model agent. Adaptive Behavior **11**(4), 209–243 (2003)
3. Beer, R.D., et al.: Toward the evolution of dynamical neural networks for minimally cognitive behavior. In: From Animals to Animats, vol. 4, pp. 421–429 (1996)
4. Bengio, Y., Frasconi, P.: An input output hmm architecture. In: Advances in neural information processing systems, pp. 427–434 (1995)
5. Bohm, C., CG, N., Hintze, A.: MABE (modular agent based evolver): A framework for digital evolution research. In: Proceedings of the European Conference of Artificial Life (2017)
6. Brooks, R.A.: Intelligence without representation. Artificial intelligence **47**(1-3), 139–159 (1991)
7. Clune, J., Stanley, K.O., Pennock, R.T., Ofria, C.: On the performance of indirect encoding across the continuum of regularity. IEEE Transactions on Evolutionary Computation **15**(3), 346–367 (2011)
8. Deb, K.: Multi-objective optimization using evolutionary algorithms. John Wiley & Sons (2001)
9. Floreano, D., Dürr, P., Mattiussi, C.: Neuroevolution: From architectures to learning. Evolutionary Intelligence **1**(1), 47–62 (2008)
10. Handley, S.G.: The automatic generations of plans for a mobile robot via genetic programming with automatically defined functions. In: Advances in Genetic Programming, vol. 18, pp. 391–407. MIT Press (1994)
11. Hintze, A., Edlund, J.A., Olson, R.S., Knoester, D.B., Schossau, J., Albantakis, L., Tehrani-Saleh, A., Kvam, P., Sheneman, L., Goldsby, H., Bohm, C., Adami, C.: Markov brains: A technical introduction. arXiv preprint arXiv:1709.05601 (2017)
12. Hintze, A., Kirkpatrick, D., Adami, C.: The structure of evolved representations across different substrates for artificial intelligence. In: Artificial Life Conference Proceedings, pp. 388–395. MIT Press (2018)
13. Hintze, A., Schossau, J., Bohm, C.: The evolutionary Buffet method. In: Genetic Programming Theory and Practice XVI, pp. 17–36. Springer (2019). https://link.springer.com/chapter/10.1007/978-3-030-04735-1_2
14. Hochreiter, S., Schmidhuber, J.: Long short-term memory. Neural Computation **9**(8), 1735–1780 (1997)
15. Kirkpatrick, D., Hintze, A.: The role of ambient noise in the evolution of robust mental representations in cognitive systems. Artif. Life Conf. Proc. (31), 432–439 (2019). https://doi.org/10.1162/isal_a_00198

16. Koza, J.R.: Genetic programming as a means for programming computers by natural selection. Statistics and Computing **4**(2), 87–112 (1994)
17. Koza, J.R., Rice, J.P.: Automatic programming of robots using genetic programming. In: AAAI, vol. 92, pp. 194–207 (1992)
18. Langdon, W.B.: Evolving data structures with genetic programming. In: Int. Conference on Genetic Algorithms, pp. 295–302 (1995)
19. Marstaller, L., Hintze, A., Adami, C.: The evolution of representation in simple cognitive networks. Neural Computation **25**(8), 2079–2107 (2013)
20. Merritt, D.J., Brannon, E.M.: Nothing to it: Precursors to a zero concept in preschoolers. Behavioural Processes **93**, 91–97 (2013)
21. Merritt, D.J., Rugani, R., Brannon, E.M.: Empty sets as part of the numerical continuum: conceptual precursors to the zero concept in rhesus monkeys. Journal of Experimental Psychology: General **138**(2), 258 (2009)
22. Miller, J.F.: Cartesian Genetic Programming. Springer (2011)
23. Miller, J.F.: Cartesian genetic programming. In: Cartesian Genetic Programming, pp. 17–34. Springer (2011)
24. Nieder, A.: Honey bees zero in on the empty set. Science **360**(6393), 1069–1070 (2018)
25. Nordin, P.: A compiling genetic programming system that directly manipulates the machine code. In: Advances in Genetic Programming, vol. 1, pp. 311–331. MIT Press (1994)
26. Nordin, P., Banzhaf, W.: Genetic programming controlling a miniature robot. In: Working Notes for the AAAI Symposium on Genetic Programming, vol. 61, p. 67. MIT, Cambridge, MA, USA, AAAI (1995)
27. Nordin, P., Banzhaf, W.: An on-line method to evolve behavior and to control a miniature robot in real time with genetic programming. Adaptive Behavior **5**(2), 107–140 (1997)
28. Nordin, P., Banzhaf, W.: Real time control of a khepera robot using genetic programming. Control and Cybernetics **26**, 533–562 (1997)
29. Reynolds, C.W.: An evolved, vision-based behavioral model of coordinated group motion. In: Proc From Animals to Animats, vol. 2, pp. 384–392 (1993)
30. Reynolds, C.W.: Evolution of obstacle avoidance behavior: using noise to promote robust solutions. In: Advances in Genetic Programming, vol. 1, pp. 221–241. Cambridge, MA: MIT Press (1994)
31. Russell, S.J., Norvig, P., Canny, J.F., Malik, J.M., Edwards, D.D.: Artificial Intelligence: A Modern Approach. Prentice Hall, Upper Saddle River (2003)
32. Schossau, J., Adami, C., Hintze, A.: Information-theoretic neuro-correlates boost evolution of cognitive systems. Entropy **18**(1), 6 (2015)
33. Spector, L., Robinson, A.: Genetic programming and autoconstructive evolution with the push programming language. Genetic Programming and Evolvable Machines **3**(1), 7–40 (2002)
34. Srivastava, N., Hinton, G., Krizhevsky, A., Sutskever, I., Salakhutdinov, R.: Dropout: a simple way to prevent neural networks from overfitting. The Journal of Machine Learning Research **15**(1), 1929–1958 (2014)
35. Stanley, K.O., Clune, J., Lehman, J., Miikkulainen, R.: Designing neural networks through neuroevolution. Nature Machine Intelligence **1**(1), 24–35 (2019)
36. Stanley, K.O., Miikkulainen, R.: Evolving neural networks through augmenting topologies. Evolutionary Computation **10**(2), 99–127 (2002)
37. Teller, A.: The evolution of mental models. In: Advances in Genetic Programming, pp. 199–220. MIT Press (1994)
38. Thomas, B.: Evolutionary algorithms in theory and practice. Oxford University Press, New York (1996)
39. Yao, X.: Evolving artificial neural networks. Proceedings of the IEEE **87**(9), 1423–1447 (1999)
40. Zhou, A., Qu, B.Y., Li, H., Zhao, S.Z., Suganthan, P.N., Zhang, Q.: Multiobjective evolutionary algorithms: A survey of the state of the art. Swarm and Evolutionary Computation **1**(1), 32–49 (2011)

Chapter 8
How Competitive Is Genetic Programming in Business Data Science Applications?

Arthur Kordon, Theresa Kotanchek, and Mark Kotanchek

8.1 Introduction

Data Science is a very complex field that incorporates mathematics, statistics, computer science and programming, math modeling, database technologies, data visualization, data analytics, and so on. From a business point of view, Data Science is a three-legged chair to create high value that combines business acumen, data wrangling and analytics. Focusing on the hard science skills such as machine learning is not enough. Knowing how to generate complex algorithms is important, however, it is even more important to understand what insights these mathematical models reveal for the business, and what actions to take based on these insights. Of special interest to business is AI-based Data Science that uses AI-driven approaches to turn data into insight and actions. The specific features of this type of AI-based Data Science are: (1) leading role of AI-driven methods, (2) a broad set of technologies, such as machine/deep learning, evolutionary computation, decision trees/random forest, SVM, swarm intelligence, and intelligent agents/chatbots, (3) broad applicability, and (4) advanced skillset requirement [11].

Genetic Programming (GP) is one of the AI-based Data Science approaches with very high potential for business applications. It has shown already an impressive list of industrial applications in different areas in manufacturing and business [1, 14]. However, the fast progress of some AI-driven methods, such as machine/deep learning and random forest have grabbed the attention of both the academic and business communities. One of the reasons for this growth is that these approaches

A. Kordon (✉)
Kordon Consulting LLC, Fort Lauderdale, FL, USA
e-mail: arthur@kordon-consulting.com

T. Kotanchek · M. Kotanchek
Evolved Analytics LLC, Midland, MI, USA
e-mail: theresa@evolved-analytics.com; mark@evolved-analytics.com

© Springer Nature Switzerland AG 2020

145

W. Banzhaf et al. (eds.), *Genetic Programming Theory and Practice XVII*, Genetic and Evolutionary Computation, https://doi.org/10.1007/978-3-030-39958-0_8

have contributed to increased efficiency and profitability of technological giants, such as Google, Amazon, and Facebook, whose business models strongly depend on image and natural language processing and all of which allocate tremendous resources for algorithmic improvement and infrastructure development. As a result, these approaches dominate the academic world and the media at the expense of the other methods. Net-net, the generic perception is that this limited subset is equivalent to AI or Data Science.

This paper's objective is to evaluate GP's competitiveness in business Data Science-driven applications and suggest the necessary steps to increase its reach, impact and competitiveness. The GP community at large needs to understand the urgency of taking appropriate actions to increase GP's presence as a relevant method of applied Data Science as a counteract to the aggressive penetration of the fashionable machine/deep learning technologies into science and industry.

8.2 Business Needs for Data Science

Globalization and new technologies triggered significant changes in the way businesses operate. Extracting value from data becomes a core capability and part of business competitiveness. The decisive factor in this process is the technologies of AI-based Data Science. The key business needs for transferring data into value by AI-driven methods are discussed shortly below.

8.2.1 Business Forecasting

One of the lessons from the great recession in 2008–2009 was that the businesses need more quantitative and trustworthy predictions about the future. The ultimate objective of business forecasting is to fill this need and deliver to the key business decision-makers a reliable forecast on specific economic variables, such as product demand, raw materials price, labor cost, etc. Some key types of business forecasting are discussed shortly below.

- Price forecasting—Price forecasting is a generic problem with growing popularity to many businesses. Fortunately, two factors facilitate the solution to this problem: (a) many different economic time series have become readily available, and (b) software capabilities for multivariate forecasting have become popular. Optimum pricing takes into account available and predicted inventory, production costs, prices from competitors, and profit margins. Price elasticity models are often used to determine how high prices can be increased before reaching strong resistance.
- Demand forecasting—Demand forecasting is based on quantitative methods for predicting future demand for products sold by a company. Such predictions are

critical for modeling and optimizing product distribution. They are necessary for total cost minimization of product prices, product sourcing locations, and supply chain expenses. The standard approach for demand forecasting is by using statistical time-series analysis, which is mostly based on linear methods. AI-based Data Science has the capability to broaden the forecast modeling options with nonlinear time-series models.

- Long-term forecasting—Developing and updating business strategy is another area where forecasting capability is needed. Selection of economic drivers and discovery of long-term trends and patterns is critical for building reliable 5–10 year future predictions. Of special importance is the nonlinear forecasting approach of integrating GP and ARIMAX (Auto-Regressive Integrated Moving-Average with eXplanatory variables) method in developing such models.

8.2.2 Effective Operation

Operating the business with low cost in a variety of economic conditions is in constant need for every enterprise. AI-driven methods offer many opportunities for cost reduction and productivity improvement. The business needs of the key segments of process operations are discussed below.

- Optimal manufacturing—Process operation optimization is mostly accomplished by optimal control and accurate planning. Optimal control requires sophisticated nonlinear or adaptive control algorithms to satisfy the growing complexity of manufacturing processes. Optimal production planning needs accurate and reliable demand forecasting. An additional requirement for successful business process optimization is high process observability with sensors or math models for parameter estimation (inferential sensors).
- Predictive maintenance—Predictive maintenance predicts equipment break-down, and the recommendations could include corrective actions, equipment replacement, or even planned failure. The objective of predictive maintenance is to recognize system faults in advance due to significant differences of certain parameters or changed behaviors. The sensor networks and statistical/machine learning patterns enable operators to recognize if there is any deviation from the normal operating regime that is a symptom of a future break.
- Supply chain optimization—A company's supply chain comprises geographically dispersed facilities where raw materials, intermediate products, or finished products are acquired, transformed, stored, or sold, and transportation links that connect facilities along with product flow. A supply chain network includes vendors, manufacturing plants, distribution centers and markets. AI-driven methods may contribute to improved efficiency and reduced cost of supply chain operations by more accurate demand estimation and optimal scheduling.

8.2.3 Growth Opportunities

The most valuable business need is for achieving sustainable growth. One key strategy for accomplishing this goal is fast innovation with higher speed than the competition. The three driving forces of fast innovation are: (1) differentiation from the competition, (2) fast time to market, and (3) disruptive innovation. The purpose of differentiation is to provide an offering which the customer believes delivers superior performance per unit of cost. The objective of fast market penetration is to earn high margins, as a temporary monopoly, and quickly create a new innovation to counter the gradual commoditization of the old one. Disruptive innovations is the most substantial driving force since the disruptive technologies redefine the competitive landscape, by making any previous competitive advantages obsolete.

The key directions in exploring the business growth opportunities are briefly discussed below.

- Exploring new markets—Market exploration is a broad category which includes improved search techniques for finding market needs and gaps in existing knowledge in patents and the available literature. Complex data analysis based on market surveys, future demand assessment, social networks, and the identification of technology gaps to assess the potential business impact of innovation are required as well. Of critical importance for the success of this task are the capabilities to develop financial and market penetration models.
- Developing new products—Rapid exploration of new ideas and fast commercialization are the key requirements for a successful new product development. Modeling capabilities are critical for exploring the technical features of innovations. Initially, fundamental knowledge and human expertise about a new invention are at a very low level, and the development of first-principles models is very time consuming and expensive. On top of that, there is very limited data available either in the literature or from laboratory experiments. Often it is also difficult to satisfy the assumptions of classical statistics to develop statistical models with a limited number of observations. As a result, AI-based methods for accurate and reliable model generation from a small number of data points are needed.
- Selecting scale-up options—The key decision in fast commercialization of a new product is in scale-up activities, i.e., either to build pilot plant facilities or to take the risk and directly jump to full-scale operation. The cost and time savings if the pilot plant phase is avoided are significant. Two capabilities are important in this critical decision—robust modeling and advanced design of experiments to validate the hypothesis of reliable performance after scale-up.
- Attracting new customers—For many businesses the key source for growth is attracting new customers. It requires better understanding of current customers and their behaviors and preferences. Companies are keen to expand their traditional data sets with social media data, browser logs as well as text analytics

and sensor data to get a more complete picture of their customers. The gained insight and customer segmentation are the basis for approaching potential new customers.

8.2.4 Multi-Objective Optimization and Decision Making

Measuring and monitoring business performance is critical. Focusing on the wrong metrics can be extremely detrimental, as firms poorly invest money and time measuring, monitoring and optimizing metrics that don't matter. Equally detrimental are key performance indicators (KPIs) which are poorly structured or that are too difficult and/or too costly to obtain or to monitor on a regular basis [24].

The most effective KPIs marry business/financial objectives with product market fit. Although important, doing so effectively is often complicated by the fact that most businesses have short, medium and long-term objectives which need to be simultaneously achieved. KPIs should keep everyone on the same page and moving in the same direction. If a KPI is too high level or too vague, it can be interpreted in many different ways, and actioned in many different ways across an organization. Hence, business KPIs should be accurate and specific enough to inform specific actions.

Currently, in many organizations, optimization has mainly been applied to silo functions. Considering what optimization does—help users make the best decisions considering a set of constraints or trade-offs—then the value is always greater when focused on higher value opportunities. Looking at the problem from a business angle, it's where the user can optimize the enterprise—simultaneously optimizing supply and demand, and in turn physical, human and financial assets. [2]

Hence, the ability of the enterprise to make decisions about which demand to pursue, with what products/services and with the right resource configuration while considering all relevant market, regulatory, physical and policy constraints is a differentiator. Those organizations who know how to effectively integrate internal and external data streams to simultaneously optimize multiple KPIs and manage trade-offs are in a position to gain greater market share, growth, and financial returns.

8.3 Data Science Competitive Landscape

8.3.1 Defining Key Competitors for Data Science Applications

The competitive advantage of a Data Science approach, is defined as technical superiority that cannot be reproduced by other technologies and can be translated with minimal efforts into a position of competitive advantage in the marketplace [13]. One of the most important steps in defining competitive advantages of AI-

Business Need	Linear Regression/ Logistic Regression	Neural Networks	Deep Learning	Random Forests	Support Vector Machines	Genetic Programming
Multi-Objective Decisions	Good	Weak	Weak	Weak	Weak	Good
Financial Optimization	Average	Weak	Weak	Weak	Weak	Average
Business Forecasting	Good	Weak	Weak	Weak	Weak	Good
Effective Operations	Average	Weak	Weak	Average	Average	Good
Growth Opportunities	Weak	Good	Average	Average	Average	Good
Limitations	Assumes known model form & uncorrelated variables	Requires large training datasets	Requires massive datasets	Overfitting of small datasets	Difficulty handling missing records	Computationally intensive
Total Cost of Ownership	Low	Average	High	High	High	Low
Training Cost	Low	Average	High	Average	High	Low

Fig. 8.1 Estimating business needs satisfaction and related application cost for selected competitive Data Science methods

driven Data Science approaches is identifying the key technological rivals in the race, such as statistics (represented by linear and logistic regression), machine learning (represented by neural networks), deep learning, random forest, GP as a representative of evolutionary computation, and SVM. The selection is based on a method's capability, popularity, and applicability.

8.3.2 Comparison on Business Needs Satisfaction

From a business application point of view, the comparison should include an estimate on how these technologies contribute to solving the key business needs, discussed in the previous section. Of interest are the expected cost and key limitations. A qualitative estimate of these factors, based on the authors' rich industrial experience, is shown in Fig. 8.1.

8.3.3 How Popular Is GP in the Data Science Community?

The popularity of GP relative to key competitive approaches, such as logistic regression, neural networks, and random forest, based on appropriate queries on the Internet was estimated in 2014 by [7]. The queries on Kaggle.com indicated the use of new methods by data scientists; on site:edu—the queries estimated potential academic research; on site:syllabus—the queries estimated potential classes; on

Approach	Kaggle	site:edu	site:edu syllabus	site:goy	site: linkedin
logistic regression	1310	25100	392	15400	102000
machine learning	4510	39400	1590	5890	880000
deep learning	1450	65700	4880	9270	728000
random forest	1320	17,000	348	5190	55200
GP	24	8520	224	1790	2650
GP% 2019	0.3	5.5	3.0	4.8	0.1
GP% 2014	0.46	1.2	2.7	11.2	1.7

Fig. 8.2 Popularity of GP and selected competitive Data Science methods

site:gov—the queries estimated the potential funding or applications in the Government, and on linkedIn.com—the queries estimated the generic popularity across professionals. The conclusion was that GP lagged behind all other methods and was seen as an outsider in the mainstream machine learning and AI community.

Following the method described in [7], a more recent query has been done in 2019 and the results are shown in Fig. 8.2.

One difference with the GP popularity comparison study in 2014 is that deep learning has been added as a new competitor. The percentage of GP in the corresponding application areas for both 2014 and 2019 studies are added as well. As it is shown in Fig. 8.2, the relative popularity of GP is very low on the sites—Kaggle.com and LinkedIn.com—most populated by data scientists. Of deep concern is also the declining trend of GP% in 5 years for these two important application areas. The conclusion that could be made from both studies is that GP is not very popular in the Data Science community and is virtually unknown to the majority of data scientists.

8.4   Current State-of-the-Art of Genetic Programming as Business Application Method

8.4.1   Competitive Advantages of GP

The key competitive advantages of GP are shown in the mind-map in Fig. 8.3 and discussed shortly below.

- Transparent interpretable models—Pareto GP allows the simulated evolution and model selection to be directed toward structures based on an optimal balance between accuracy and expressional complexity [22]. A survey of industrial applications in The Dow Chemical Company shows that the selected models, generated by Pareto GP, are simple and accurate [11]. In addition, a very important factor in favor of symbolic regression is that engineers and scientists

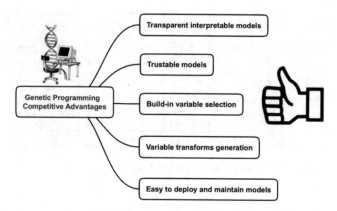

Fig. 8.3 Key GP competitive advantages

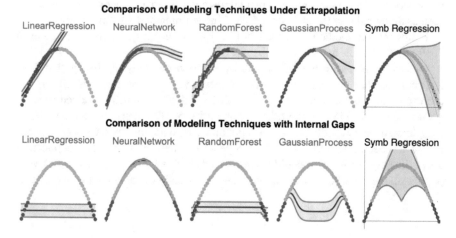

Fig. 8.4 Comparison of various modeling techniques under extrapolation and interpolation with internal gaps in data

    prefer mathematical expressions and very often can find an appropriate physical interpretation.

- Trustable models—The plethora of developed models evolved can be exploited by combining a diverse subset into an ensemble. This is, effectively, a super-model where the aggregate provides a robust prediction and the divergence of the constituent models a trust metric which can detect extrapolation as well as fundamental changes in the targeted system. This is illustrated in the Fig. 8.4 below which shows the extrapolation performance of a simple ball trajectory as well as the situation of handling internal sampling gaps. Note that while GP predicts well, it also flags the lack of trust in those circumstances whereas conventional modeling techniques cannot identify the lack of observational support.

- Built-in variable selection—One of the unique features of GP is its built-in mechanism to select variables related to the target response during the simulated evolution process and to gradually ignore variables that are not [21]. An inputs ranking is based on the variable presence in its equations during the simulated evolution. An advantage of this type of variable selection is that it implicitly includes nonlinear relationships between the target and the inputs.
- Variable transforms generation—A side effect from the severe fight for high fitness of the competing equations in simulated evolution, is the generation of simple transforms that survive the competition. Some of them appear as building blocks in many equations, and as such, can be defined, analyzed, and select as features. Selecting the proper transforms is not trivial, however, and is limited to simple equations of limited number of original variables. Validating the proper physical dimensionality of the transform is highly recommended. Domain experts do not accept features with incorrect physical dimensions.
- Easy to deploy and maintain models—Evolutionary computation has a clear advantage in marketing the technology to potential users. The scientific principles are easy to explain to almost any audience. We also find that process engineers are much more open to implement symbolic regression models in manufacturing plants. Most of the alternative approaches are expensive, especially in real-time process monitoring and control systems. In addition, the simple symbolic regression models require minimal maintenance.

8.4.2   Key Weaknesses of GP

The key weaknesses of GP are shown in the mind-map in Fig. 8.5 and discussed shortly below.

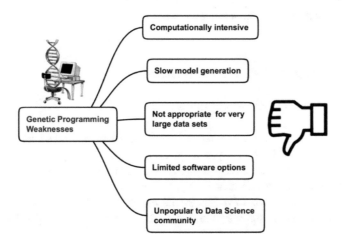

Fig. 8.5 Key GP weaknesses

- Computationally intensive—Simulated evolution requires substantial number-crunching power. Fortunately, the continuous growth of computational power, along with parallel computing mitigates this issue. Additional gains in productivity are made by improved algorithms. The third way to reduce computational time is shrinking down the data dimensionality. For example, only the most informative data can be selected for the simulated evolution by other methods like random forest (by variable selection) and support vector machines (by record selection).
- Slow model development—An inevitable effect of the computationally intensive simulated evolution is the slow model development. However, the slow speed of model generation does not significantly raise the development cost. The lion's share of this cost is taken by the model developer's time which is relatively low and limited to a small number of potential models on the Pareto front for selection and validation. With automatic ensemble generation this cost is minimal.
- Not appropriate for huge numbers of variables—An implicit assumption in Pareto GP is that relatively few features (variables) matter and, as a result, GP is not very efficient when the search space is tremendously large. It is preferable to limit the number of variables to up to one thousand and the number of records to up to a million records in order to get results in an acceptable time to model development. Additionally, waveforms or images are not good inputs; in this case, processing such into candidate features is preferred as well as being more insightful. Iterative reduction in candidate features is useful to reduce the size of the haystack in these needle-in-a-haystack search problems.
- Limited software options—In comparison to other Data Science methods, the choice for available programming packages is very limited. Of special importance to business applications is the availability of professional user-friendly software. A short survey of current software options and an example of professional software is given in Section 8.5.3.
- Unpopular to Data Science community—As was clearly shown in Section 8.3.3, the Data Science academic community and data scientists at large are unaware about the capabilities and applicability of GP.

8.4.3   Successful Genetic Programming Applications

8.4.3.1   Examples of GP Applications

The big potential of GP for solving real-world problems, based on its advantages, has been recognized by industry since the early 2000s. The pioneer in this effort was Dow Chemical where various application areas, such as inferential sensors, empirical emulators, accelerated first-principles model building, effective design of experiments, etc., have solved industrial problems and demonstrated long-term value creation (see a summary in [12]). A more recent survey of real-world

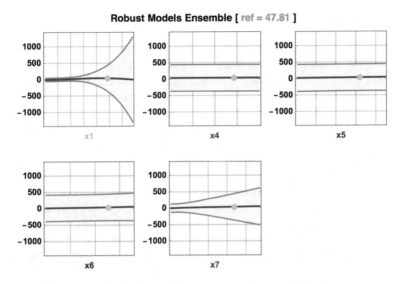

Fig. 8.6 Ensemble response plot (blockchain currency fees)—from [25]

applications further demonstrates its growing importance and financial impact in manufacturing [6, 15, 16, 20, 23], the health sector [4], the financial sector [25] and education [3].

Due to its requirement for transparent models, GP has significant advantage in the financial sector. An interesting application of GP is the study of behavior and causation in cryptocurrency markets. Of special importance is the ensemble generation option of GP that creates a trust metric of the models. Figure 8.6 shows how an intelligent agent perceives the trustability of the prediction, what may happen in regions of unknown parameter space (when it is exposed to unseen data) or when the underlying system changes [25].

Another interesting application is improving evaluating techniques of instructors by using symbolic regression based non-experimental data from previous scores collected by the university as input. The generated models can predict 60–70% of variation in students' exam scores and performance [3].

8.4.3.2 GP Applications with High Value Creation

Some GP business implementations have matured into established technologies with sustainable value generation. The applied models have been used long-term (some longer than 15 years) with consistent profit flow. Two examples, robust inferential sensors and nonlinear business forecasting, are discussed shortly below.

8.4.3.3 Robust Inferential Sensors

Soft or inferential sensors deduce the critical parameters in manufacturing processes from easy-to-measure variables such as temperatures, pressures and flows. The dominant technology in the market, based on neural networks, has several flaws, such as poor extrapolation, unreliable estimation, and high maintenance cost. The robust inferential sensors are in the form of explicit algebraic equations automatically generated by Pareto GP with an optimal tradeoff between accurate predictions and simple expressions. As a result, they offer more robust performance in the face of minor process changes in comparison to neural networks. We will illustrate this technology with two examples from robust inferential sensors with long-term operation at Dow Chemical [11].

The first example relates to distillation tower control. Obtaining an accurate and fast prediction of a process quality variable (in this case, propylene concentration) can enable better control of a distillation column. The current analytical technique allows measurement of propylene every 10 min, which is not sufficient for control purposes. The solution was a robust inferential sensor that provides a prediction of propylene concentration every minute. It was an ensemble of equation-based models with an optimal balance between performance and low complexity. Satisfying the requirement for robustness to sensor failure favors models with different inputs.

The second example relates to the important issue of NOx emissions monitoring. The discussed technology has been successfully applied for NOx emission monitoring of two gas turbines, GT1 and GT2 at Dow Chemical. The sensors have performed excellently after deployment in 2004. During more than 10 years of operation, they passed the annual relative accuracy test audits (RATAs) (the longest period allowed by law) with the single exception of a 6-month schedule case. To achieve an annual schedule, a sensor must pass with better than 7.5% relative accuracy. Both turbines were mechanically rebuilt in 2008 but the models still passed RATA with annual schedule without the need to fit a different model structure. Robust inferential sensors continue to gain popularity in industry. In addition to Dow Chemical (with more than 100 applied inferential sensors) it has been explored at GE [7] and recently applied at Georgia Pacific LLC. The total value creation from the deployed models due to improved control, process optimization, reduced shutdowns and lab analysis is estimated to tens of millions of dollars.

8.4.3.4 Nonlinear Business Forecasting

A hybrid modeling approach of integrating GP and the most-used in industry forecasting method—ARIMAX (Auto-Regressive Integrated Moving-Average with eXplanatory variables) has a good potential for nonlinear business forecasting. The idea is to combine the advantages of both methods—nonlinear transform generation by GP and forecasting by an established technology, such as ARIMAX.

The nonlinearity is represented with transforms, generated by GP and used as explanatory variables in ARIMAX models. The final forecast is generated by these ARIMAX models [10].

For some manufacturing companies, especially in processing industry, forecasting raw materials prices is of critical importance. As the largest US chemical company with a broad range of products, Dow Chemical is using different raw materials. Their total cost varies in the range of billions of dollars. Reducing this cost with reliable forecasting models can deliver significant value and directly influence the company's profit.

In order to accomplish this goal, a large-scale forecasting project for development and deployment of predictive models for the top 51 costlier raw materials prices was initiated with the support of the purchasing department. Some of these models are nonlinear, based on the hybrid GP-ARIMAX approach. All models are using the most important economic drivers, some of them—leading indicators. GP is also used in this variable selection [19]. The progress of forecasting performance is tracked on a monthly basis. The average performance after the first 6 months in operation was in the range of the Mean Absolute Percentage Error (MAPEs) between 1.7 and 28% with a mean error of 10.3% (which is in the borderline of a good forecast definition.) Taking into account the big volume of the raw materials cost, even 1% reduction due to more accurate price forecast is in the range of 50–70 million dollars [9].

8.5 How to Increase Competitive Impact of Genetic Programming in Data Science Applications?

The defined advantages of GP relative to identified key Data Science competitors and its impressive application record in industry are the basis for defining an action plan for increasing the visibility and impact of this technology. The objective is to reverse the growing relative dominance of machine/deep learning in the Data Science community and especially for future business applications. The key steps to accomplish this goal and to promote effectively the great potential of GP are discussed below.

8.5.1 Develop a Successful Marketing Strategy

8.5.1.1 Data Science Marketing

The most critical issue in the proposed action plan from a practical point of view is the lack of professional marketing that communicates the true capabilities of different Data Science technologies. Unfortunately, it is often replaced by hype delivered by vendors with promised voodoo from an avalanche of buzzwords. An alternative could be a more realistic but limited form of marketing, called research marketing [8].

Relative to generic marketing with its broad advertisement efforts, research marketing is limited to direct interactions with targeted potential customers either by social networks, publications or by different types of presentations. In this case, the marketing cost is often significantly lower then broad advertising efforts. On the other hand, the size of the audience is very limited. The situation is changing, however, with the growing influence of Web-based marketing and social networks. In this section, we give some ideas on applying this marketing strategy in interacting with three key Data Science communities, related to industrial GP applications who are in the business of applying Data Science: statisticians, machine/deep learning data scientists, and business analysts.

8.5.1.2 Marketing to Statistical Community

Statistics is still the most widespread Data Science approach in practice. Most engineers, biologists, and economists have some statistical background. Very often, however, this popularity is at the expense of mass-scale misuse and abuse of the approach. It also shapes a mindset that all problems related to data must be resolved by statistics and all other methods are statistically not credible. As a result of the recent AI-based Data Science invasion, however, this community is in a defensive position switching the mode from directly challenging the machine/deep learning methods to gradually adapting to the new reality. This opens opportunities to promote and advance GP as an alternative to other machine/deep learning AI-based Data Science approaches, as GP generated models are transparent algebraic equations which can be statistically evaluated versus non-transparent machine/ deep learning black boxes.

The key benefit from the synergy of GP and statistical model building in industrial model development is broadening the modeling capabilities of both approaches. On the one hand, GP allows model building in cases where it would be very costly or physically unrealistic to develop a linear model. On the other hand, statistical modeling with its well-established metrics gives GP models all the necessary measures of statistical performance.

8.5.1.3 Marketing to Machine Learning Community

The target audience is the academic research community related to Data Science, data scientists with a degree in this discipline, and the growing army of semi-educated citizen data scientists. The typical behavior that has to be taken into account is their religious belief in technical dominance of machine/deep learning combined with an obsession with open source Python-based software.

The suggested marketing strategy is focused on explaining the unique capabilities of GP and the benefits of integrating this approach with machine/deep learning. An important element is also emphasizing the broad nature of AI-based Data Science beyond the narrow machine/deep learning view. The unique features of evolutionary computation and its related approaches should be explained in a popular manner with an emphasis on describing the differences between GA and GP. The impressive technical capabilities of GP and its solid business application record, discussed in this paper, can be used in the technology introduction. Examples of marketing slides and elevator pitches for AI-based Data Science methods, including GP, are given in [11].

The key driving force behind integrating modeling methods is balancing the weaknesses of one method with the strengths of another one. As a result, the integrated solution has better structure, higher credibility, broader acceptability, and reduced cost. The key attractive GP feature of interest to machine/deep learning community is the generation of transparent interpretable models with optimal trade-off of accuracy and complexity, with known trust metrics enabling robust interpolation and extrapolation. This may compensate for the key weaknesses of machine/deep learning—the lack of understanding and interpretability of the derived black-box models.

8.5.1.4 Marketing to Business Community

Marketing to the business audience should be organized in a different way. The focus must be on business problems solved by GP, such as the big value creation opportunities of GP, demonstrated in robust inferential sensors and nonlinear business forecasting applications in large enterprises, such as Dow and Georgia Pacific LLC. (see detailed examples in [1, 11]).

It is recommended that for this audience the GP approach is explained in plain English with minimal jargon and technical details. A critical point is clarifying that GP is part of AI-driven technologies as is machine/deep learning. Another topic is showing GP's competitive advantages, especially the comparison with machine/deep learning. Contrasting the low total cost of ownership of GP developed models versus the high cost of developing, deploying, and maintenance of machine/deep learning systems is a good selling point as well.

An important part of marketing strategy is to explain the current hype about AI at the center of which are machine/deep learning technologies, advertised with inflated capabilities by the vendors and the media. The key reason for the fast rising status of these two technologies is their importance to the business model of the key technical giants, such as Google, Facebook, Amazon, etc. Whereas, the specific business needs of an enterprise are different. Often there is no need for image and natural language processing or voice recognition, which are the key application areas of deep learning. In fact many enterprises are seeking to glean critical insights from highly valuable, sparse datasets which GP is well-positioned to satisfy whereas machine-learning/ deep learning technologies are not equipped to do. As shown in Fig. 8.1, GP is well-positioned to satisfy most of the business needs of an enterprise.

8.5.2 Broaden Application Areas

Demonstrating new application areas is the best opportunity for increasing GP competitiveness in business Data Science applications. A generic area of high potential is explainable and interpretable models. It is driven by the demand for model interpretability in some business applications in finance and health care. In addition, DARPA started the Explainable Artificial Intelligence (XAI) initiative for creating a suite of machine learning techniques that (1) Produce more explainable models, while maintaining a high level of learning performance (e.g., prediction accuracy) and (2) Enable human users to understand, appropriately trust, and effectively manage the emerging generation of artificially intelligent partners [5]. The current approach of solving this problem with generating model explanation by natural language processing of the black-box models is not very effective from business perspective. Using GP could be a viable alternative. Explainable models support causal reasoning, which is another area for future business applications [17].

8.5.3 Improved Professional Development Tools

Several software tools for exploring GP have been developed by different academic groups in:

1. Matlab (GP Matlab Toolbox- developed by Sara Silva (http://gplab.sourceforge.net))
2. Python (Karoo GP package—developed by Kai Staats with support for symbolic regression and classification (http://kstaats.github.io/karoo_gp/)), and
3. R (RGP package, developed by Thomas Bartz-Beielstein's group (https://rdrr.io/cran/rgp/man/rgp-package.html)).
4. HeuristicLab (https://dev.heuristiclab.com/) integrates GP with a broad range of machine learning, evolutionary computation, and swarm intelligence algorithms for prediction, classification, and optimization.
5. DataModeler (www.evolved-analytics.com) is the only commercial tool which has survived and continues to be available and supported.

8.5.4 Increase GP Visibility and Teaching in Data Science Classes

The following action items are recommended to reverse the declining trend of GP in Data Science community at large:

- Writing a new reference book for the current state of the art GP—it should be a popular book to introduce the broad Data Science community to the field. The last similar book was written more than 10 years ago and doesn't represent the current state of the art [18]. It is very difficult to raise the interest of data scientists on this technology without such book, particularly taking into account the hundreds of available books on machine/deep learning. It is very important to discuss GP in the broad context of AI-based Data Science as well.

- Publications with comparative studies—contrasting GP with machine/deep learning methods is needed especially in popular or machine/deep learning journals.
- Popularizing GP in general AI or machine learning conferences—introducing the approach by tutorials, workshops, and presentations at these events with high participation is the first step to attract the attention of many data scientists.
- Publication in high-visibility journals—spreading the technology in popular technical journals and emphasizing the integration capabilities between GP and machine/deep learning could raise the attention of data scientists. Improving model explainability and neuroevolution are two potential topics of interest to the broad Data Science community.
- Including evolutionary computation in general and GP in particular in Data Science classes—the current narrow-focused machine/deep learning courses that dominate academia has to be challenged. The future data scientists should be taught on a broad Data Science basis that includes both EC and GP. A good reference with such balanced approach and focus on business application, described in [11], can be used for teaching such classes.

8.6   Conclusions

The current invasion of machine/deep learning methods in the academic circles, industry, and the media has significantly reduced the interest on other relevant AI-driven approaches, such as evolutionary computation and swarm intelligence. Recent surveys show that the popularity of GP is declining, and it is virtually unknown to the Data Science and business communities. At the same time, analyzing the competitive advantages and weaknesses of the approach show that it is well-positioned for business applications. GP has demonstrated an impressive application record in different business and real-world areas. Of special importance for the credibility of this approach in future business applications are two established GP-based modeling technologies—robust inferential sensors and nonlinear business forecasting—with demonstrated consistent value creation of tens of millions of dollars.

This analysis of the current state of the art of GP as a business application method shows that the root cause for the declining popularity is not in its technical capabilities and potential for future business applications but in insufficient research marketing efforts. Several steps for increasing GP visibility and competitiveness, such as: focused marketing to statistical, machine learning, and business communities; broaden application areas; improved professional software tools; and more aggressive academic presence and teaching are recommended. This multi-pronged approach is a starting point to counteract the current dominance of machine/deep learning technologies in science and business.

References

1. Amir, G., Amir, A., Rayon, C. (eds.): Handbook of Genetic Programming Applications. Springer (2015)
2. Centurion, C.: The power of optimizing supply and demand simultaneously. https://blog.riverlogic.com/optimizing-supply-and-demand-simultaneously (2014)
3. Duzhin, F., Gustafson, A.: Machine learning-based app for self-evaluation of teacher-specific instructional style and tools. Educ. Sci **8**, 2–15 (2018)
4. Dwarshuis, N.J., Song, H.W., Patel, A., Kotanchek, T., Roy, K.: Functionalized microcarriers improve t cell manufacturing by facilitating migratory memory t cell production and increasing cd4/cd8 ratio. bioRxiv p. 646760 (2019)
5. Gunning, D.: Explainable Artificial Intelligence (XAI). https://www.darpa.mil/attachments/XAIProgramUpdate.pdf (2017)
6. Gusel, L., Brezornik, M.: Application of genetic programming for modelling of material characteristics. Expert Systems with Applications **38**, 15,014–15,019 (2011)
7. Gustafson, S., Narasimhan, R., Palla, R., Yousuf, A.: Using genetic programming for data science: lessons learned. In: R. Riolo, W. Worzel, M. Kotanchek, A. Kordon (eds.) Genetic Programming Theory and Practice XIII, Genetic and Evolutionary Computation, pp. 117–135. Springer (2016)
8. Kordon, A.: Marketing computational intelligence in industry. In: Proceedings of CEC 2010, Barcelona, pp. 3185–3192 (2010)
9. Kordon, A.: Applying data mining in raw materials forecasting. In: SAS Analytics 2012 Conference, Las Vegas (2012)
10. Kordon, A.: Applying genetic programming in business forecasting. In: R. Riolo, T. McConaghy, E. Vladislavleva (eds.) Genetic Programming Theory and Practice XI, Springer Series on Genetic and Evolutionary Computation, pp. 101–117. Springer (2014)
11. Kordon, A.: Applying data science: How to create value with artificial intelligence. Springer (2020)
12. Kordon, A., Castillo, F., Smits, G., Kotanchek, M.: Application issues of genetic programming in industry. In: T. Yu, R. Riolo, B. Worzel (eds.) Genetic Programming Theory and Practice III, Springer Series on Genetic and Evolutionary Computation, pp. 241–258. Springer (2005)
13. Kordon, A., Jordaan, E., Castillo, F., Kalos, A., Smits, G., Kotanchek, M.: Competitive advantages of evolutionary computation for industrial applications. In: Proceedings of CEC 2005, Edinburgh, UK, pp. 166–173 (2005)
14. Kotanchek, M., Smits, G., Kordon, A.: Industrial strength genetic programming. In: R. Riolo, B. Worzel (eds.) Genetic Programming Theory and Practice I, pp. 239–256. Kluwer (2003)
15. Kovacic, M., Sarler, M.: Application of the genetic programming for increasing the soft annealing productivity in steel industry. Materials and Manufacturing Processes **24**, 369–374 (2009)
16. Okhovat, A., Mousavi, M.: Modeling of arsenic, chromium and cadmium removal by nanofiltration process using genetic programming. Applied Soft Computing **12**, 793–799 (2012)
17. Pearl, J., Mackenzie, D.: The book of Why: The new science of cause and effect. Basic Books, NY (2018)
18. Poli, R., Langdon, W., McPhee, N.: A field guide to genetic programming. Lulu Selfpublishing (2008)

19. Rey, T., Kordon, A., Wells, C.: Applied data mining for forecasting using SAS. SAS Press (2012)
20. Rostami, A., Arabloo, M., Ebadi, H.: Genetic programming approach for prediction of supercritical co2 thermal conductivity. Chemical Engineering Research & Design **122**, 164–175 (2017)
21. Smits, G., Kordon, A., Jordaan, E., Vladislavleva, C., Kotanchek, M.: Variable selection in industrial data sets using pareto genetic programming. In: T. Yu, R. Riolo, B. Worzel (eds.) Genetic Programming Theory and Practice III,, pp. 79–92. Springer (2006)
22. Smits, G., Kotanchek, M.: Pareto front exploitation in symbolic regression. In: U. O'Reilly, T. Yu, R. Riolo, B. Worzel (eds.) Genetic Programming Theory and Practice II, pp. 283–300. Springer (2004)
23. Sonebi, M., Cevic, A.: Genetic programming based formulation for fresh and hardened properties of self-compacting concrete containing pulverized fuel ash. Construction and Building Materials **23**, 2614–2622 (2009)
24. Taylor, J.: Best practices for picking the right kpis for your business. https://www.klipfolio.com/blog/best-practices-for-picking-business-kpis (2016)
25. Venegas, P.: Asymmetric trust and causal reasoning in blockchain-based ais. In: Proceedings of IXth International Conference on Complex Systems, Cambridge, USA, pp. 44–55 (2018)

Chapter 9
Using Modularity Metrics as Design Features to Guide Evolution in Genetic Programming

Anil Kumar Saini and Lee Spector

9.1 Introduction

Modularity is ubiquitous in nature. Most of the systems around us—from our brains to almost every software program ever written—are modular. In biological systems, modularity contributes to the evolvability of an organism, which is its ability to adapt to new surroundings over evolutionary time [1]. In software engineering, it is advantageous in more obvious ways; parts of the program can be changed without affecting the whole program, and modules can act as building blocks that can be arranged in different orders or added on top of other modules to form entirely new programs.

Although genetic programming can easily solve simple problems like the ones found in introductory programming textbooks [2], evolving significantly more complex programs that are nonetheless routinely written by humans is still well beyond the state of the art. How do humans do it? Among other heuristics, humans frequently use modularity to write complex programs; they organize their code into units that can be reused again and again. Since modularity is pervasive in almost all complex systems, it might be the ingredient that genetic programming needs to evolve complex software.

A. K. Saini (✉)
College of Information and Computer Sciences, University of Massachusetts, Amherst, MA, USA
e-mail: aks@cs.umass.edu

L. Spector
Department of Computer Science, Amherst College, Amherst, MA, USA

School of Cognitive Science, Hampshire College, Amherst, MA, USA

College of Information and Computer Sciences, University of Massachusetts, Amherst, MA, USA
e-mail: lspector@hampshire.edu

© Springer Nature Switzerland AG 2020
W. Banzhaf et al. (eds.), *Genetic Programming Theory and Practice XVII*, Genetic and Evolutionary Computation, https://doi.org/10.1007/978-3-030-39958-0_9

In this chapter, we describe two modularity metrics—Reuse and Repetition—that were inspired by similar metrics in software engineering. We show how these metrics can be used to define design features that influence parent selection to incentivize the evolution of modular programs. Our experiments demonstrate that the use of this technique can significantly decrease the average size of successful programs.

In the following section, we briefly present the history of prior work on modularity in genetic programming. We then present our modularity metrics and the method by which they can be calculated. This is followed by discussions of how the metrics can be used during evolution. We then present our experiments and their results, and finish with a summary of our conclusions and suggestions for future work.

9.2 Modularity in Genetic Programming

The concept of modularity has long been discussed in genetic programming. John Koza introduced the concepts of Automatically Defined Functions (ADFs) [3] and Architecture Altering Operations [4] in tree-based genetic programming that can be used to evolve reusable modules in the evolving programs. Many other researchers [12] used these concepts to build much more flexible ways of inducing modularity. Module Acquisition [5], for example, designates a group of nodes in a tree as a module and protects it from manipulation by genetic operators. Automatically Defined Macros (ADMs) [6] allows for the evolution of macros capable of altering program architectures. Hierarchical Locally Defined Modules (HLDM) [7] introduces a hierarchical way of defining and using modules in genetic programming. These and other related methods often provide function-like templates to the evolving programs to modify and use during the course of evolution.

Other efforts include designing the genetic programming systems in such a way that modules can be defined and used within the evolvable code itself, instead of being constructed from the templates provided prior to the start of evolution. For example, in PushGP [13], "code" itself is type, which can be manipulated using specific instructions. SignalGP [17] is another technique where the program is a collection of modules. The concept of tags [14], which provides a procedure-calling mechanism during evolution, has also been introduced to encourage the adoption of modules.

Although the importance of modules has long been felt and efforts have been made to help programs acquire modules during the process of evolution, a very few attempts have been made to measure modularity. One such attempt is Functional Modularity [18], which considers modules as functional units and calculates modularity based on their performance on a set of test cases. There are certain other measures of modularity, such as Q-metric [19, 20], which can be used for programs that can be represented as networks.

Despite these efforts, evolving modularity is still termed as one of the open issues in genetic programming [11]. Most of the above-mentioned strategies to evolve modular programs try to give evolution access to the tools it may need to develop modules without offering any real incentive to actually use them. Additionally, these strategies make certain assumptions about the structure of the program, and consequently, some parameters like the number, types, or other information about modules need to be specified in advance before the evolution can start.

One of the motivations to push genetic programming systems to produce modular programs is that evolution often prefers non-modular programs. To understand this, let us look at a hypothetical scenario. Imagine there are two programs with the same error vectors solving the same problem: one with a module reused multiple times, and another without any module. During mutation in the first program, if any change occurs to this module, we would notice a huge change in the error vector, but the same might not happen in the case of the non-modular program.

Although modularity as a concept is not new to the field of software engineering, there have not been many attempts at defining modularity in the context of automatic programming, where instead of humans, machines write the programs. This chapter discusses the modularity metrics introduced in [9] and proposes the ways in which they can be used during evolution to incentivize the development of modules.

9.3 Modularity Metrics

Depending on the field of study, there can be different types of modularity. For example, developmental, morphological, evolutionary, and other kinds of modularity have been discussed in the literature [10]. And there are corresponding metrics to measure them.

In this chapter, we focus on the concept of modularity used in software engineering. Programmers often employ modules to solve parts of a given problem and reuse them multiple times to avoid code repetition. They also use many metrics to calculate the modularity of a given software. Coupling, for example, measures the interdependence between different modules; cohesion, on the other hand, measures the amount of interaction among components of a particular module [8]. A modular software design, therefore, is characterized by low coupling and high cohesion.

General-purpose programming languages like Python, Java, etc. contain constructs in the form of functions, classes, etc. that can be used to identify a module. Genetic programming systems, on the other hand, often do not have such standard constructs and generally use different ways of achieving modularity. Due to this, it becomes difficult to detect modules in different representations in genetic programming, and the metrics used in software engineering might not work for evolved programs.

Nevertheless, the metrics presented in this chapter are inspired by some of the concepts used in software engineering like code reusability, component-based development, etc. Reuse and Repetition—two metrics presented in this chapter—

are also based on the heuristics humans use to write programs; in order to have less repetition, they try to reuse their code using features like functions, procedures, etc.

Since we want a formulation that could be used irrespective of the underlying genetic programming system, we use the information contained in the execution trace of a program to calculate the metrics. Moreover, we maintain that instead of one single metric, there can be multiple metrics measuring different aspects of modularity.

In this section, we first define a module in the context of automatic programming. And then after going over the design principles used to formulate the metrics, we provide the exact equations to calculate those metrics.

9.3.1 Module

A module may be defined as a part of the program, which can be used and modified independently of other parts [9]. For the purpose of this chapter, a single token as well as a collection of tokens—keywords, variables, constants, etc. can be considered as modules, provided the following conditions are met:

1. The order in which the tokens appear in a module should be the same as the order in which they appear in the program. For example, if we have a set of instructions ABC in the execution trace, for it to be considered a module, the same set of instructions should be present in the program in the same order.
2. The module should have a definite beginning and an end. For a group of instructions, as in most of the programming languages, this is indicated by brackets, keywords, or indentations. In the case of single instructions, however, no such construct is needed.

9.3.2 Design Principles for Modularity Metrics

In order to come up with exact formulations for Reuse and Repetition, some design principles were used. The formulations which were in direct conflict with any of these principles were rejected, and those which complied with all of them were taken up for consideration. And the final set of metrics we present in Sect. 9.3.4 is one of them. The design principles are:

1. The frequency with which a module gets executed should contribute more than its size towards the metrics.
2. All consecutive sub-sequences of a module are considered as modules. For example, if ABC is a module, A, B, C, AB, and BC are also modules.
3. Since we are extracting modules from execution traces, there might be some instructions in the trace which were not there in the program. This can happen when an instruction from the program calls instructions from other programs

or libraries. To keep things simple, we will not use such instructions in our calculation of the metrics.

4. The modularity metrics measure the structural properties of a program. Therefore, the formulations of these metrics may not take into account the usefulness or other functional aspects of modules.

5. The modules should appear or be used at least twice for them to be considered in the calculation. This helps in identifying the boundaries of modules.

6. Since both Reuse and Repetition consider the frequency and size of modules, and they measure similar properties of modules, they can have similar formulations. They should, however, be independent of each other. In other words, a program can have a high value of Reuse with a low value of Repetition and vice-versa.

9.3.3 Reuse and Repetition

Reuse measures how frequently a copy of a module gets executed. Repetition, on the other hand, measures how frequently a given module appears in the program. The main difference between the two is that in the former one, one copy gets executed multiple times, whereas, in the latter, there are already multiple copies present in the program, each of which gets executed once.

A program will have high reuse if it has a function that is called repeatedly or a block of code that gets iterated multiple times. The program will have a high value of Repetition if the same set of instructions are written multiple times and executed separately.

9.3.4 Reuse and Repetition from Execution Trace

An execution trace is a sequence of instructions arranged by the order in which they are executed. This sequence is often different from the actual program since the instructions at a particular position (say, a line number) in the program can call instructions at other locations.

As previously described in [9], the procedure to calculate the modularity metrics from execution traces is as follows.

1. Assign an identifier to every token in the program. For simplicity, let this identifier be the position of the token in the program.

2. Execute the program. Maintain two traces, one with instructions (called the execution trace) and the other with identifiers (called the metadata trace). During the course of execution, if we come across an instruction that was not present in the program, we give it a special identifier. Such instructions will not be considered while computing the metrics.

Usually, in order to calculate the error vector of a given program, it is executed on a set of test cases. Consequently, we may have multiple executions traces for a single program. And since the metrics are calculated on the execution traces, we will have a list of values instead of a single number for each metric. To simplify this, we can use one of the two approaches: use summary statistics (for example, mean, maximum, etc.), or choose a value randomly from the list.

As mentioned above, we have two types of traces after the program gets executed. The exact procedure to extract modules from these traces depends on the language in which they are represented.

For the sake of simplicity, a module is any group of instructions or identifiers that is repeated at least twice in the respective traces. Now, a simple and naive way to quantify the amount of reuse and repetition is to compute the proportion of execution trace under reuse or repetition. The corresponding formulation for Reuse (U) and Repetition (P) would be:

$$U = \frac{\sum_{i=1}^{m} l_i \cdot f_i}{l} \tag{9.1}$$

and

$$P = \frac{\sum_{i=1}^{n} l_i \cdot f_i}{l}, \tag{9.2}$$

where there are m modules being reused, n modules being repeated, l_i and f_i are respectively the length and size of a given module i, and l is the length of the trace. The length of the trace is simply the number of instructions present in it and the lengths of both the traces are equal. The main issue with this formulation is that it gives equal weight to the size and frequency of a module, which is in direct contradiction to the design principles given in Sect. 9.3.2.

After increasing the weight of the frequency and normalizing it accordingly, another possible formulation can be:

$$U = \frac{\sum_{i=1}^{m} l_i \cdot 2^{f_i}}{2^l} \tag{9.3}$$

and

$$P = \frac{\sum_{i=1}^{n} l_i \cdot 2^{f_i}}{2^l}. \tag{9.4}$$

This formulation is also problematic. Since the measures give more importance to the frequency of usage (exponentiation) than to the length of a module (simple multiplication), we get higher value of Reuse when we have more frequent modules in an execution trace of a given length. In other words, as the length of the trace increases, the reuse measure starts to prefer short and frequent modules over bigger and less frequent ones. Table 9.1 illustrates this point. Although intuitively reuse

Table 9.1 Reuse values of some toy examples calculated using Eq. 9.3

Execution trace	Reuse value
ABCABCDEF	0.078125
ABCABCABCDEF	0.01953125
ABCABCABCABCDEF	0.0048828125
ABCABCABCABCABCDEF	0.001220703125

Each letter denotes an instruction in the execution trace. We assume there is no repetition

should increase since the module ABC is getting reused more often, the actual reuse calculated from Eq. 9.3 is decreasing. Similar issues occur during Repetition calculations.

The final formulations change the denominator to remedy this. The following equations are simplified versions of the equations given in [9]:

$$U = \frac{\sum_{i=1}^{m} l_i \cdot 2^{f_i}}{2^u} \qquad (9.5)$$

and

$$P = \frac{\sum_{i=1}^{n} l_i \cdot 2^{f_i}}{2^v}. \qquad (9.6)$$

where u is the number of unique identifiers used in the metadata trace, and v is the total number of instructions of the program used in the execution trace. In other words, in normalization, we use the number of unique identifiers or instructions used in the traces, instead of the length of the traces. These numbers are also often different from the length of the program as some instructions get executed multiple times, and some might not get executed at all due to conditional operations. Again, while calculating u and v, we do not include the instructions that are not present in the program and appear only in the traces.

9.4   Using Modularity Metrics to Guide Evolution

Every program generated by genetic programming has certain 'Software Quality Features.' One of these features is correctness (errors on test cases). Although correctness—which focuses only on the 'output' of a program, not its 'structure'— might be the most important feature, it is not the only one. There are many other features that focus on the structure of the generated programs. In this chapter, we will term these features 'design features' and differentiate them from 'correctness features.' We will use errors on test cases as correctness feature and modularity metrics as design features.

In this section, we will explore ways to combine modularity metrics with errors of individuals to guide the evolution of programs.

9.4.1   Using Design Features During Parent Selection

The procedure to combine design features with error values will be different for different parent selection methods. Here, we describe the procedure with lexicase selection [15]. What we essentially aim to do is to have additional pressure to prefer more modular programs during parent selection.

In each iteration of lexicase selection, the test cases (or, the error values on test cases) are shuffled in random order. The design features can be sorted among these error values in one of the following ways:

1. The first option is to shuffle error values and design features together.
2. The second option is to consider design features after error values. What this means is that out of two individuals, one with higher values of design features would be preferred over the other only if they both have the same error values.
3. The third option is to consider design features before error values. What this means is that out of two individuals, one with higher values of design features would always be preferred over the other, irrespective of error values.

We note that modularity metrics are used during selection just like normal error values, but they are not taken into consideration when determining whether a program is a solution to a problem or not.

Another method of using modularity metrics to guide the evolution of modular programs is to filter out individuals with low values of modularity metrics from the population before the parent selection commences.

9.4.2   Using Design Features During Variation

Modularity metrics can also be used to guide the mutation and crossover operators. For example, the values of reuse and repetition can serve as inputs to a given variation operator, and the rate and other parameters of variation can be decided accordingly. Additionally, instead of applying a variation operator uniformly, we can use the reuse and repetition values of the parts of an individual to decide the location of variation.

9.5   Experiments and Results

In this section, we describe the experiments that were conducted to investigate the effects of using modularity metrics as design features. We run our experiments on Clojush[1] (Clojure implementation of PushGP) which evolves programs in a stack-based programming language called Push. Since calculating modularity metrics is time-intensive, we decided to use the problem of symbolic regression because of its faster runtime per generation compared to other more complex problems. Specifically, we try to evolve programs to compute the mathematical expression of $(x^3 + 1)^3 + 1$.

Before presenting our experimental set-up, we first describe the procedure to extract from Push programs the modules to be used to compute the metrics. We also introduce the concept of autosimplification as a way to remove unnecessary instructions from the programs prior to calculating the metrics.

9.5.1   Extracting Modules from Push Programs

As mentioned in Sect. 9.3.4, the procedure to retrieve modules from the execution trace of a given program depends on the language in which the program is represented. Since we are using PushGP in our experiments, we present the algorithm to extract modules from Push programs running on that system. The same procedure can be used for other languages with minimal changes.

Push is a stack-based programming language with separate stacks for every data type [13]. During execution, the instructions can take their inputs from and place their outputs on different stacks. In each iteration, the top element of the execution stack gets executed. Hence, the sequence of the top elements on the execution stack after every iteration becomes the execution trace.

Algorithm in Fig. 9.1, which describes the procedure to calculate the metrics for Push programs, can also be used for programs written in similar stack-based programming languages. Table 9.2 gives an example of a Push program and the corresponding execution and metadata traces.

We first convert the metadata trace into a different representation in the following way. A single instruction is represented as [a:a] where a is the identifier of that instruction. A group of instructions inside parentheses, with the identifier of the first instruction being a and that of the last one being b, is represented as [a:b]. Additionally, modules in an execution trace are called instruction-modules (for example, (exec_dup (exec_swap 1 2))), and those in metadata trace are called identifier-modules (for example, [1:4]).

---

[1]https://github.com/lspector/Clojush.

Input: execution trace of the program containing instructions;
   metadata trace of the program containing corresponding identifiers;
Result: Reuse and Repetition values
mTrace := modified metadata trace (see Table 9.2 for an example);
len := length of *mTrace*;
seqs := an empty set;
for $i = 1,2,3...len$ **do**
   for *module [m:n] in mTrace* **do**
      if $n - m == i$ **then**
         temp_items := take the next $(i+1)$ items $[m_j,n_j]$ from *mTrace* provided
         $[m_j,n_j] \neq [m,n]$ and $m_j \geq m$, and $n_j \leq n$;
         temp_items := sort the items by the first element of each pair;
         seqs := seqs + **AllContinuousSeqs**(temp_items);
         delete these items from *mTrace*;
      end
   end
end
seqs_for_reuse = seqs after removing the identifier-modules appearing only once;
use Equation 9.5 to calculate Reuse;
seqs_for_repetition = unique identifier-modules from seqs;
get the corresponding instruction-modules and their frequencies;|
use Equation 9.6 to calculate Repetition;

Fig. 9.1 Algorithm 1—Calculating metrics for Push programs

Table 9.2 A Push program example and the corresponding execution and metadata traces

Program	`(exec_dup (exec_swap 1 2))`
Execution trace	`((exec_dup (exec_swap 1 2)) exec_dup (exec_swap 1 2) exec_swap 2 1 (exec_swap 1 2) exec_swap 2 1)`
Metadata trace	`((1 (2 3 4)) 1 (2 3 4) 2 4 3 (2 3 4) 2 4 3)`
Metadata trace (modified)	`([1:4] [1:1] [2:4] [2:2] [4:4] [3:3] [2:4] [2:2] [4:4] [3:3])`

To extract modules for Reuse, we use the metadata trace. In each iteration of the outer loop, we search the metadata trace for modules of a certain size. Whenever we find a module of size s, from the next $(s + 1)$ modules in the trace, we collect the ones which are 'included' in the bigger module. For example, if $[1:4]$ is the module under consideration, the modules $[1:1]$, $[1:3]$, $[2:4]$, etc. are considered to be 'included' in the bigger module. Before deleting these smaller modules from the metadata trace, we use them to find modules containing consecutive identifiers as follows. **AllContinuousSeqs()** takes a set of identifier-modules, and output those proper subsets which contain consecutive identifiers. For example, if the input is $\{[2:2],[3:3],[4:4], [6:8]\}$, it will output $\{([2:2]), ([3:3]), ([4:4]), ([2:2], [3:3]), ([3:3],[4:4]), ([2:2],[3:3],[4:4]),([6:8])\}$. This is in tune with the design principles described in Sect. 9.3.2, which allows the subsets of modules to be considered as modules as well. Following this procedure, we will have a list of modules at the end of the outer loop. And the modules which appear at least twice in this list will we considered for computing Reuse.

Now, to get instruction-modules for Repetition calculation, we first get unique identifier-modules from the list described above, and look for the corresponding instruction-modules in the execution trace. From these modules, the ones which are repeated at least twice are considered for computing Repetition. Note that only the modules containing the same set of instructions with different identifiers are considered for calculating Repetition.

9.5.2   Autosimplification

Sometimes, a program evolved by a given genetic programming system contains many instructions that do not contribute to its performance on the test cases. In other words, some code elements can be removed from the program without affecting its performance on the test cases. While these unnecessary instructions do not affect the overall error vector of the program, they can reduce the accuracy of modularity metrics. Hence, we perform two sets of experiments, one where we calculate the metrics on non-simplified programs, and the other where we calculate them on the simplified ones.

In genetic programming, automatic simplification of evolved programs has been shown to improve generalization [22]. It has also been used as a genetic operator [22, 23]. Algorithm in Fig. 9.2 gives a simple procedure to automatically simplify a given Push program with certain number of steps. Although the procedure can produce different simplified programs each time it is run, the prior work shows that it often produces consistent results in practice. For programs written in other languages, similar algorithms can be developed.

9.5.3   Experimental Set-up and Results

As discussed in Sect. 9.4, there are multiple ways of using the modularity metrics to incentivize the development of modules in evolving programs. In our experiments, we focus on using the metrics during lexicase selection.

We follow the procedure laid down in Sect. 9.4.1. Prior to parent selection, we first compute the errors on a set of test cases. Then to calculate the metrics, we choose a test case randomly from the list of test cases, execute the program on that test case, and then use the execution trace so obtained to compute the metrics. During lexicase selection, we shuffle the errors and metrics together for every selection event. In addition to individuals with low errors, we prefer the ones with high values of Reuse and low values of Repetition.

Input: *prog*, the program to be simplified;
      numSteps, number of simplification steps;
      ErrorFunction(), the function used for calculating errors;

errVector := *ErrorFunction(prog)* ;
repeat
    | rand := a random number between 0 and 1;
    | **if** *rand < 0.5* **then**
    | | *newProg* := remove a small number (typically 1 or 2) of random instructions from
    | | *prog*;
    | **else**
    | | *newProg* := remove a random parenthesis pair from *prog*;
    | **end**
    | *newErrVector* := *ErrorFunction(newProg)* ;
    | **if** *newErrVector == errVector* **then**
    | | *prog* := *newProg*;
    | **end**
until *the numSteps limit is reached*;
return *prog*

Fig. 9.2 Algorithm 2—Auto-simplification

Table 9.3 Genetic programming parameters

Parameter	Value
Population size	1000
Number of generations	500
Parent selection algorithm	ϵ-lexicase
Mutation operator	Uniform Mutation by Addition and Deletion
Mutation rate	0.09
Number of runs per condition	30

We perform experiments on the same problem under four different conditions with 30 runs for each condition: without using any metrics, using only reuse metric, using only repetition metric, and using both the metrics. In these experiments, the metrics are calculated on non-simplified programs. The parameters used during the runs are listed in Table 9.3. To deal with floating-point numbers, we use a version of lexicase selection called ϵ-lexicase selection [16]. We do not perform crossover and instead use UMAD (Uniform Mutation by Addition and Deletion)[21] mutation operator. During evolution, the programs have access to all the instructions that operate on float and execution stacks.

The results are given in Table 9.4. The number of successes is the number of runs out of 30 that evolved a successful program passing all the test cases. The size of a push program is the total number of instructions and parenthesis pairs present in it. In addition to presenting the average size of successful programs, we also show the average size of successful programs after simplification (number of simplification steps being 5000) so as give a sense of the proportion of nonessential instructions in them. To test statistically significant differences, we used the pairwise comparisons for proportions for success rates and the Mann-Whitney-Wilcoxon Test for the sizes of successful programs. From the table, it is clear that using reuse as a design feature makes the successful programs smaller, without significantly affecting the success rate. However, when we use repetition metric either separately or with reuse, we get larger programs without any significant difference in success rates. This might be due to the fact that, in order to have a low value of repetition, the programs can often have useless instructions, which basically increases the denominator in Eq. 9.6. This is well supported by the last two rows of the table, where the difference between the sizes of programs before and after simplification is very large.

To restrain the modularity metrics from preferring programs with more nonessential instructions, we used the concept of autosimplification as discussed in Sect. 9.5.2. In other words, prior to calculating the metrics, we simplified the program with the number of simplification steps being 50. Accordingly, we performed 30 runs for each of the conditions mentioned earlier. The results are given in Table 9.5. Although we do see a slight improvement in the success rate while using the reuse metric, the results are very similar to Table 9.4. We still get large programs while using the repetition metric as we did before. As a future work, therefore, we can either explore increasing the number of simplification steps or use some other mechanism to rein in the increase in the program size when using the repetition metric.

Although the average size of the successful programs after simplification in the last row is significantly less than the corresponding value for no-intervention case, the performance in terms of the number of successes and the average size of the non-simplified programs is actually worse.

From the results described in Tables 9.4 and 9.5, it can be concluded that the best results are obtained by computing the reuse metric on simplified programs and using it as a design feature in lexicase selection. However, these results are preliminary in nature and are mainly intended to show how the modularity metrics can be used in the evolutionary framework.

Table 9.4 Using metrics as design features

	Number of successes	Average size of successful programs	Average size of successful programs after simplification
No intervention	12	43.08	20.42
Reuse only	12	<u>31.00</u>	14.67
Repetition only	5	60.80	26.8
Both metrics	7	58.71	11.86

The metrics have been calculated on non-simplified programs. The underline indicates that the value is significantly smaller than the corresponding value in the no-intervention case

Table 9.5 Using metrics as design features

	Number of successes	Average size of successful programs	Average size of successful programs after simplification
No intervention	12	43.08	20.42
Reuse only	17	34.12	17.71
Repetition only	5	67.00	23.8
Both metrics	5	62.80	<u>8.6</u>

The metrics have been calculated on simplified programs, with the number of simplification steps being 50. The underline indicates that the value is significantly smaller than the corresponding value in the no-intervention case

9.6 Conclusions and Future Work

In this chapter, we first discussed a set of modularity metrics—Reuse and Repetition—that measure different aspects of modularity. We then presented multiple schemes of using these metrics in the framework of evolving programs. We demonstrated one of these methods by computing the metrics on Push programs and using them as design features in lexicase selection. We presented some preliminary results which show that using the reuse metric gives us more compact successful programs.

We use symbolic regression in our experiments, which is very easy to solve and might not even benefit from having modules. Therefore, as a future work, we need to run the same set of experiments on more complex problems like the ones from the benchmark suite of [2]. We can further optimize the whole procedure to calculate the metrics and also explore other methods of using them as mentioned in Sects. 9.4.1 and 9.4.2.

While the metrics presented in this chapter focus only on the structure of modules, new metrics that take into account the usefulness of those modules in solving different test cases can also be added to the metrics suite.

Considering almost all complex systems around us are modular in nature, we might benefit substantially from encouraging modularity in genetic programming systems if we want to synthesize programs as complex as the ones written by humans. We also believe the lessons learned from employing modularity metrics in the evolutionary framework can help us develop the tools to tackle some of the most difficult and unsolved problems in genetic programming. Further experiments to investigate the utility of modularity in the evolution of programs may also shed some light on the role of modularity in the evolution of biological systems.

Acknowledgements We would like to thank other members of Hampshire College Institute for Computational Intelligence for their valuable inputs.

This material is based upon work supported by the National Science Foundation under Grant No. 1617087. Any opinions, findings, and conclusions or recommendations expressed in this publication are those of the authors and do not necessarily reflect the views of the National Science Foundation.

References

1. Clune, J., Mouret, J. B., & Lipson, H. (2013). The evolutionary origins of modularity. Proceedings of the Royal Society B: Biological Sciences, 280(1755), 20122863.
2. Helmuth, T., & Spector, L. (2015, July). General program synthesis benchmark suite. In Proceedings of the 2015 Annual Conference on Genetic and Evolutionary Computation (pp. 1039–1046). ACM.
3. Koza, J. R. (1992). Genetic programming: on the programming of computers by means of natural selection (Vol. 1). MIT press.
4. Koza, J. R. (1994). Architecture-altering operations for evolving the architecture of a multi-part program in genetic programming.
5. Angeline, P. J., & Pollack, J. (1993, February). Evolutionary module acquisition. In Proceedings of the second annual conference on evolutionary programming (pp. 154–163).
6. Spector, L. (1995). Evolving Control Structures with Automatically Defined Macros. Submitted to the 1995 AAAI Fall Symposium on Genetic Programming.
7. Banzhaf, W., Banscherus, D., & Dittrich, P. (1999). Hierarchical genetic programming using local modules. Secretary of the SFB 531.
8. Dhama, H. (1995). Quantitative models of cohesion and coupling in software. Journal of Systems and Software, 29(1), 65–74.
9. Saini, A. K., & Spector, L. (2019). Modularity Metrics for Genetic Programming. In Genetic and Evolutionary Computation Conference Companion (GECCO 2019 Companion), July 13–17, 2019, Prague, Czech Republic. ACM, New York, NY, USA, 4 pages. https://doi.org/10.1145/3319619.3326908
10. Callebaut, W., Rasskin-Gutman, D., & Simon, H. A. (Eds.). (2005). Modularity: understanding the development and evolution of natural complex systems. MIT press.
11. O'Neill, M., Vanneschi, L., Gustafson, S., & Banzhaf, W. (2010). Open issues in genetic programming. Genetic Programming and Evolvable Machines, 11(3–4), 339–363.
12. Gerules, G., Janikow, C. (2016, July). A survey of modularity in genetic programming. In 2016 IEEE Congress on Evolutionary Computation (CEC) (pp. 5034–5043). IEEE.
13. Lee Spector. 2001. Autoconstructive evolution: Push, pushGP, and pushpop. In Proceedings of the Genetic and Evolutionary Computation Conference (GECCO- 2001), Vol. 137.
14. Spector, L., Martin, B., Harrington, K., Helmuth, T. (2011, July). Tag-based modules in genetic programming. In Proceedings of the 13th annual conference on Genetic and evolutionary computation (pp. 1419–1426). ACM.
15. Helmuth, T., Spector, L., & Matheson, J. (2015). Solving uncompromising problems with lexicase selection. IEEE Transactions on Evolutionary Computation, 19(5), 630–643.
16. La Cava, W., Spector, L., & Danai, K. (2016, July). Epsilon-lexicase selection for regression. In Proceedings of the Genetic and Evolutionary Computation Conference 2016 (pp. 741–748). ACM.
17. Lalejini, A., Ofria, C. (2018, July). Evolving event-driven programs with SignalGP. In Proceedings of the Genetic and Evolutionary Computation Conference (pp. 1135–1142). ACM
18. Krzysztof Krawiec and Bartosz Wieloch. 2009. Functional modularity for genetic programming. In Proceedings of the 11th Annual conference on Genetic and evolutionary computation. ACM, 995–1002.
19. Newman, M. E. (2006). Modularity and community structure in networks. Proceedings of the national academy of sciences, 103(23), 8577–8582.
20. Qin, Z., McKay, R., & Gedeon, T. (2018). Why don't the modules dominate-Investigating the Structure of a Well-Known Modularity-Inducing Problem Domain. arXiv preprint arXiv:1807.05976.
21. Helmuth, T., McPhee, N. F., & Spector, L. (2018, July). Program synthesis using uniform mutation by addition and deletion. In Proceedings of the Genetic and Evolutionary Computation Conference (pp. 1127–1134). ACM.

22. Helmuth, T., McPhee, N. F., Pantridge, E., & Spector, L. (2017, July). Improving generalization of evolved programs through automatic simplification. In Proceedings of the Genetic and Evolutionary Computation Conference (pp. 937–944). ACM.
23. Zhan, H. (2014, July). A quantitative analysis of the simplification genetic operator. In Proceedings of the Companion Publication of the 2014 Annual Conference on Genetic and Evolutionary Computation (pp. 1077–1080). ACM.

Chapter 10
Evolutionary Computation and AI Safety

Research Problems Impeding Routine and Safe Real-World Application of Evolution

Joel Lehman

10.1 Introduction

As the capabilities and pervasiveness of machine learning (ML) and artificial intelligence (AI) increasingly affect society, there is increasing concern about the *safety* of such systems, i.e. the potential of accidental harm from implementation errors and unintended consequences in ML algorithms. As a result, there has been increasing interest in the nascent field of *AI safety* [2, 9, 11, 20, 39, 75], which seeks to understand and solve the technical challenges in developing and deploying AI that does what its designer intended it to do. The purpose of this chapter is to explore how the study of AI safety intersects with that of evolutionary computation (EC), to both highlight an exciting and important set of safety problems within EC, and to suggest that evolution and EC have important insights that could benefit the general study of AI safety.

To frame the problem of AI safety, we adopt the framework of [2], which defines AI safety as concerned with accidents in ML systems, and defines five problems within three broad categories of issues: (1) specifying the wrong objective function, (2) making safe and efficient use of a true but expensive objective (e.g. human feedback), and (3) how to improve or adapt safely while interacting with the real world. A running example in that paper, which we adopt here, describes a robot with the task of cleaning an office using common tools; we modify the example to assume that the controller for this robot has been evolved, i.e. with an EC technique like neuroevolution [36, 74] or genetic programming (GP; [5, 29]) in the setting of evolutionary robotics (ER; [42, 47]). While this running example is posed in the reinforcement learning (RL) setting of ER, similar issues can arise whenever

J. Lehman (✉)
Uber AI, San Francisco, CA, USA
e-mail: joel.lehman@uber.com

© Springer Nature Switzerland AG 2020

W. Banzhaf et al. (eds.), *Genetic Programming Theory and Practice XVII*, Genetic and Evolutionary Computation, https://doi.org/10.1007/978-3-030-39958-0_10

an EC-trained artifact interacts with the real world; for example, a credit-scoring system trained with GP symbolic regression (e.g. as in [48]) when deployed might enact unintended consequences on the real-world borrowers its decisions affect, e.g. by basing decisions on ethically- and legally-problematic borrower traits (e.g. race).

One motivation for this chapter is to draw attention within EC to a selection of interesting and important concrete research problems (as introduced by [2]), in hopes of encouraging progress towards one of EC's aspirations: to provide mature and reliably safe solutions for real-world AI problems. If EC systems are increasingly trained, refined, and applied in the real-world, it becomes necessary to deal with real-world complications that are often side-stepped in closed-world research benchmarks; grappling with these issues is thus necessary for EC to transition into a reliable approach for safely solving real-world problems. For example, if evolution is occurring in an environment alongside humans (e.g. evolving a robot controller that interacts with people in an office setting) much care is needed to design an appropriate fitness function that at least does not cause harm in its early incarnations; in contrast, fitness functions in more traditional closed-world ER simulations often undergo many iterations of free-form debugging, with no real danger or cost (beyond wasted time and computation), where initial attempts often create highly-unexpected outcomes [35]. To enable reliable real-world deployment of EC, it may be useful to come up with new automated design procedures, to import tools from AI safety in statistical ML, or to perform new and directed EC research on solving technical safety problems.

A complementary motivation is to highlight AI safety problems for which EC techniques might be particularly well-suited to make significant contributions. For example, the subfields of quality diversity (QD; [37, 50]) and open-ended evolution [60, 66] might provide a natural mechanism to create a diverse set of test-scenarios to illuminate rare but important potential failures modes of ML systems (that might otherwise go unidentified). For example, the fooling images work of [46] shows how EC can automatically identify diverse visual patterns that a deep neural network will confidently misidentify). Overall, while most current AI safety work is conducted with traditional statistical ML (e.g. gradient-based deep learning approaches), EC might bring new ideas, perspectives, and techniques to bear on such problems.

A final motivation is to consider if and how natural evolution solved problems similar to those tackled by AI safety researchers. For example, evolution has designed various means of collaboration among social animals and between mutualistic species, that in effect minimize negative side-effects to other agents (an important topic in AI safety). Additionally, evolution has uncovered ways to explore more safely both across an evolutionary timescale (i.e. through the evolution of evolvability [27, 68], whereby evolution favors improved variation) and an individual organism's lifetime (i.e. through the complementary instincts of curiosity and fear [7]). The hope is that biological inspiration might point the way towards potential solutions to these kinds of safety problems in EC or in ML at large.

The conclusion is that AI safety is likely to be a growing field of interest in coming years that offers a range of interesting technical challenges, and that EC may both have important insights to offer and benefits to gain from research in that community.

10.2 Background

The next sections describe the field of AI safety, and how EC is applied in the real world, which helps to understand safety concerns from an EC perspective.

10.2.1 AI Safety

The field of *AI safety* [2, 11] seeks to pose and solve technical challenges involved in developing AI that in practice does what its designer intends it to do. The hope is to help foresee and avoid harmful accidents that might result from good-intentioned AI gone astray, for example, through misspecified fitness functions or differences between the training and testing environments. While the name "AI safety" naturally evokes ideas of direct physical safety (e.g. how to make sure there are sufficient guard-rails that prevent a robotic arm from accidentally hitting a human), the problems studied in AI safety also encompass more abstract and broad concerns. Such concerns include immediate and short-term ones, like how a mobile robot driven by RL can continually improve its policy by exploring, without taking any catastrophic actions (such as those that cause harm to itself, to the environment, or humans); they also include more speculative concerns about the future (e.g. how to make sure an AI that surpassed human intelligence would still be controllable and aligned with our interests).

One central challenge in AI safety, relevant both to short and long-term concerns, is known as the *value alignment* problem: How to align what a computational agent values with what we value. This problem might appear at first simple, because as designers of agents we have complete control over their incentives. However, such alignment remains an unsolved technical challenge. Currently we do not know how in practice to algorithmically specify (or learn from data) the complexity of what humans care about, e.g. our moral intuitions, common-sense knowledge, and cultural norms, all of which can potentially come to bear upon what we intend for a computational agent to do. In other words, EC as of yet lacks a procedure to specify a correct and complete fitness function that encompasses all the background context that could be important for a system that interacts appropriately with humans and society.

More concretely, even for an AI system that interacts with the real world in very limited ways, it is still often a challenge to design a fitness function that truly measures or incentivizes correct behavior [35]. Indeed, the typical paradigm in AI remains to specify a fixed and relatively simple objective function (e.g. a fitness function in EC) that is then optimized through search; however, as practitioners in

EC are well-aware, an intuitive fitness function can often be optimized in unexpected ways [35]. While there exist candidate approaches to value alignment [20, 39, 75], the problem at core currently remains unsolved.

Interestingly, even if incentives are aligned, i.e. the learning system is provided with the correct objective function, how to successfully (and safely) optimize that objective function is still a difficult and unsolved problem in its own right. For example, an RL agent that is given the correct objective to optimize can still make mistakes *while it is being optimized* (e.g. it can make harmful mistakes while exploring how to improve its policy); or, the objective might be challenging to optimize (e.g. it might instantiate a fitness landscape with many local optima), and the locally-optimal policies found by search in practice might not be value-aligned.

One useful framework for categorizing technical challenges in AI safety comes from [2], which divides safety problems into five categories: avoiding negative side effects, reward hacking, scalable oversight, safe exploration, and robustness to distributional shift (see Table 10.1 for short descriptions of each). We adopt this framework in this paper for relating AI safety problems to EC and evolution, and later in this paper describe each of these problems in detail and how they emerge in EC.

One general consideration for AI safety is that it is most relevant when considering applying AI algorithms to real-world situations, where human well-being, broadly speaking (e.g. including not only physical safety, but also social harm from biased high-stakes decisions [71] or offense from insensitive classifications [1]), might be at stake. Thus the next section reviews common paradigms for applying EC to the real world.

Table 10.1 The table describes five categories of technical challenges in AI safety, as identified by [2]

Avoiding negative side effects	Negative side effects result from a fitness function that correctly specifies how to *narrowly* achieve a goal, but does not penalize possible harms to the environment or other agents
Reward hacking	Reward hacking is when a fitness function fails to well-specify how to achieve a goal; evolution can therefore maximize fitness in an unexpected and undesirable way
Scalable oversight	Scalable oversight requires effectively and efficiently balancing the use of cheap proxy fitness functions (e.g. a simple heuristic) with expensive but more accurate fitness evaluations (e.g. human assessment)
Safe exploration	Safe exploration studies how evolution can learn effective behavior while minimizing catastrophic actions taking during learning
Robustness to distributional shift	Robustness to distributional shift requires real-world applications of evolution to safely deal with situations not seen during training

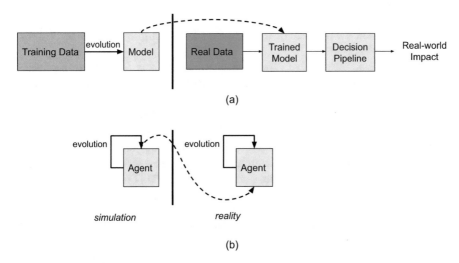

Fig. 10.1 Common paradigms for EC to impact the real world. In (**a**) supervised learning, a model is evolved by fitting training data, and then is deployed into a larger decision pipeline that involves real data; the decisions resulting from the pipeline (that are influenced by the trained model) translate into real world impact. In (**b**) evolutionary robotics or RL, an agent is often first trained in simulation, and then transferred across the reality gap, where it can potentially be further evolved. Alternatively, in embodied evolution, an agent is trained from the onset in the real world

10.2.2   EC and the Real World

There are many different motivations for studying EC. While one entirely legitimate such motivation is to understand the creative potential of algorithms inspired by biological evolution for its own sake, researchers in EC often explicitly aim towards real-world applications of their ideas, or at least paint a viable path towards how their ideas might be translated into beneficial real-world impact. Below we describe how such translation often happens in both supervised and reinforcement learning problems (see Fig. 10.1 for a high-level summary).

10.2.2.1   Supervised Learning

When EC is applied to supervised learning, i.e. training where the task is to predict or classify over a labeled training set, it is important to recall that supervised training performance is rarely an end in itself. While improved accuracy on a benchmark is often critical for publishing a paper about symbolic regression [30] or neural classification [52], such accuracy is only practically important insofar as it feeds into the downstream task the model is *applied to*. For example, a classification model of credit-worthiness might be applied to decide whether a loan should be granted or not. While improved accuracy will likely contribute to such use cases, it will not take into account the nuances of the domain, like the differing impact of different kinds of mistakes [1].

Thus, while applications of supervised learning might at first not seem relevant to AI safety, the objective that a supervised-learning EC model is trained towards (e.g. classification accuracy) nearly always serves only as a *proxy* for the "true" downstream objective in the real world (e.g. efficient loan allocations that abide by legal and moral norms). Notably, limitations of such proxies are well-known; for example, the fairness, accountability, and transparency community within ML has highlighted how maximizing training accuracy can result in models that base decisions on societally-unacceptable criteria [71, 76]. This pervasive type of gap between the proxy and the true objective can be seen as a manifestation of the general value alignment problem, and techniques for minimizing such a gap highlight how AI safety research can be relevant to EC-based supervised learning.

10.2.2.2   Reinforcement Learning

When EC is instead applied to robotics or RL, evolution most often first occurs within simulated environments. The idea is that policies trained in simulation can subsequently be transferred to reality [21, 28], and potentially further evolved in the real world. The reasons for training in simulation include that real-world evaluations can be slow, tedious, and expensive, often risking damage to hardware (like a robot) and to the broader environment (like humans coexisting with the robot). Simulation enables more convenient large-scale experimentation (given sufficient computation), although both how to design accurate simulations for complicated domains and how to successfully transfer policies from simulation to the real world remain challenging areas of research [21, 28, 49].

Safety concerns in this paradigm can emerge from simulations that do not reveal safety-critical edge-cases later encountered when models are deployed in the real-world, or from changing circumstances in reality (i.e. distributional shift) that are not captured for in simulation. Another paradigm in EC is embodied evolution [70], wherein evolution is conducted in the real world, to circumvent the challenges of building accurate simulators and crossing the reality gap. In this setting, to the extent that evolved policies interact with humans or can damage their robotic body or their environment, there may be the need for potentially expensive supervision (an AI safety issue discussed in more detail later). In general, because there is never the protective buffer of simulation between a policy and the real-world, safety considerations in embodied evolution may be more challenging than in other settings.

The conclusion is that as EC strives and achieves greater real-world impact, there will likely be a corresponding increased risk (albeit still potentially minor in many domains) of unintentional harm, independent of the specific paradigm by which EC models are trained and deployed.

10.3   EC and Concrete AI Safety Problems

This section explores more concretely how ideas from EC intersect with those from AI safety. We adopt the framework of [2], which identifies five classes of concrete problems that can cause AI accidents: avoiding negative side effects, reward hacking, scalable oversight, safe exploration, and robustness to distributional shift (described from a high level in Table 10.1). For each of these five problems we introduce the problem, describe how it can arise in EC, how it relates to various research areas in EC, and suggest directions for potential solutions to such problems. Note that our main aim here is to frame AI safety for EC researchers and practitioners, and as a result, we will not comprehensively survey the broader study of AI safety within ML; for more comprehensive surveys, see [2] or [11].

10.3.1   Avoiding Negative Side Effects

The problem of negative side effects is that a well-specified fitness function must not only reward achieving a desired goal narrowly, but should also penalize possible negative consequences on the broader environment. That is, a fitness function is often under-specified in practice, even if the conditions of achieving the desired goal are correctly described. The reason is that there are many ways to short-sightedly accomplish a goal that humans would nonetheless find unacceptable. For example, borrowing from [2], a robot might knock over an expensive vase en route to its destination; even if the robot arrives successfully at its destination (its goal), the damage to the vase is an unacceptable negative byproduct of the robot pursuing its goal.

If the fitness function does not penalize for breaking the vase, the resulting negative side effects could be viewed as a failure of the researcher to express the correct fitness function. However, while one could attempt to anticipate and hard code into the fitness function every negative contingency, such exhaustive anticipation is often unrealistic, and at best tedious. Ideally, there would be a way to automatically (or with minimal supervision) augment a goal-directed fitness function to penalize such undesired impacts. The challenge in designing such an automated method relates to the value alignment problem in AI safety, in that there is much background context (e.g. about what objects in the environment are fragile or important) that a human brings to their understanding of what an acceptable solution is; such context is difficult to effectively and exhaustively translate into a fitness function (although some projects do aim to distill such background knowledge [41]).

Interestingly, most EC and ER environments are constructed such that there is little potential for negative side effects; the reason is that richer environments are more challenging to model, more computationally demanding to simulate, and such complications are often orthogonal to the research questions under study. In practice, simulated environments in ER are nearly always closed-world and spartan, containing only elements directly relevant to the task at hand. For example, a common variety of ER task involves simulated wheeled robots navigating through an enclosed environment containing only walls and artifacts directly related to the task (e.g. a light switch that can be triggered, or tokens that can be collected). Negative side-effects are often impossible by definition: The robot can not damage itself or anything of importance in its environment.

In ER and EC experiments that involve the real world, or interacting with humans, there is more potential for negative side-effects, although experimenters nearly always a priori minimize that possibility by design. For example, when transferring policies evolved in simulation to the real world, the real world environment is often engineered to mimic the spartan simulated one, and often such transfers are one-off experiments (i.e. the robot will not then be operating in an ongoing way) under intensive supervision. However, despite the minimization by design of negative side effects, the conclusion is that as (or if) EC and ER progresses, we likely will want or need evolved agents to be deployed in complex open-world or human-coinhabited environments; in such situations, the problem of negative side effects can no longer be avoided. Thus, when aiming toward the real world, simulated environments may need to be augmented to include the potential for negative side-effects (and for learning to avoid them), or automated techniques for mitigating side-effects from real-world deployment may need to be developed.

So far, the problem of negative side-effects appears to be an under-studied aspect of how to scale EC, one that may provide exciting future research directions. One possible paradigm for minimizing negative-side effects is to train EC agents through interactive evolutionary computation (IEC; [64]), i.e. to involve humans directly in the breeding process. Due to the problem of user fatigue in IEC [64], i.e. that the task of breeding can become monotonous and exhausting, it is difficult to scale IEC, which necessitates learning surrogate models [22] or applying distributed IEC [55], i.e. systems that involve many humans breeding in potentially uncoordinated ways. Overall, the interaction of IEC with embodied evolution or ER in general (as in [73]) could benefit from greater study from a safety perspective. Current research directions in ML that address negative side effects include penalizing for changes to the environment [4], or algorithms that *satisfice* instead of optimize unboundedly [65] (motivated by the idea that side-effects may often result from extreme optimization). Both such approaches could potentially be adapted for EC.

10.3.2 Reward Hacking

The problem of reward hacking, like that of negative side-effects, is caused by an incompletely- or incorrectly-specified fitness function. While negative side-effects are collateral damage incurred while successfully achieving the desired objective, reward hacking is when optimization uncovers unexpected ways to maximize the fitness function *without* achieving the desired objective. For example, if the true objective of a cleaning robot is to clean the office, but its fitness function rewards for each individual mess the robot cleans, the robot may discover that it maximizes fitness by creating new messes that it can subsequently clean [2].

The phenomenon of reward hacking is familiar to most EC practitioners; nearly all of us have encountered situations where an intuitive fitness function is maximized by counter-intuitive (and undesirable) behavior. Indeed, that so many illuminating (and funny) anecdotes of reward hacking existed in the EC community was one main inspiration behind the crowd-sourced documentation effort of [35], which describes many reward-hacking examples. A representative example is found in Karl Sims' seminal virtual creatures work [57]. In early attempts to evolve locomotion gaits by rewarding forward motion, the result was not locomotion, but morphological evolution towards tall rigid bodies that could exploit their potential energy by falling or somersaulting forward.

Beyond EC, the challenge of constructing incentives for agents (like fitness functions) that cannot be undermined is well known in other fields. For example, in economics, Goodhart's law [17] states that "when a measure becomes a target, it ceases to be a good measure." Similar understanding goes by the name of the principle-agent problem in economics and political science [53], and similar challenges exist in designing contracts in law [18]. Further, there are many historical examples of perverse incentives, where an incentive to solve one problem instead exacerbates it; for example, a French colonial program in Hanoi paid citizens for turning in rat tails, in hopes of exterminating rats, but it instead led to *farming* rats [67]. This consilience of evidence suggests that designing incentives is generally difficult, and that humans are habitually overconfident about their ability to skillfully do so, often failing to anticipate subtle loopholes instantiated by intuitive reward structures. In this way, reward hacking in EC and ML is one manifestation of a broader problem.

In practice, reward hacking in EC is often solved through iteration. First, an intuitive fitness function leads to surprising and undesirable outcomes, that are understandable only in hindsight. The experimenter then attempts to modify the fitness function to patch the problem, which potentially may lead to a different kind of exploit that must also be patched. Interestingly, because these failed incentives can be viewed as failures of the experimenter, and happen within the loop of scientific experimentation that precedes a polished experimental setup, they are often not reported scientifically [35]; as a result, the prevalence and importance of reward hacking in EC may be under-appreciated and understudied.

While frustrating, when evolution occurs in simulation such reward hacking may not cause harm much beyond wasted experimenter effort and time. However, the ability for EC practitioners to quickly and safely explore new tasks, especially in settings such as embodied evolution or reality-gap crossing, is undercut by the expertise and trial-and-error needed to construct reliable fitness functions.

As in negative side-effects, IEC is one avenue for helping to overcome reward hacking, by involving human judgment to assess quality during evolution rather than by crafting fixed heuristics. Beyond directed human breeding, humans may also supply other (potentially richer) forms of guidance to further constrain or replace traditional fitness functions, like demonstrations of acceptable behavior or heuristic advice, as in [25].

Such EC research directions can be seen as connected to similar potential solutions in traditional ML, such as imitation learning [3], wherein an agent learns how to imitate expert demonstrations of behavior; cooperative inverse reinforcement learning [18], where a reinforcement learning agent cooperates with a human to discover and optimize the human's preferences; or reward modeling [39], wherein a machine learning model is trained to predict human preferences (similar to surrogate models used in IEC [22]. Exploring if and how such ML methods could apply to EC (e.g. evolutionary imitation learning, or applying deep learning models to learn models of human preferences to drive evolution) may be a productive area of future research.

10.3.3 Scalable Oversight

The problem of scalable oversight is that in EC and learning systems in general, there is often expensive-to-gather information that accurately reflects how acceptable a solution is, but such guidance is too expensive to be applied as the primary driver of search. For example, a very accurate measure of fitness for a cleaning robot might require expensive manual testing of how dirty a carpet is before and after the robot is deployed within a room. Other proxy measures may be more cheaply available, such as a human giving a quick glance to a room, or by the robot measuring how much dirt it is picking up. However, such proxies might exacerbate problems such as negative side effects or reward hacking [2]. For example, a robot maximizing dirt picked up might knock over a plant to gain access to more dirt, or a robot maximizing human approval after a quick glance might hide messes under a rug. The issue is how to efficiently and effectively apply combinations of cheap proxy signals with occasional expensive feedback, to produce a practical (and well-behaved) learning system.

One way the issue of scalable oversight emerges in EC is through the practical construction of real-world fitness functions (e.g. fitness functions for fine-tuning policies in reality that were first learned in simulation, or fitness functions applied in embodied evolution). In other words, when applying evolution in a real-world situation, what sensors are available on a robot, what a human can easily evaluate, or how the environment can be augmented with automated sensors to evaluate aspects of behavior (e.g. motion capture equipment or ceiling-mounted cameras) will affect what fitness functions are possible to automate, and the overall cost-effectiveness of executing different experiments.

However, scalable oversight, like other AI safety issues, is often eliminated by design from simulated EC domains. Experiments in which cheap proxy fitness evaluations are not possible or in which they fail (due to reward hacking or negative side-effects) are unlikely to be pursued or published. However, if progress could be made on enabling more scalable oversight, it might extend the range of what kinds of embodied evolution or real-world fine-tuning could be performed. In this way, scalable oversight is an interesting avenue of research not only for safety reasons, but because it may help expand the complexity of domains for which real-world EC can be applied.

The area of EC research most similar to scalable oversight is that of surrogate-assisted EC [22], wherein expensive-to-calculate fitness functions are approximated with a learned model; of particular interest (for their potential efficiency) are surrogate models that intelligently choose which points in the search space to subject to expensive ground-truth fitness queries. For example, [14] applies Bayesian optimization to enable a data-efficient quality diversity [50] algorithm. Another related area of EC study are methods that estimate which genomes are likely to successfully cross the reality gap [28]; the reason is that there is an analogy between simulations (and their relation to reality) and proxy fitness measures (and their relation to ground-truth fitness).

From a ML perspective, [2] propose that a semi-supervised formulation of reinforcement learning may be a productive paradigm for tackling scalable oversight. The idea is that an agent only receives reward information on a small subset of its experience (as opposed to the traditional RL setting where reward is observed for each action taken in the environment). In particular, the agent must learn *when* to request expensive reward information, and is incentivized to learn cheap proxy measures that correlate with the expensive reward. Because EC uses fitness functions that operate over an individual's entire evaluation, rather than the per-timestep rewards of traditional RL, it may not be easy to translate such a paradigm to EC (although it could be an interesting direction for research). One potential way of framing semi-supervised RL for evolutionary RL is to learn a semi-supervised reward predictor (with ML) that could assign fitness to individuals by observing their sensory-motor stream.

10.3.4   Safe Exploration

The problem of safe exploration is how evolution (or individuals capable of life-time learning) can explore new solutions without ever (or only very rarely) taking catastrophic actions, i.e. ones that harm valuable aspects of the environment, including humans or expensive equipment such as robots. Note that safe exploration remains a problem even if objectives are correctly specified: Even if a fitness function correctly identifies all unacceptable negative side-effects, and a *properly-trained* agent would thus avoid such effects, *during learning* an agent might still undertake catastrophic actions. For example, the cleaning robot may suffer a fitness penalty for breaking a vase, but it still needs to experience that penalty during training to learn to avoid breaking it. A related problem is that given a robotic controller that behaves safely, there is no guarantee that an arbitrary mutation of it will also be safe. The danger of exploration is a deep philosophical problem, in that the very act of exploration seems inherently to be about stepping into the unknown. However, humans can often successfully explore new possibilities and emerge relatively unscathed (sometimes using mental models to predict whether a new strategy would be catastrophic before trying it, somewhat similarly to model-based RL [63]), suggesting that practical solutions may be possible.

There are two main ways that real-world accidents from safe exploration can emerge in EC. First, take the case of learning a plastic policy (e.g. a policy that learns from experience *during its lifetime* [58, 59]). For example, a robot might be trained to explore any environment it is embedded within, in search of a particular goal. In effect, such an agent must learn *how to explore*, and if the deployment plan involves the real world (through embodied evolution, or crossing the reality gap), then there are risks from unsafe exploration. For example, in a new environment, a learned exploratory strategy might lead the robot to damage itself. Second, there is the case where a learned (non-plastic) policy is either trained in the real world (embodied evolution), or is fine-tuned in the real world after being trained in simulation. In this case, exploring the space of policies (through mutations of existing policies) may result in unsafe policies. For example, in some robotics domains solutions are known to be fragile, i.e. that most mutations result in degenerate (possibly damaging) behavior [33, 38]. For concreteness, a robot trained to walk successfully in simulation may lose some performance when transferred across the reality gap, and there is no guarantee that perturbations of the transferred policy (explored in hopes they will improve the walking policy) will not cause the robot to fall and harm itself.

Overall, it may be impossible to solve the issue of safe exploration without involving some form of human oversight. The reason is that learning what is unsafe seemingly requires either: (1) an accurate model of the world that includes robust identification of catastrophes, (2) labelled data of all possible causes of unsafe scenarios in a domain, or (3) active experience in the domain with feedback from an overseer that prevents unsafe actions from being taken. All three require either extensive domain knowledge, e.g. (1) or (2), or direct human intervention (3). In this way, the problem of safe exploration may be intrinsically tied (like some of the other problems) to that of scalable oversight: Given that potentially expensive human feedback is needed, how can it be gathered and exploited in an efficient way to enable reliable real-world exploration?

Interestingly, like other problems mentioned here, often the issue of safe exploration in EC currently arises *outside* the formal scientific process: Domains are constructed that intrinsically minimize risk (e.g. through spartan closed-world design), and guard-rails to minimize damage to real-world robots and their environment are engineered on a robot-by-robot or domain-by-domain basis by experimenters; failure modes (e.g. robot damage) encountered in such experiments are unlikely to be deemed of enough scientific import to be published. Thus, one contribution to studying safe exploration in EC would be to introduce a variant of common ER benchmarks that simulate the idea of safe embodied evolution; for example, a maze navigation task could include deep holes that would endanger a robot, or fragile and valuable aspects of the environment.

Another possible avenue of research for contributing to safe evolutionary exploration is to improve the robustness and evolvability of genomes. For example, some EC methods find parts of the search space that are more robust to mutation [33], or adapt variation operators to increase robustness or evolvability [38, 72], or attempt to enforce small changes to an evolved policy [34]. While not initially motivated by safe exploration, it may be possible to adapt such techniques towards that end. The idea is that with well-tuned variation, parent policies that are safe may be more likely to produce safe children policies, under the assumption that larger policy changes are more likely to be degenerate.

EC could also attempt to solve existing safe exploration benchmarks from the RL community, e.g. the safe exploration grid-world of [40] or domains explored by [45]. Potential safe exploration techniques could also be imported or adapted from studies of safe exploration in RL [15]. Promising such techniques include the approach of [54], wherein human oversight is used to train a supervised learning model that blocks unsafe actions, or [43], wherein catastrophic actions are explicitly stored and rehearsed to endow a RL agent with an intrinsic sense of fear. Similar models could be trained to block unsafe actions for ER or in embodied evolution.

10.3.5 Robustness to Distributional Drift

The problem of robustness to distributional shift is that when AI systems are deployed, they may encounter situations that deviate from the ones it was trained upon. In such situations, a naively trained agent may demonstrate arbitrarily inappropriate behavior, because extrapolating to novel circumstances is challenging. Accidents can thus result in this paradigm if an agent's policy results in ill-suited actions when encountering new situations.

In some EC communities, such as ER, experiments may not always explore how well a learned behavior generalizes to situations other than the exact ones experienced in training; i.e. in the language of statistical ML, the training set doubles as the testing set. As a result, there may be little understanding of how a policy would generalize, and how pathological a robot's behavior would be if it encountered a novel situation. Note that interestingly, the issue of poor generalization is a topic of recent interest in deep RL as well [10, 23, 77].

While this paradigm may not be intrinsically problematic, i.e. if the research question does not involve generalization or real-world deployment, graceful degradation of out-of-training-distribution performance becomes critical as policies are deployed in the real-world (especially open-world scenarios where it is well-understood that all possible situations cannot be anticipated, and that circumstances will likely shift over time).

Several EC communities study partial solutions to this problem. For example, one subfield of EC studies dynamic fitness landscapes [6, 51], wherein evolution continues as circumstances change, which could continually align the policy to the current distribution of scenarios. Further, such fluid adaptation may favor (or be enabled by mechanisms that encourage) more *evolvable* representations, i.e. representations offering diverse and adaptive variation, another important and related field of EC study [26, 68]. Complementarily, others in EC study meta-learning [58], i.e. evolutionary approaches to learning *how to learn*, which may enable a policy to quickly learn online from its own mistakes.

While these research communities provide important insights for tackling distributional shift, new benchmark tasks may be needed to ground out the risks from real-world distributional shift and to determine which (or which combinations) of these techniques would help ameliorate such risks in practice. For example, an ER domain could be introduced in which environments are produced through procedural content generation (PCG; [56]), but where the distribution of PCG parameters changes over evolutionary time; different approaches could be compared by how many catastrophic failures are encountered across evolutionary time.

Solutions could also take inspiration from the study of distributional shift within ML. For example, the insight in Inverse Reward Design [19] is that the fitness function encountered during training should only be trusted insofar as it reflects situations that occur during training (i.e. the human designer of the fitness function designed it explicitly to solve such training situations). An agent should thus have uncertainty over what such a fitness function implies for for situations that never appear in training environments. It may be possible to export such an insight to an evolutionary context, perhaps by querying a human for guidance or forcing a known safe policy to take over when anomalous circumstances are encountered (e.g. as studied by the fields of novelty/anomoly detection [8, 44] or uncertainty-aware RL [13, 24]).

10.4 Discussion

One interesting question is if EC has unique contributions to make to the general study of AI safety. A potential benefit of evolution relative to traditional ML is its divergent creative potential—evolution seems well-suited to discovering a great diversity of well-adapted artifacts. Subfields of EC that study artificial life [32], open-endedness [60], and quality diversity [50] focus on this facet of evolution, which may be of use for helping in particular with the problem of robustness to distributional shift. That is, evolution could be driven to discover a wide range of new training situations to discover latent flaws in learned policies or models, to augment a limited training set that might not cover the diversity of situations that could later be encountered. For example, the work of [46] applies a QD algorithm to find, in a single evolutionary run, a set of diverse images that reliably fool a deep neural network vision model; following work has shown that these kinds of adversarial images can provide safety hazards for real-world use cases of such vision models [12, 31]. Similar QD approaches might also be used to evolve scenarios to stress-test robotic policies. Work in this spirit includes [16], wherein novelty search and GP are used to probe latent behavior of a robotic navigation system and an automobile door locking control system. Similarly, the environments evolved by open-ended systems like POET [69] could be adapted as a testing suite for fixed policies.

A related question is to consider what lessons biological evolution has for AI safety. Many problems faced by AI safety have been solved, at least in some abstract sense, by biology. For example, the problem of negative side effects in AI safety is related to the evolution of cooperation and sociality in biology, in that cooperation often entails considering other agents and their goals in addition to one's own goal (whether through behavioral convention, as in bees, or deliberative thought, as in humans). From this perspective, the negative side effects of a robot pursuing its own limited agenda result from not understanding or taking into account the broader preferences of outside agents (e.g. that a vase is a valuable artifact and should not be broken while cleaning a room). Humans have evolved moral instincts, the ability to empathize with others, and verbal and written language, all of which enables us to understand the gestalt of a task another human might ask us to perform, thereby helping us avoid reward hacking and negative side effects. Similarly, the robustness of our genetic architecture to random mutations and the natural instincts of curiosity and fear are nature's hard-won solution to the problem of safe exploration on a genetic and individual level, respectively. In the same way that evolution (and EC) have a privileged position in the study and understanding of human-level AI (because evolution is the only algorithm to so-far produce human-level intelligence), evolution and EC may also have a privileged position in understanding the AI safety challenges that biology has in some sense solved.

An important question for future study is if methods in EC manifest different kinds of AI safety concerns than those considered within traditional ML, e.g. due to their lack of formal gradient-following or because some EAs produce AI as the result of a divergent creative process (as opposed to optimizing an explicit objective function as common in most ML). Because this question is yet unanswered, it is unclear whether the long-term safety agendas currently popular in ML [9, 20, 39] are applicable to AI produced by paradigms such as evolutionary artificial life or open-ended evolution, which in their grandest aspirations (just as in traditional ML or AI) include producing agents with human-level intelligence [62]. If current safety agendas do not apply to ambitious forms of EC, then formulating new agendas that targeting them may be a valuable pursuit.

A final discussion topic is to draw together some of the recurring themes from considering each AI safety problem separately, in hopes of highlighting promising research questions and paradigms. One theme is the potential need for modifications of EC benchmarks to include safety considerations or the adoption of existing AI safety benchmarks within EC. Benchmarks, for better or worse, help draw researcher attention, and can render seemingly nebulous problems more concrete. Because existing EC domains and benchmarks minimize safety concerns by design (because researchers most often are pursuing research questions orthogonal to safety), new benchmarks may help to catalyze safety research, especially if they are variants of domains familiar to EC researchers. For example, EC techniques could be applied to the AI safety grid-worlds of [40]. Alternatively, existing ER domains (such as maze navigation or ball-gathering) could be augmented with catastrophic actions (for investigating safe exploration), or could include held-out test environments that could test for robustness to distributional shift. Another overarching theme is the potential for some form of IEC to help in the solution to nearly all of the reviewed problems; this is not surprising, because many AI safety problems emerge precisely because human insight is relegated to constructing a fixed setup (i.e. in EC the genetic encoding and the fitness function), and IEC is a framework for allowing human choice to intervene during evolution. Safety considerations may drive more efficient ways to perform IEC (through improved surrogate models), as well as the construction of new forms of IEC. For example, IEC most often helps steer what individuals reproduce, but IEC solutions to problems such as safe exploration may require humans to interact more directly with policies *as they execute*, i.e. to intervene to prevent unsafe actions. One source of inspiration may be systems such as the neuroevolution-based game NERO [61], in which a human experimenter can interact in real time to dynamically change the environment, parameters of the fitness function, and even embody a virtual agent to probe learned agent behaviors.

10.5   Conclusion

AI safety is an important research topic for enabling EC to reach one of its aspirations, which is to maximize its beneficial real-world impact. At first glance, such research might seem uninteresting, because it can evoke sentiments of domain-specific engineering, rather than the pursuit of grand scientific questions; however, AI safety enfolds interesting and philosophically deep unsolved technical challenges, including how to avoid catastrophe while learning about the world, and how to create fitness functions that incentivize agents that abide by the spirit rather than the letter of the law. As ML and AI grow in import, we can expect funding and interest in AI safety to similarly grow, and the hope of this paper is to advocate for EC researchers to both contribute and take note of advances in this developing field.

References

1. Google photos labeled black people 'gorillas'. https://www.usatoday.com/story/tech/2015/07/01/google-apologizes-after-photos-identify-black-people-as-gorillas/29567465/. Accessed: 2019-05-01
2. Amodei, D., Olah, C., Steinhardt, J., Christiano, P., Schulman, J., Mané, D.: Concrete problems in AI safety. arXiv preprint arXiv:1606.06565 (2016)
3. Argall, B.D., Chernova, S., Veloso, M., Browning, B.: A survey of robot learning from demonstration. Robotics and Autonomous Systems **57**(5), 469–483 (2009)
4. Armstrong, S., Levinstein, B.: Low impact artificial intelligences. arXiv preprint arXiv:1705.10720 (2017)

5. Banzhaf, W., Nordin, P., Keller, R.E., Francone, F.D.: Genetic Programming — An Introduction. Morgan Kaufmann, San Francisco (1998)
6. Branke, J., Schmeck, H.: Designing evolutionary algorithms for dynamic optimization problems. In: Advances in Evolutionary Computing, pp. 239–262. Springer (2003)
7. Buss, D.: Evolutionary psychology: The new science of the mind. Psychology Press (2015)
8. Chandola, V., Banerjee, A., Kumar, V.: Anomaly detection: A survey. ACM computing Surveys (CSUR) **41**(3), 15 (2009)
9. Christiano, P., Shlegeris, B., Amodei, D.: Supervising strong learners by amplifying weak experts. arXiv preprint arXiv:1810.08575 (2018)
10. Cobbe, K., Klimov, O., Hesse, C., Kim, T., Schulman, J.: Quantifying generalization in reinforcement learning. arXiv preprint arXiv:1812.02341 (2018)
11. Everitt, T., Lea, G., Hutter, M.: Agi safety literature review. arXiv preprint arXiv:1805.01109 (2018)
12. Eykholt, K., Evtimov, I., Fernandes, E., Li, B., Rahmati, A., Xiao, C., Prakash, A., Kohno, T., Song, D.: Robust physical-world attacks on deep learning models.   arXiv preprint arXiv:1707.08945 (2017)
13. Eysenbach, B., Gu, S., Ibarz, J., Levine, S.: Leave no trace: Learning to reset for safe and autonomous reinforcement learning. arXiv preprint arXiv:1711.06782 (2017)
14. Gaier, A., Asteroth, A., Mouret, J.B.: Data-efficient design exploration through surrogate-assisted illumination. Evolutionary Computation **26**(3), 381–410 (2018)
15. García, J., Fernández, F.: A comprehensive survey on safe reinforcement learning. Journal of Machine Learning Research **16**(1), 1437–1480 (2015)
16. Goldsby, H.J., Cheng, B.H.: Automatically discovering properties that specify the latent behavior of UML models.   In: International Conference on Model Driven Engineering Languages and Systems, pp. 316–330. Springer (2010)
17. Goodhart, C.A.: Problems of monetary management: The UK experience. In: Monetary Theory and Practice, pp. 91–121. Springer (1984)
18. Hadfield-Menell, D., Hadfield, G.K.: Incomplete contracting and ai alignment. In: Proceedings of the 2019 AAAI/ACM Conference on AI, Ethics, and Society, pp. 417–422. ACM (2019)
19. Hadfield-Menell, D., Milli, S., Abbeel, P., Russell, S.J., Dragan, A.: Inverse reward design. In: Advances in Neural Information Processing Systems, pp. 6765–6774 (2017)
20. Irving, G., Christiano, P., Amodei, D.: AI safety via debate. arXiv preprint arXiv:1805.00899 (2018)
21. Jakobi, N., Husbands, P., Harvey, I.: Noise and the reality gap: The use of simulation in evolutionary robotics.   In: European Conference on Artificial Life, pp. 704–720. Springer (1995)
22. Jin, Y.: Surrogate-assisted evolutionary computation: Recent advances and future challenges. Swarm and Evolutionary Computation **1**(2), 61–70 (2011)
23. Justesen, N., Torrado, R.R., Bontrager, P., Khalifa, A., Togelius, J., Risi, S.: Procedural level generation improves generality of deep reinforcement learning.   arXiv preprint arXiv:1806.10729 (2018)
24. Kahn, G., Villaflor, A., Pong, V., Abbeel, P., Levine, S.: Uncertainty-aware reinforcement learning for collision avoidance. arXiv preprint arXiv:1702.01182 (2017)
25. Karpov, I.V., Valsalam, V.K., Miikkulainen, R.: Human-assisted neuroevolution through shaping, advice and examples.   In: Proceedings of the 13th annual Conference on Genetic and evolutionary computation, pp. 371–378. ACM (2011)
26. Kashtan, N., Noor, E., Alon, U.: Varying environments can speed up evolution. Proceedings of the National Academy of Sciences **104**(34), 13,711–13,716 (2007)
27. Kirschner, M., Gerhart, J.: Evolvability.   Proceedings of the National Academy of Sciences **95**(15), 8420–8427 (1998)
28. Koos, S., Mouret, J.B., Doncieux, S.: The transferability approach: Crossing the reality gap in evolutionary robotics. IEEE Transactions on Evolutionary Computation **17**(1), 122–145 (2013)
29. Koza, J.R.: Genetic programming: On the programming of computers by means of natural selection. MIT press (1992)

30. Koza, J.R.: Genetic programming II: Automatic discovery of reusable subprograms. Cambridge, MA, USA **13**(8), 32 (1994)
31. Kurakin, A., Goodfellow, I., Bengio, S.: Adversarial examples in the physical world. arXiv preprint arXiv:1607.02533 (2016)
32. Langton, C.G.: Artificial life: An overview. Mit Press (1997)
33. Lehman, J., Chen, J., Clune, J., Stanley, K.O.: ES is more than just a traditional finite-difference approximator. In: Proceedings of the Genetic and Evolutionary Computation Conference, pp. 450–457. ACM (2018)
34. Lehman, J., Chen, J., Clune, J., Stanley, K.O.: Safe mutations for deep and recurrent neural networks through output gradients. In: Proceedings of the Genetic and Evolutionary Computation Conference, pp. 117–124. ACM (2018)
35. Lehman, J., Clune, J., Misevic, D., Adami, C., Altenberg, L., Beaulieu, J., Bentley, P.J., Bernard, S., Beslon, G., Bryson, D.M., et al.: The surprising creativity of digital evolution: A collection of anecdotes from the evolutionary computation and artificial life research communities. arXiv preprint arXiv:1803.03453 (2018)
36. Lehman, J., Miikkulainen, R.: Neuroevolution. Scholarpedia **8**(6), 30,977 (2013)
37. Lehman, J., Stanley, K.O.: Evolving a diversity of virtual creatures through novelty search and local competition. In: Proceedings of the 13th annual Conference on Genetic and Evolutionary Computation, pp. 211–218. ACM (2011)
38. Lehman, J., Stanley, K.O.: Improving evolvability through novelty search and self-adaptation. In: 2011 IEEE Congress of Evolutionary Computation (CEC), pp. 2693–2700. IEEE (2011)
39. Leike, J., Krueger, D., Everitt, T., Martic, M., Maini, V., Legg, S.: Scalable agent alignment via reward modeling: A research direction. arXiv preprint arXiv:1811.07871 (2018)
40. Leike, J., Martic, M., Krakovna, V., Ortega, P.A., Everitt, T., Lefrancq, A., Orseau, L., Legg, S.: AI safety gridworlds. arXiv preprint arXiv:1711.09883 (2017)
41. Lenat, D.B.: Cyc: A large-scale investment in knowledge infrastructure. Communications of the ACM **38**(11), 33–38 (1995)
42. Lewis, M.A., Fagg, A.H., Solidum, A.: Genetic programming approach to the construction of a neural network for control of a walking robot. In: Proceedings 1992 IEEE International Conference on Robotics and Automation, pp. 2618–2623. IEEE (1992)
43. Lipton, Z.C., Azizzadenesheli, K., Kumar, A., Li, L., Gao, J., Deng, L.: Combating reinforcement learning's sisyphean curse with intrinsic fear. arXiv preprint arXiv:1611.01211 (2016)
44. Markou, M., Singh, S.: Novelty detection: A review - part 1: Statistical approaches. Signal Processing **83**(12), 2481–2497 (2003)
45. Moldovan, T.M., Abbeel, P.: Safe exploration in Markov decision processes. arXiv preprint arXiv:1205.4810 (2012)
46. Nguyen, A., Yosinski, J., Clune, J.: Deep neural networks are easily fooled: High confidence predictions for unrecognizable images. In: Proceedings of the IEEE Conference on Computer Vision and Pattern Recognition, pp. 427–436 (2015)
47. Nolfi, S., Floreano, D., Floreano, D.D.: Evolutionary robotics: The biology, intelligence, and technology of self-organizing machines. MIT press (2000)
48. Ong, C.S., Huang, J.J., Tzeng, G.H.: Building credit scoring models using genetic programming. Expert Systems with Applications **29**(1), 41–47 (2005)
49. Pollack, J.B., Lipson, H., Ficici, S., Funes, P., Hornby, G., Watson, R.A.: Evolutionary techniques in physical robotics. In: International Conference on Evolvable Systems, pp. 175–186. Springer (2000)
50. Pugh, J.K., Soros, L.B., Stanley, K.O.: Quality diversity: A new frontier for evolutionary computation. Frontiers in Robotics and AI **3**, 40 (2016)
51. Richter, H.: Detecting change in dynamic fitness landscapes. In: 2009 IEEE Congress on Evolutionary Computation, pp. 1613–1620. IEEE (2009)
52. Rocha, M., Cortez, P., Neves, J.: Evolution of neural networks for classification and regression. Neurocomputing **70**(16–18), 2809–2816 (2007)

53. Ross, S.A.: The economic theory of agency: The principal's problem. The American Economic Review **63**(2), 134–139 (1973)
54. Saunders, W., Sastry, G., Stuhlmueller, A., Evans, O.: Trial without error: Towards safe reinforcement learning via human intervention. In: Proceedings of the 17th International Conference on Autonomous Agents and MultiAgent Systems, pp. 2067–2069. International Foundation for Autonomous Agents and Multiagent Systems (2018)
55. Secretan, J., Beato, N., D Ambrosio, D.B., Rodriguez, A., Campbell, A., Stanley, K.O.: Picbreeder: Evolving pictures collaboratively online. In: Proceedings of the SIGCHI Conference on Human Factors in Computing Systems, pp. 1759–1768. ACM (2008)
56. Shaker, N., Togelius, J., Nelson, M.J.: Procedural content generation in games. Springer (2016)
57. Sims, K.: Evolving virtual creatures. In: Proceedings of the 21st annual conference on Computer graphics and interactive techniques, pp. 15–22. ACM (1994)
58. Soltoggio, A., Bullinaria, J.A., Mattiussi, C., Dürr, P., Floreano, D.: Evolutionary advantages of neuromodulated plasticity in dynamic, reward-based scenarios. In: Proceedings of the 11th international Conference on Artificial Life (Alife XI), CONF, pp. 569–576. MIT Press (2008)
59. Soltoggio, A., Stanley, K.O., Risi, S.: Born to learn: the inspiration, progress, and future of evolved plastic artificial neural networks. Neural Networks **108**, 48–67 (2018)
60. Standish, R.K.: Open-ended artificial evolution. International Journal of Computational Intelligence and Applications **3**(02), 167–175 (2003)
61. Stanley, K.O., Bryant, B.D., Miikkulainen, R.: Real-time neuroevolution in the NERO video game. IEEE Transactions on Evolutionary Computation **9**(6), 653–668 (2005)
62. Stanley, K.O., Lehman, J., Soros, L.: Open-endedness: The last grand challenge you've never heard of. While open-endedness could be a force for discovering intelligence, it could also be a component of AI itself (2017)
63. Sutton, R.S.: Integrated architectures for learning, planning, and reacting based on approximating dynamic programming. In: Machine Learning Proceedings 1990, pp. 216–224. Elsevier (1990)
64. Takagi, H.: Interactive evolutionary computation: Fusion of the capabilities of EC optimization and human evaluation. Proceedings of the IEEE **89**(9), 1275–1296 (2001)
65. Taylor, J.: Quantilizers: A safer alternative to maximizers for limited optimization. In: Workshops at the Thirtieth AAAI Conference on Artificial Intelligence (2016)
66. Taylor, T., Bedau, M., Channon, A., Ackley, D., Banzhaf, W., Beslon, G., Dolson, E., Froese, T., Hickinbotham, S., Ikegami, T., et al.: Open-ended evolution: Perspectives from the OEE workshop in York. Artificial Life **22**(3), 408–423 (2016)
67. Vann, M.G.: Of rats, rice, and race: The great Hanoi rat massacre, an episode in French colonial history. French Colonial History **4**(1), 191–203 (2003)
68. Wagner, G.P., Altenberg, L.: Perspective: Complex adaptations and the evolution of evolvability. Evolution **50**(3), 967–976 (1996)
69. Wang, R., Lehman, J., Clune, J., Stanley, K.O.: Paired Open-Ended Trailblazer (POET): Endlessly Generating Increasingly Complex and Diverse Learning Environments and Their Solutions. arXiv preprint arXiv:1901.01753 (2019)
70. Watson, R.A., Ficici, S.G., Pollack, J.B.: Embodied evolution: Distributing an evolutionary algorithm in a population of robots. Robotics and Autonomous Systems **39**(1), 1–18 (2002)
71. Whittaker, M., Crawford, K., Dobbe, R., Fried, G., Kaziunas, E., Mathur, V., West, S.M., Richardson, R., Schultz, J., Schwartz, O.: AI now report 2018. AI Now Institute at New York University (2018)
72. Wierstra, D., Schaul, T., Peters, J., Schmidhuber, J.: Natural evolution strategies. In: 2008 IEEE Congress on Evolutionary Computation (IEEE World Congress on Computational Intelligence), pp. 3381–3387. IEEE (2008)
73. Woolley, B.G., Stanley, K.O.: A novel human-computer collaboration: Combining novelty search with interactive evolution. In: Proceedings of the 2014 annual Conference on Genetic and Evolutionary Computation, pp. 233–240. ACM (2014)

74. Yao, X.: Evolving artificial neural networks. Proceedings of the IEEE **87**(9), 1423–1447 (1999)
75. Yudkowsky, E.: Coherent extrapolated volition. Singularity Institute for Artificial Intelligence (2004)
76. Zafar, M.B., Valera, I., Gomez Rodriguez, M., Gummadi, K.P.: Fairness beyond disparate treatment & disparate impact: Learning classification without disparate mistreatment. In: Proceedings of the 26th International Conference on World Wide Web, pp. 1171–1180. International World Wide Web Conferences Steering Committee (2017)
77. Zhang, C., Vinyals, O., Munos, R., Bengio, S.: A study on overfitting in deep reinforcement learning. arXiv preprint arXiv:1804.06893 (2018)

Chapter 11
Genetic Programming Symbolic Regression: What Is the Prior on the Prediction?

Miguel Nicolau and James McDermott

11.1 Introduction

In Genetic Programming Symbolic Regression (GPSR), what is the prior on the prediction $\hat{y}$? This fundamental question does not seem to have been asked in the GP literature.

The prior is a core concept in a Bayesian understanding of statistics, and is thought of as one's belief concerning the value of some variable of interest, prior to observing evidence, based on assumptions or on domain knowledge. Sometimes the variable whose prior is under discussion is an explicit model parameter which we are trying to infer from data. In the context of neural networks, for example, "weight decay" (L_2 regularisation of weights) is equivalent to a prior on weights—specifically, a prior belief that good weights will be normal and small in magnitude [16]. For another example [9] proposes a weakly informative prior on the learnable parameters of logistic regression. In Bayesian linear regression, the model is $\hat{y} \sim \mathcal{N}(\beta^T X, \sigma^2 I)$ and domain knowledge can be injected through priors on β and σ^2. In the context of inference it is most common to discuss priors on a model's internal parameters, as in this example. However, the prior on the Bayesian linear regression output $\hat{y}$ also exists and in this case is normal.

M. Nicolau (✉)
University College Dublin, Quinn School of Business, Belfield, Dublin, Ireland
e-mail: miguel.nicolau@ucd.ie

J. McDermott
National University of Ireland, Galway, Ireland
e-mail: james.mcdermott@nuigalway.ie

© Springer Nature Switzerland AG 2020   201
W. Banzhaf et al. (eds.), *Genetic Programming Theory and Practice XVII*, Genetic
and Evolutionary Computation, https://doi.org/10.1007/978-3-030-39958-0_11

In our context, the variable of interest is again $\hat{y}$, the output of our model, as opposed to an internal parameter or hyperparameter. Whatever prior exists on $\hat{y}$ amounts to an assumption concerning the likely values of y, but rather than being based on an explicit assumption or on domain knowledge, it is an implicit and sometimes unintended result of the language and mechanisms of GP. Any algorithm for regression exhibits some bias. Concerning the GP initialisation algorithm and search algorithm we aim to ask: is it the *right* bias?

We will call the distribution of $\hat{y}$ after the GP initialisation step the *Initialisation Prior*, and the distribution of $\hat{y}$ after an evolution *not* driven by fitness the *GPSR Prior*. It is appropriate to call these *priors* because the initialisation prior is a distribution of values produced directly at initialisation, later to be shifted by evolution; and the GPSR prior is the distribution of values which the GPSR search process would produce even in the absence of training data. In both cases, the prior is a distribution which is "easy" to achieve. Other distributions are not ruled out but require evidence and search effort to be achieved. These priors are called *prior predictive distributions*. In reality, there is not a single initialisation prior or GPSR prior, but rather a family of each, parameterised by algorithm configuration and independent variables. We will thus achieve families of results.

In recent thinking on semantic GP, the *semantics* of a program f is often defined as a vector $[f(X_0), f(X_1), \ldots, f(X_{m-1})] = [\hat{y}_0, \hat{y}_1, \ldots, \hat{y}_{m-1}]$ obtained by applying f to each fitness case X_i in a data set, $0 \leq i < m$. Thus, the distribution of $\hat{y}_i$ (over all trees generated by some GP procedure) is one dimension of the prior on GPSR semantics. Saying the same thing in another way, the prior is the distribution of outputs of trees created by some procedure, in semantic space.

There are several reasons for studying these priors, including:

- Understanding algorithm behaviour as the result of distribution mismatch: if there is a strong mismatch between the y distribution for a particular problem and the $\hat{y}$ prior, performance may be degraded and algorithm dynamics may be affected.
- As an application of this, it may be possible to improve performance by choosing algorithm parameters which match the resulting $\hat{y}$ prior to the y distribution.
- Understanding mutation behaviour: several forms of mutation generate random trees using an initialisation method (e.g. standard subtree mutation [15] and geometric semantic mutation [24] often use the GROW algorithm [15]), so understanding their behaviour requires understanding the $\hat{y}$ initialisation prior.

In the following section (Sect. 11.2) we discuss these motivations in more detail. Following that, we present related work (Sect. 11.3), methodology, experiments and results (Sect. 11.4), applications and consequences (Sect. 11.5) and conclusions (Sect. 11.6).

11.2 Motivation

11.2.1 Distribution Mismatch, Problem Difficulty, and Performance

For any problem and algorithm configuration, we can observe the distribution of y and compare it to the prior on $\hat{y}$. If they are quite different we will say that there is distribution mismatch. This is important as our intuition is that if the true distribution of y for the problem is far from the GP prior, then GP has to work much harder to match the true distribution. The problem becomes harder. We might guess that if a lot of the search effort goes towards matching the distribution, then less is available for matching the individual training cases. More concretely, we adapt and expand the following explanation from Keijzer [12]. If our prior does not match the true y distribution, then early in a GP run, most individuals will be making predictions which are not even in the correct range. The best individual may be one which *predicts a good constant*, that is, a constant that is close to the true y distribution. Such an individual would be a strong attractor and could be a local pseudo-optimum.[1] A large number of the budget of fitness evaluations may be spent on neighbours of such individuals, and improvements in fitness will come about mostly by improving the constant. Eventually, this effect will come to an end as an individual is found which predicts a near-optimal constant, i.e. near the mean of observed y. By this time, diversity may have dropped (and much of the search effort wasted), so there is little ability to evolve from this individual to also match the individual training cases. Keijzer [12] verified this explanation with an experiment:

> Consider for example the two target functions $t = x^2$ and $t = 100 + x^2$. When using standard symbolic regression to find these functions, a large difference in search efficiency can be observed. Whereas the first target function is readily found (often even in the initial generation), the second target function is usually not found at all. [In experiments on the second target function,] genetic programming routinely converges on the average of the training data: a value of 100.37. Only in 8 cases out of the 50 runs that were performed did the particular genetic programming system find something that had a better performance than this average. [...] Once found, diversity has dropped to such a point that the additional square of the inputs is no longer found.

Thus, a mismatch between the y distribution and the $\hat{y}$ prior really matters for problem difficulty. The same experiment could be carried out with a real-world rather than synthetic dataset, by adding (e.g.) $+100$ to the y values in the training and test sets, and observing the change in GP performance.

Keijzer's reasoning also helps us to understand algorithm dynamics: it explains an observation that the best individual during a run on such a problem is often a very small one.

---

[1]In many GP configurations, there are no true local optima since the mutation operator can jump to anywhere in the search space in single step. We can informally define a local pseudo-optimum as a point where improving steps are not impossible but highly unlikely.

11.2.2 Algorithm Configuration

As argued above, an incorrect prior can affect performance. One possibility for researchers is to configure their algorithms to avoid incorrect priors. Choosing algorithm parameters to give a $\hat{y}$ which matches the y distribution on a given problem can be done without any true domain knowledge. Our experiments will aim to give some insight into the effect of GPSR parameters including tree depth and function sets on the prior.

In particular, we will continue some recent work which investigated the effect of GP function sets on the robustness of GPSR [27, 28]. This included both performance on training set and ability to generalise beyond the training set. The present paper will also investigate the effect of GP function sets, but on a different topic—$\hat{y}$ priors. The close relationship between the two is that the robustness of GP is partly explained by the match between its $\hat{y}$ prior and the true y, and partly by other factors. Thus, the present paper attempts to disentangle these effects identified in the previous work and to further justify choosing function sets even without any knowledge that particular functions are present in the true underlying data-generating process.

Recent software provides default function sets such as (add, sub, mul, div) [37] or (add, sub, mul, pdiv, neg, cos, sin) [8], where pdiv is Koza's protected division operator. A review of recent literature in GPSR highlighted (add, sub, mul, pdiv, sin, cos, exp, log) as the most commonly used default function set [28]. Other commonly used functions include (inv, ppow, psqrt, tanh), with ppow and psqrt being Koza's protected versions of x^y and $\sqrt{x}$. Recently, the analytic quotient [26] (aq) operator has seen more application.

One immediate observation on the effect of function set on the GPSR prior is that some function sets are biased-positive and others not, primarily due to the presence of protected operators such as the protected square root. Is this bias beneficial? An initial answer is that it depends on whether the observed y distribution has the same bias. Many problems may have positive y distributions, for example when y occurs as a count or a measurement of a physical quantity such as mass in kilograms or length in metres.

There is a further complication. If the true y distribution is biased-positive, as in the examples above, we may be motivated to choose a function set which is biased-positive, e.g. including a protected square root. However, if our X distribution is all positive, then we may achieve a biased-positive $\hat{y}$ even with a function set which is not biased-positive. The same arguments apply again: X variables measured as counts, or values of mass or length, etc., will be non-negative. Thus, we have an empirical question concerning how strongly the independent variables affect the GPSR prior, relative to the strength of the effect of the function set.

Other parameters such as a limit on tree depth will also affect the initialisation and GPSR priors, and so we can ask whether they should be manipulated in order to achieve a match. Our experiments will address these questions.

11.2.3 Understanding the Behaviour of Search Operators

Several mutation operators in GP generate a random tree as part of their workings, often using an initialisation operator such as the GROW algorithm to do so. Subtree mutation generates one, usually using GROW [31], and pastes it in place of a randomly-selected subtree of the parent individual. Geometric Semantic GP (GSGP [24]) mutation for regression problems generates two random trees and uses a linear combination with the parent. In order to understand the behaviour of such operators, it is necessary to understand the behaviour of the GROW component. This is an application of our initialisation prior. We will address this in Sect. 11.5.3.

11.3 Previous Work on GP Biases

Our work is related to the long stream of literature on GP bias. Much research has focussed on bloat, i.e. the tendency for GP runs to produce larger and larger individuals without improvements in fitness, e.g. [6, 29, 32], with excellent reviews by [19, 33]. Several researchers in the context of bloat studied the effect of mutation and crossover on code growth. A related strand of research aimed to control this bias, e.g. [11, 34, 35].

Further research concentrates on other biases associated with GP operators. Keijzer and Foster [13] pointed out that the crossover operator introduces a bias towards unbalanced trees. McDermott [21] characterised the bias of mutation operators towards exploration or exploitation.

It is also common to study the bias of GP systems in terms of their prior on how the GP language (terminals and functions) is used. For example, Mauceri et al. [20] studied the frequency of use of each function in the Grammatical Evolution (GE) function set in time series feature extraction problems, comparing them to the prior probability of their use in random generation of programs; they also studied the frequency of use of each time-point in the time-series, comparing this to the prior probability in random programs. In both cases, it was found that the evolutionary process in the presence of problem-specific fitness pushed the distribution far from the prior. Whigham [41] proposed a distinction between two forms of bias on the distribution of programs formed by GP: language bias, which depends on e.g. the choice of grammar in a GE system; and search bias, which depends on variation operators and evolutionary dynamics. Our distinction between the initialisation prior and GPSR prior on $\hat{y}$ echoes Whigham's distinction. He also proposed a system [40] by which the language bias could be dynamically updated.

Our work can be seen as a contribution to the GP bias literature in the sense that we are trying to characterise the bias of (1) the GP language and (2) the search process, but focussing on a different variable (in contrast to the above work on tree depth, number of nodes, choice of primitive functions, etc.): the $\hat{y}$ outputs of the GP trees.

Recent semantically-oriented thinking about GP is also closely related to our work. Beadle and Johnson sought to characterise and control the semantic diversity of their GP initialisation procedure [1]. Here, "semantic diversity" means diversity in the $\hat{y}$ outputs of initialisation. In our terms, they aimed to avoid creating too low a variance in their $\hat{y}$ distribution.

11.4 Methodology, Experiments, and Results

In this section we first set out what we know about priors from first principles. We then describe our experimental setup, and give results.

11.4.1 Reasoning from First Principles

We can expect the GP prior to be quite diffuse or "uninformative", with long tails and many extreme values [27, 28]. It is easy to see that the support of the GP prior and initialisation prior are infinite, that is $\hat{y}$ could take on values in any range, even infinite values (depending on the function set).

When we consider the initialisation prior, the question we are asking—what is the prior on the output of an arithmetic expression in variables X_i, constructed according to a given non-deterministic tree-sampling procedure and for a given distribution on X—is a very general one, not specific to GP. The GPSR prior is GP-specific.

For the initialisation prior, the mathematics needed is the mechanisms of calculating distributions of sums, products, and other combinations and transformations of random variables. For example, given two normally-distributed random variables x_1 and x_2 with means μ_1 and μ_2 and variances σ_1^2 and σ_2^2, their sum $x_1 + x_2$ has normal density, with mean $\mu_1 + \mu_2$ and variance $\sigma_1^2 + \sigma_2^2$ [10] (Chapter 7). Using this and similar standard statistical methods [10, 36], it would in principle be possible to calculate the distribution of the output of a GROW-initialised tree. Similarly, drawing on previous research into GP search biases (e.g. [34] and many more), we could characterise the shapes of trees after evolution, and then in principle calculate the distribution of their output. In practice we would not be able to go very far, since the calculations become difficult even for some common cases of combining two variables in well-known distributions [10, 36].

Instead, we will proceed experimentally, by taking the X values from some common GPSR problems and from some constructed distributions, and generating many trees (by an initialisation algorithm alone for the initialisation prior, or by running many GPSR runs for the GPSR prior), and observing the resulting $\hat{y}$ values. We can then visualise the PMF of distributions using histograms, and measure minimum, median, maximum, mean, standard deviation, and so on. All these values are reported in Appendix A, Table 11.3.

In most cases we will regard the distinction between different training cases on the same problem as of no interest, and simply concatenate $\hat{y}$ values corresponding to distinct training cases, in order to study an overall distribution for the problem. However, in Sect. 11.5.3 we will consider the distinction.

We have considered Shapiro–Wilks tests for normality, but in most cases our histograms make it immediately clear that $\hat{y}$ results are not normal, so we think such tests unnecessary. We have also considered using Kolmogorov–Smirnov tests, and methods such as Kullback–Leibler divergence and Wasserstein distance to measure the dissimilarity between a pair of distributions, for example between y and $\hat{y}$ for a given problem. However, in practice our results show large differences, and many empirical distributions include extreme values and large areas with no support. Therefore we expect that the "signal"—results from all of these tests—would be overwhelmed by the "noise"—our choices of how to deal with the unusual distributions we are observing, in particular how to smooth, and what bin size and lower and upper bounds to use when discretising data. Again, we consider that our histograms are sufficiently clear-cut to allow conclusions without statistical tests.

Although our empirically-observed distributions have probability zero in large areas, these areas are not impossible outputs for GP. The GP initialisation and search process are not Bayesian updates, so an observed probability of zero in the prior does not imply a probability of zero after observation of data.

11.4.2   Setup

For all experiments performed, we used a standard GP implementation [8], with commonly used parameters, as shown in Table 11.1. We compare four function sets: set 1 contains the standard four arithmetic operators (with protected division); set 2 replaces pdiv with analytic quotient (aq); set 3 is the most commonly encountered in the literature; and set 4 has been shown to be very efficient at both approximation (train) and generalisation (test) [27].

11.4.3   Initialisation Prior

The initialisation procedure used with a GP run (typically ramped half-and-half) has a direct effect on the $\hat{y}$ values generated. To observe this, we devised a *vanilla* dataset, with a single predictor, drawn from a normal distribution with mean zero and standard deviation 1, $\mathcal{N}(0, 1)$, and 500 observations. We then ran initialisation with each function set with the parameters shown in Table 11.1, for 5 distinct runs, thus generating a set of $500 \times 500 \times 5 = 1{,}250{,}000$ $\hat{y}$ observations. Histograms of these are shown in Fig. 11.1.

Some $\hat{y}$ values generated are "extreme outliers", so in order to be able to visualise the data, we resorted to Winsorization (also known as clipping): we calculate

Table 11.1 GP parameters

Replacement	Elitist
Elitism	1% pop. size
Population size	500
Generations	50
Initialisation	Ramped h'n'h
Init. tree depth	2–6
Max. tree depth	20
Sub-tree crossover prob.	90%
Sub-tree mutation prob.	10%
Func. set 1	$(+, -, *, \text{pdiv})$
Func. set 2	$(+, -, *, \text{aq})$
Func. set 3	$(+, -, *, \text{pdiv}, \sin, \cos, \exp, \log)$
Func. set 4	$(+, -, *, \text{aq}, \sin, \tanh)$

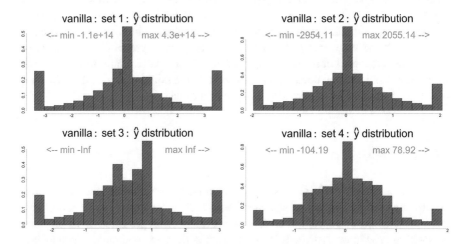

Fig. 11.1 $\hat{y}$ density histograms for all function sets, using the vanilla dataset, at generation 0

standard whisker intervals[2] for each distribution as the min/max values to visualise, and clip any values outside that range to the min/max values. In the following we will also refer to "extreme" values, meaning values which are several orders of magnitude farther from zero than these lower and upper bounds. Finally, we give the lowest/highest $\hat{y}$ values observed, which are infinite in some cases, as text on the plots.

We can observe that, although the inter-whisker range for all function sets is similar, the shape of the distributions is substantially different, as is the range of the outliers generated. Function sets 1 and 2 generate almost symmetric distributions,

---

[2]We adopt the whisker definition of 3rd-quantile $+ 1.5 * $ IQR for the upper whisker, and inversely for the lower whisker.

with the pdiv operator creating an asymmetry with more $\hat{y} = 1$ values generated. That operator is also responsible for generating much larger $\hat{y}$ values, with a range $[-1.1 \times 10^{14}, 4.3 \times 10^{14}]$, versus $[-2914.11, 2055.14]$ for set 2.

Set 3 generates a fundamentally asymmetric distribution, with a large proportion of values within the range $[0, 1]$, mostly due to the combination of pdiv with the trigonometric operators. It also generates some infinities.

Finally, set 4 generates a distribution somewhat similar to that of set 2, but with a higher proportion of values within the range $[-1, 1]$, due to its use of trigonometric operators. It also generates the least extreme values, with an observed range of $[-104.19, 78.92]$.

11.4.4 GPSR Prior

As defined above, by GPSR Prior we mean the prior of expressions initialised in a standard way, and then subjected to an (unguided) evolutionary process. To observe these, we used the same setup as in Sect. 11.4.3, but let each population evolve for 50 generations, with the same constant fitness value assigned to all individuals. We then compared the $\hat{y}$ distributions obtained at generation 50 to those from generation 0; Fig. 11.2 shows these, using the same visualisation techniques as described in Sect. 11.4.3.

Overall, there are only minor changes to the observed distributions, with sets 1 and 2 losing some of their central tendency, and no noticeable changes to sets 3 and 4. These changes are related to the constituent elements of the function sets, and their tendency to generate code growth (bloat): sets 1 and 2, equipped with only 2-arity functions, have a tendency to generate larger trees, given that our GP

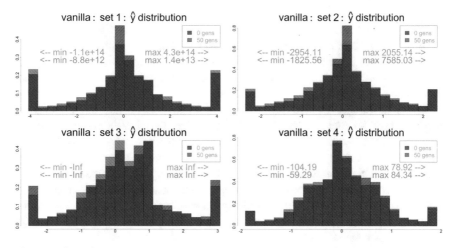

Fig. 11.2 $\hat{y}$ density histograms for all function sets, at generation 0 vs. generation 50

Fig. 11.3 Mean tree size per generation, for all four function sets. Sets 1 and 2 overlap exactly here, as they contain the same number of 2-arity functions, and runs for all function sets were executed with the same sets of random seeds

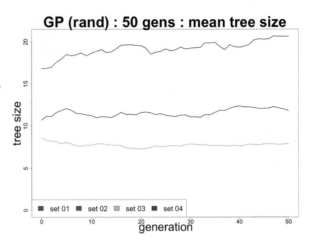

trees are bounded by depth but not by size. Sets 3 and 4 (particularly the former) have more 1-arity functions, which generate smaller trees at initialisation and during our unguided search. Figure 11.3 illustrates this phenomenon. Incidentally, these findings are further validation of the hypothesis that fitness causes bloat [17].

11.4.5 Effect of Tree Depth on Initialisation Prior

To further analyse whether the minor differences between initialisation prior and GPSR prior are caused by differences in tree size, we devised another experiment, taking tree size differences to the extreme. We again created an initial GP population using the parameters from Table 11.1, but used initialisation depth of either 2 or 15, and compared the resulting distributions. These are shown in Fig. 11.4.

We see that the overall effect, across function sets, is that for larger trees we have wider distributions and more extreme values. There are no observable extreme outliers at depth 2, for all function sets, and the ranges of values observed are quite small. They are the same for sets 2 and 4, as the extra functions from set 4 do not contribute to $\hat{y}$ growth at such shallow depths.

Depth 15 changes the distributions to a very large extent, confirming the findings from Sect. 11.4.4. Set 1 generates 35% of outliers (i.e. $\hat{y}$ values outside of our designated range), and a range of observed values extending to $[-1.6 \times 10^{104}, 5.3 \times 10^{107}]$. More extreme values can be expected, because a larger tree gives more locations at which something extreme can happen (e.g. an asymptote caused by divide-by-near-zero), and something extreme often only needs to happen once in a tree to cause an extreme output. This change in distribution is also observed with set 2, but with a smaller proportion of outliers (18%), and a substantially smaller range of $\hat{y}$ values generated.

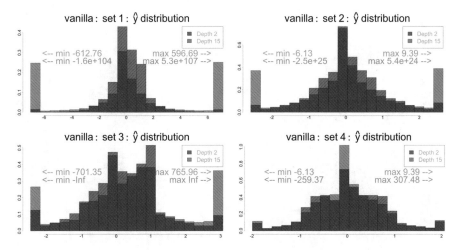

Fig. 11.4 $\hat{y}$ density histograms for all function sets at generation 0, with initialisation depth 2 vs. depth 15

Set 3 exhibits similar changes, with 16% of all $\hat{y}$ values generated being outliers. These are quite asymmetric, and biased towards positive values. Its range of observed values essentially extends to $[-\infty, +\infty]$.

$\hat{y}$ distributions generated by set 4, on the other hand, seem to be only mildly influenced by larger tree sizes. There is no quantifiable increase in the number of outliers generated, and the main difference seems to be a reduction in the proportion of values generated around the range $[0, 1]$.

11.4.6   Effect of Problem Dimension on Initialisation Prior

Another factor that might influence the initialisation prior is the dimension of the problem being solved. To study this effect, we generated a second vanilla dataset, with 10 independent variables, all sampled from $\mathcal{N}(0, 1)$. All GP settings reverted to Table 11.1, including initialisation depth. The differences between 1 and 10 independent variables are shown in Fig. 11.5.

There is a small but noticeable difference, particularly with sets 1, 2 and 4: a small diminution of central tendency is observed. This is likely due to a decrease of "cancelling" constructs such as sub(x,x), pdiv(x,x) or aq(x,x) in the generated trees. Interestingly, this does not increase the range of $\hat{y}$ values observed, which in fact seems to be reduced. This is not easily observed with set 3, the reason likely being that the sub and pdiv operators are used much less in the resulting trees, due to the presence of many other functions.

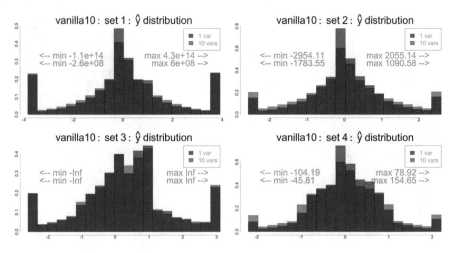

Fig. 11.5 $\hat{y}$ density histograms for all function sets at generation 0, using a dataset with one predictor vs. ten predictors

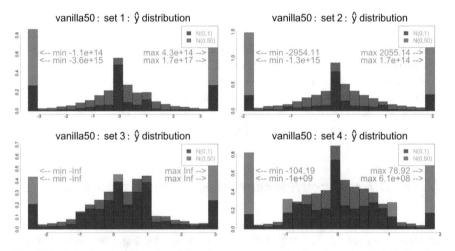

Fig. 11.6 $\hat{y}$ density histograms for all function sets at generation 0, using a dataset with X values drawn from $N(0, 1)$ vs. $N(0, 50)$

11.4.7 Effect of X Range on Initialisation Prior

Although [0, 1] is a common range observed in X values, particularly if the problem has been normalised, many real-world problems are not or cannot be normalised. To investigate the effect of a wider X range, we generated another "vanilla" dataset, now with values drawn from $\mathcal{N}(0, 50)$, and compared the resulting $\hat{y}$ distributions to those of the original $\mathcal{N}(0, 1)$ dataset. The results are shown in Fig. 11.6.

There is a large change in the resulting $\hat{y}$ distributions, due to the larger X range. All function sets generate a much larger proportion of outliers. Sets 1 and 2 keep a large proportion of values at $[-1, 1]$, but distribute the remaining values across a much wider range. Interestingly, the range of $\hat{y}$ values does not change much, only their distribution. Set 3 sees a similar effect, but far less pronounced. Set 4 is the odd one out. While generating a smaller proportion of values in the range $[-1, 1]$, the shape of the distribution remains similar. However, the range of generated $\hat{y}$ values increases by many orders of magnitude. Further results on the effect of X on $\hat{y}$ are presented in Sect. 11.4.8.

11.4.8   Comparing the y and $\hat{y}$ Distributions Across Problems

We have seen what the GP initialisation and GPSR priors look like, and how these are influenced by several parameters, both evolutionary parameters and data parameters. But how similar are these to distributions encountered in typical GP benchmarks?

To investigate this, we analyse the distribution of y for four problems commonly used in the GPSR literature: Keijzer-5 [12], Korns-8 [14], Vladislavleva-4 [39] and the real-world Housing dataset [18]. These are summarised in Table 11.2, and their y-value distributions are plotted in Fig. 11.7. These plots give a glimpse at the widely different y distributions that can be observed in different datasets.

Based on these results, we can predict that on the Housing and Korns datasets, with a mean y value far from zero, GP will likely start by predicting good constants, and may lose diversity before it can try to match individual training cases, as explained in Sect. 11.2.

We next compare the initialisation prior $\hat{y}$ to y across two common problems, Housing and Vladislavleva-4. The y-value distributions have already been shown in Fig. 11.7. The $\hat{y}$ distributions are shown in Fig. 11.8

These results show that the y distribution for specific problems can be similar in both shape and range to the initialisation or GPSR prior, as observed with the Keijzer-5 dataset, or substantially different in both, as observed in the Korns-

Table 11.2 Benchmarks analysed

Benchmark	Dimension	Data source	Training set
Keijzer-5	3	$\dfrac{30 * x_0 * x_2}{((x_0 - 10) * x_1{}^2)}$	$U[-1, 1, 1000] \times$
			$U[1, 2, 1000] \times$
			$U[-1, 1, 1000]$
4-4 Korns-8	5	$6.87 + (11 * \sqrt{7.23 * x_0 * x_3 * x_4})$	$U[-50, 50, 10000]$
Vladislavleva-4	5	$\dfrac{10}{5 + \sum_{i=0}^{4}(x_i - 3)^2}$	$U[0.05, 6.05, 1024]$
Housing	13	Housing values	Observations

Independent variables ranges are defined as $U[a, b, c]$, meaning c uniform random samples drawn from a to b inclusive. Both Korns-8 and Vladislavleva-4 draw all input variables from the same distribution (with replacement)

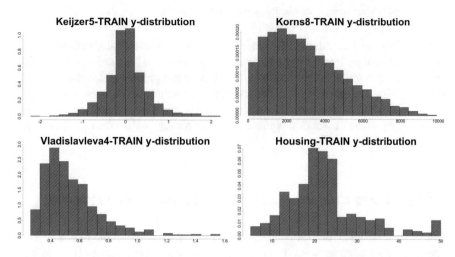

Fig. 11.7 y density histograms for four datasets commonly encountered in the GP literature (training data only): Keijzer-5, Korns-8, Vladislavleva-4, and the Housing dataset

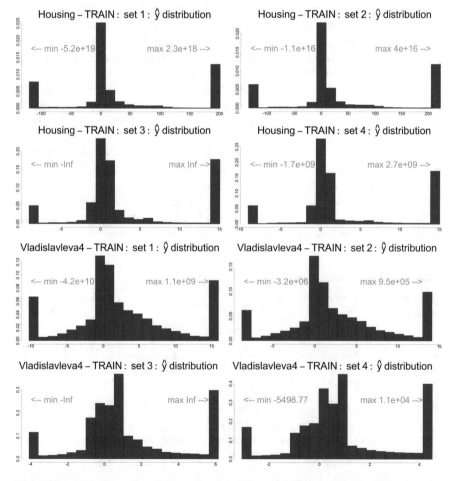

Fig. 11.8 $\hat{y}$ density histograms for two datasets (training only): Vladislavleva-4 and Housing, with all function sets

8 dataset. In the Housing and Vladislavleva-4 datasets (Fig. 11.8) there is also a mismatch between y and $\hat{y}$.

Comparing next across problems, we see that the $\hat{y}$ distribution is much wider (larger variance, more extreme values) for Housing than for Vladislavleva-4. In our context this effect is due only to the difference in their X distributions. This effect seems comparable in size to the effect of the function set (compare function sets 01 and 04 in Fig. 11.8).

We can therefore also infer that the $\hat{y}$ distribution can differ substantially between individual training cases, as it will be substantially affected by the training cases' X values.

Standard statistical measures for all of the distributions which have been discussed in this section are reported in Appendix A, Table 11.3. Where necessary, NaN and infinite values have been removed to allow these calculations. Overall, we observe that many $\hat{y}$ distributions have high variance, are highly skewed and are fat-tailed (large positive excess kurtosis).

11.5 Applications

In this section we consider some applications of our results.

11.5.1 Algorithm Behaviour and Performance

We have found that large distribution mismatches occur in many situations: across X ranges and different real-world problems; for different function sets and maximum tree depths; at initialisation and after evolutionary search. We have seen a plausible explanation, originating with Keijzer, for evolutionary dynamics in such cases. In summary, it seems sensible for practitioners with real-world problems to try out initial runs, and where performance is poor, to investigate whether distribution mismatch could be the culprit by looking for best-of-generation individuals which merely predict good constants.

Another possible application of our work arises as a special case, in binary classification by GPSR. In many papers (see e.g. the survey [7, p. 131]), the condition $f(X) > 0$ is used as the discriminant for binary classification of the query point X. If the GPSR prior is not symmetric about 0, then this introduces a bias towards predicting the positive (or negative) class which may not match the true distribution of binary labels. Again, the search algorithm would have to expend search effort to reach a part of the search space with a correct bias. This is separate from the well-known problem of the accuracy metric being misleading on unbalanced data [2]. This observation may have real-world consequences since with several function sets and in many real-world datasets, the initialisation and GPSR priors will be biased positive.

11.5.2 Algorithm Configuration

What function set should we choose for a particular problem? Note that several of the problems we have investigated are among the GP benchmarks commonly used in recent years. Some of them, when proposed as benchmarks, are accompanied by specific function sets. For strict comparison of algorithm performance it may be necessary to use the specified function sets. However, when the perspective is that of maximising real-world performance, it is clear that the choice of function set matters [27, 28]. Our results support this previous work. For example, the good results achieved by certain function sets [27, 28] can now be explained as the result of a good match between the GPSR prior and the true y distribution (both distributions are biased positive, with relatively low variance, the true y has no extreme values and the prior has few $|\hat{y}| > 3$).

In practice, in many forms of regression it is common to normalise or standardise data, either X or y or both, before modelling. This is expected to improve the performance of some models, such as kernel regression, and make no difference to others, such as linear regression. For cases where the observed y distribution is itself strangely distributed, e.g. taking on both extreme positive and extreme negative values, authors have investigated suitable transformations for y. For example, Burbidge et al. [3] suggest a sigmoidal transformation, the inverse hyperbolic sine. These techniques amount to controlling distribution mismatch by changing y, as opposed to by changing the GPSR prior. They would help to address the issue described by Keijzer (see Sect. 11.2), i.e. a y distribution which is simply *shifted* far from the GPSR prior. But they may not make the distribution match the *shape* of the GPSR prior, and in particular the choice between 0–1 normalisation and standardisation amounts to a choice between positive-only and zero-centered distributions. As we have seen, GPSR priors for several function sets match neither. Also, both normalisation and standardisation will give a y distribution much narrower than the very wide GPSR prior which arises with several function sets. Thus, algorithm configuration to control the GPSR prior may still be worthwhile.

Another way to avoid the problem of distribution mismatch is to use linear scaling in the objective function [12]. It solves the problem without removing the mismatch explicitly because it means an individual which predicts a good constant is no longer a good individual in early generations: an individual which achieves any improvement on matching the correct "shape" of the true y predictions will do better. However, linear scaling has not been taken up by the majority of GP users, who seem to value the transparency of a simple objective function, with some studies questioning its generalisation ability [5]. For such cases, again, controlling the GPSR prior may be needed.

Our results on asymptotes using different function sets may be useful in finding well-behaved, robust models (i.e. models less likely to blow up on unseen data).

11.5.3   Understanding GSGP Mutation

How does geometric semantic mutation [24] behave? It is defined as $m(p) = p + ms.(r_1 - r_2)$ where p is the parent program, and r_1 and r_2 are randomly-generated trees. The method of random generation is not specified [24], but some later work is explicit and uses the GROW operator for this [4]. The effect of the small constant ms is to control the degree of exploration of the operator. The term "geometric" is defined in the original to mean that all offspring are contained in a ball of radius ϵ centred at the parent. The reason for taking $r_1 - r_2$ is not explicit, but its effect is to centre the distribution of $m(p)$ at p. The initialisation prior (i.e. the prior on r_1 and on r_2) is not in general symmetric or zero-centred, so taking $m(p) = p + ms.r_1$ would give an asymmetric mutation operator, not centred at p. One possible disadvantage is that the variance of $r_1 - r_2$ is even larger than that of a single r_1 [36].

The behaviour of the operator remains to be investigated empirically. In Fig. 11.9 we illustrate it. GSGP mutation is not an ϵ-ball. There is no radius ϵ within which all outputs are contained, because according to our previous results, the distribution is long-tailed in each dimension. Also, the variance in each dimension can differ, due to differing distributions of X values of training cases. If (e.g.) training case 0 has larger values for independent variables than those of training case 1, then the result will be an ellipse-shaped distribution rather than spherical, with a wider range for $\hat{y}_0$ than for $\hat{y}_1$. However, the ellipse is not in general axis-aligned because the output of a program $m(p)$ on training case i is correlated with that on training case j.

Thus the GSGP mutation is not distributed as an ϵ ball, but as a potentially long-tailed, non-axis-aligned elliptical distribution. In each dimension it is centered at p due to the subtraction. Moraglio and Mambrini [25] propose an alternative formulation to reshape the distribution to be not only centered but isotropic, i.e. equally distributed in all directions from p, which requires taking the Moore–Penrose inverse of a matrix derived from the training data. Another alternative was proposed by McDermott et al. [22]: $m(p) = p + ms.r$, where r is a random tree and ms is drawn from a Normal centred at 0. This achieves symmetry and somewhat reduces growth in tree size, but still does not constrain the output to be an ϵ ball, or isotropic. Some later work, e.g. [38], defines mutation as $m(p) = p + \sigma(ms.(r_1 - r_2))$, where σ is a sigmoid mapping. This constrains the distribution to an ϵ ball but is not isotropic.

In addition to the properties of (1) being constrained to an ϵ ball, (2) being symmetric per-dimension, and (3) being isotropic, we remark that two stronger conditions on mutation can also be defined: (4) a *geometric complete* mutation can be defined as one in which all points in an ϵ ball are possible results of the mutation (cf. geometric complete crossover [23, p. 304]); an even stronger condition (5) is one in which the result is uniformly distributed on the ϵ ball (implying that it is also complete and isotropic). Properties 1–3 can be achieved but no known GP mutation achieves properties 4 or 5. It is useful for researchers studying GSGP and related ideas to distinguish these five properties.

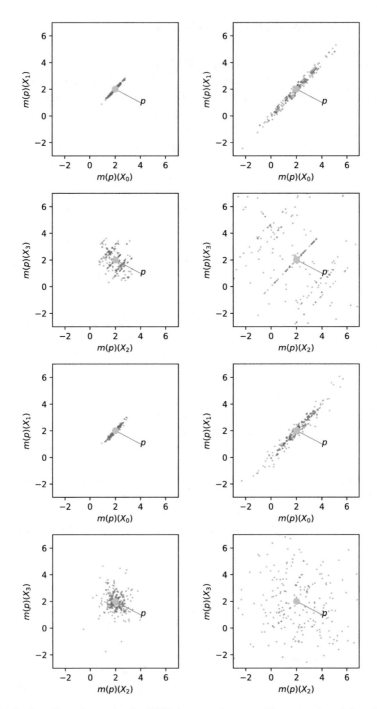

Fig. 11.9 The effect of mutation in GSGP in semantic space. The semantics of the offspring $m(p) = p + ms.(r_1 - r_2)$ is distributed centred at the parent semantics p. Here $p = (2, 2, \ldots 2)$ for

An elliptical distribution may lead to faster movement through one dimension of semantics than another, which may be helpful if the target's values in different dimensions differ markedly. On the other hand, the shape of the mutation distribution depends on X, not on y. Further, the long-tailed mutation distribution gives the potential for damaging asymptotes at every mutation. In GSGP, subtrees are amalgamated but never deleted, and so asymptotes, if they appear, tend to remain. It is therefore important for a GSGP implementation to prevent asymptotes, for example by choosing the GP language carefully. Some avenues for future work are opened up by these perspectives.

11.6 Conclusions

In this paper, we have asked the question: what is the implicit prior on $\hat{y}$ which results from our GP initialisation and search algorithms? We have compared these priors against y distributions for common problems and found mismatches which we think are important in that they impact on algorithm dynamics and performance.

One application of our results is then on how to tune algorithm parameters with distribution mismatch in mind. We have tested the impact of some algorithm parameters, including function set and maximum tree depth, on this mismatch, and found that overall the function set 4 ($+$, $-$, $*$, aq, sin, tanh) gives a better match with observed y distributions. With the other function sets, a large tree depth gives more extreme values, but with function set 4 this tendency is reversed, and so with this function set we do not need to recommend very tight limits on tree depth. We have also observed the effect of the number of variables and their range.

A second application of our results is in understanding the behaviour of operators, e.g. the GSGP mutation operator. We have demonstrated empirically the distribution achieved by this operator.

Fig. 11.9 (continued) illustration. Pairs (r_1, r_2) were generated by the GROW algorithm and their semantics illustrated on two training cases at a time. The centering is due to the subtraction $r_1 - r_2$, eliminating any offset or skew that may be present in r_1 and r_2 individually. For small mutation step $ms = 1/2$, the distribution is tight (left); for $ms = 2$, the distribution is wider (right); in either case values are not constrained to any radius. For many pairs of training cases, the distribution is wider for one than the other, leading to an elliptical rather than spherical distribution. For some pairs of training cases, there is a correlation between the semantics on those pairs (a positive correlation in rows 1 and 3, but a negative correlation could also occur); for other pairs, little or no correlation (rows 2, 4). For r_i of maximum depth 2, many offspring are simple and contain a naked variable x, and linear patterns occur (rows 1–2); for maximum depth 15 this does not occur (rows 3–4)

11.6.1 Limitations and Future Work

As argued by Silva and Costa [33], fitness is the one ingredient which, if removed, would undermine all theories of bloat. Therefore, in our experiments on GP "search" without fitness, we have achieved distributions of GP trees which are not bloated and therefore not representative of distributions in the presence of fitness. As a result our distributions of $\hat{y}$ could be inaccurate. However, running without fitness is essential if we are to discuss priors, as opposed to results from particular problems. One possible solution for future work is to use previous work on limiting distributions of GP tree sizes to characterise the eventual shapes of trees, and then to simulate such trees to study their distribution of $\hat{y}$ values. Poli et al. [30] showed that such a limiting distribution exists for crossover-only GP, the Lagrange distribution of the second kind. Extensions would be required to deal with mutation and to characterise tree shapes. The latter would be possible through studying the balancedness of trees as done by Keijzer and Foster [13].

A further possible limitation is that in our work on the GPSR prior we have taken the final generation to be representative of the output of the GP run. In reality, the final generation (in the presence of fitness) will likely include some very small trees [30], and these are unlikely to be the best of the run. It is the best of the run which is the true output of the GP run. Therefore, our GPSR priors probably over-represent small individuals. By our reasoning in Sect. 11.4.5, such individuals tend to have fewer extreme values than large individuals. The true GPSR prior— based only on the best of the run—would then have more extreme values than the distributions we have shown. In future work we will investigate whether this issue makes a difference.

Acknowledgement Thanks to Alberto Moraglio and Alexandros Agapitos for discussion.

Appendix A: Table of Distribution Statistics

See Table 11.3.

Table 11.3 Measures of spread and central tendency

Dataset	fset	Min	Max	Mean	Median	SD	Skew	Kurt
gen 0 : depth 2–6 : 1 var : $\mathcal{N}(0,1)$: $\hat{y}$								
Vanilla	01	−1.1e+14	4.3e+14	1.7e+08	0.0031	4e+11	983.51	1.1e+06
Vanilla	02	−2954.11	2055.14	−0.021	0.00	9.84	−138.22	3.5e+04
Vanilla	03	−Inf	Inf	1.4e+300	0.34	Inf	NaN	NaN
Vanilla	04	−104.19	78.92	0.052	0.008	1.07	0.93	298.37
gen 50 : depth 2–6 : 1 var : $\mathcal{N}(0,1)$: $\hat{y}$								
Vanilla	01	−8.8e+12	1.4e+13	−1.8e+07	0.0063	1.7e+10	269.44	4.6e+05
Vanilla	02	−1825.56	7585.03	0.061	0.00	12.45	278.72	1.3e+05
Vanilla	03	−Inf	Inf	2.9e+302	0.22	Inf	NaN	NaN
Vanilla	04	−59.29	84.34	0.031	0.00	1.03	3.45	204.85
gen 0 : depth 2 : 1 var : $\mathcal{N}(0,1)$: $\hat{y}$								
Vanilla	01	−612.76	596.69	0.11	0.043	8.32	−0.48	1432.52
Vanilla	02	−6.13	9.39	0.065	0.0031	1.03	0.49	5.60
Vanilla	03	−701.35	765.96	0.20	0.23	7.65	0.70	2443.81
Vanilla	04	−6.13	9.39	0.052	0.0084	0.88	0.68	7.57
gen 0 : depth 15 : 1 var : $\mathcal{N}(0,1)$: $\hat{y}$								
Vanilla	01	−1.6e+104	5.3e+107	4.2e+101	0.00	4.7e+104	NaN	NaN
Vanilla	02	−2.5e+25	5.4e+24	−1.6e+19	0.00014	2.3e+22	−1032.60	1.1e+06
Vanilla	03	−Inf	Inf	1.1e+302	0.36	Inf	NaN	NaN
Vanilla	04	−259.37	307.48	0.0057	0.00	1.60	−16.62	5582.72
gen 0 : depth 2–6 : 10 vars : $\mathcal{N}(0,1)$: $\hat{y}$								
Vanilla	01	−2.6e+08	6e+08	281.21	0.0044	6.5e+05	599.53	6e+05
Vanilla	02	−1783.55	1090.58	0.017	0.00022	3.31	−94.83	8.4e+04
Vanilla	03	−Inf	Inf	3e+301	0.32	Inf	NaN	NaN
Vanilla	04	−45.81	154.65	0.0086	0.00	1.08	4.31	524.20

(continued)

Table 11.3 (continued)

Dataset	fset	Min	Max	Mean	Median	SD	Skew	Kurt
gen 0 : depth 2–6 : 1 var : $\mathcal{N}(0, 50)$: $\hat{y}$								
Vanilla	01	-3.6e+15	1.7e+17	4.2e+11	0.0000069	1.7e+14	924.90	9.5e+05
Vanilla	02	-1.3e+15	1.7e+14	-1.3e+09	0.00	1.2e+12	-970.09	1e+06
Vanilla	03	-Inf	Inf	-1.6e+301	0.41	Inf	NaN	NaN
Vanilla	04	-1e+09	6.1e+08	-2508.95	0.0000017	1.4e+06	-264.17	2.4e+05
datasets : y								
Keijzer5	-	-2.22	2.21	0.012	-0.0026	0.50	0.35	2.21
Korns8	-	11.77	9909.16	3075.62	2746.47	2004.64	0.69	-0.16
Vladislavleva4	-	0.26	1.57	0.54	0.50	0.19	1.62	3.77
Housing	-	5.00	50.00	22.35	21.20	8.85	1.07	1.53
gen 0 : datasets : $\hat{y}$								
Keijzer5	01	-8.2e+11	4.7e+14	3.2e+08	0.081	3.3e+11	1250.16	1.7e+06
Keijzer5	02	-132.07	76.95	0.22	0.081	1.84	-4.91	300.69
Keijzer5	03	-Inf	Inf	1.3e+301	0.42	Inf	NaN	NaN
Keijzer5	04	-10.36	25.26	0.22	0.14	0.96	1.29	19.57
Korns8	01	-6.1e+20	3.2e+18	-2.5e+13	0.000021	1.2e+17	-4999.16	2.5e+07
Korns8	02	-1e+14	6.4e+13	-2.5e+07	0.00	8.2e+10	-88.61	2.3e+05
Korns8	03	-Inf	Inf	2.4e+302	0.43	Inf	NaN	NaN
Korns8	04	-5.1e+08	4.8e+08	795.17	0.00	1.4e+06	-8.05	2.2e+04
Vladislavleva4	01	-4.2e+10	1.1e+09	-1.6e+04	1.08	2.6e+07	-1597.85	2.6e+06
Vladislavleva4	02	-3.2e+06	9.5e+05	-15.14	1.19	6364.54	-242.41	9.3e+04
Vladislavleva4	03	-Inf	Inf	9.7e+301	0.68	Inf	NaN	NaN
Vladislavleva4	04	-5498.77	1.1e+04	1.66	0.61	38.56	71.22	1.6e+04
Housing	01	-5.2e+19	2.3e+18	-5.4e+13	1.22	5.5e+16	-929.70	8.7e+05
Housing	02	-1.1e+16	4e+16	-2.1e+11	2.76	9.8e+13	93.61	4.5e+04
Housing	03	-Inf	Inf	3.2e+302	0.84	Inf	NaN	NaN
Housing	04	-1.7e+09	2.7e+09	3.5e+05	0.60	2.2e+07	82.74	8138.03

References

1. Beadle, L., Johnson, C.G.: Semantic analysis of program initialisation in genetic programming. Genetic Programming and Evolvable Machines **10**(3), 307–337 (2009)
2. Bhowan, U., Johnston, M., Zhang, M., Yao, X.: Evolving diverse ensembles using genetic programming for classification with unbalanced data. IEEE Transactions on Evolutionary Computation **17**(3), 368–386 (2013)
3. Burbidge, J.B., Magee, L., Robb, A.L.: Alternative transformations to handle extreme values of the dependent variable. Journal of the American Statistical Association **83**(401), 123–127 (1988). https://doi.org/10.1080/01621459.1988.10478575. https://amstat.tandfonline.com/doi/abs/10.1080/01621459.1988.10478575
4. Castelli, M., Silva, S., Vanneschi, L.: A C++ framework for geometric semantic genetic programming. Genetic Programming and Evolvable Machines **16**(1), 73–81 (2015)
5. Costelloe, D., Ryan, C.: On improving generalisation in genetic programming. In: L. Vanneschi, S. Gustafson, A. Moraglio, I.D. Falco, M. Ebner (eds.) European Conference on Genetic Programming, EuroGP 2009, Tübingen, Germany, April 15–17, 2009, Proceedings, *Lecture Notes in Computer Science*, vol. 5481, pp. 61–72. Springer (2009)
6. Dignum, S., Poli, R.: Generalisation of the limiting distribution of program sizes in tree-based genetic programming and analysis of its effects on bloat. In: D. Thierens, H.G. Beyer, J. Bongard, J. Branke, J.A. Clark, D. Cliff, C.B. Congdon, K. Deb, B. Doerr, T. Kovacs, S. Kumar, J.F. Miller, J. Moore, F. Neumann, M. Pelikan, R. Poli, K. Sastry, K.O. Stanley, T. Stutzle, R.A. Watson, I. Wegener (eds.) GECCO '07: Proceedings of the 9th annual conference on Genetic and evolutionary computation, vol. 2, pp. 1588–1595. ACM Press, London (2007)
7. Espejo, P.G., Ventura, S., Herrera, F.: A survey on the application of genetic programming to classification. IEEE Transactions on Systems, Man, and Cybernetics, Part C (Applications and Reviews) **40**(2), 121–144 (2010)
8. Fortin, F.A., De Rainville, F.M., Gardner, M.A., Parizeau, M., Gagné, C.: DEAP: Evolutionary algorithms made easy. Journal of Machine Learning Research **13**, 2171–2175 (2012)
9. Gelman, A., Jakulin, A., Pittau, M.G., Su, Y.S.: A weakly informative default prior distribution for logistic and other regression models. Ann. Appl. Stat. **2**(4), 1360–1383 (2008). https://doi.org/10.1214/08-AOAS191
10. Grinstead, C.M., Snell, J.L.: Introduction to probability. American Mathematical Soc. (2012)
11. Iba, H., de Garis, H., Sato, T.: Genetic programming using a minimum description length principle. In: K.E. Kinnear, Jr. (ed.) Advances in Genetic Programming, chap. 12, pp. 265–284. MIT Press (1994)
12. Keijzer, M.: Improving symbolic regression with interval arithmetic and linear scaling. In: EuroGP, pp. 70–82. Springer (2003)
13. Keijzer, M., Foster, J.: Crossover bias in genetic programming. In: European Conference on Genetic Programming, pp. 33–44. Springer (2007)
14. Korns, M.F.: Accuracy in symbolic regression. In: R. Riolo, E. Vladislavleva, J.H. Moore (eds.) Genetic Programming Theory and Practice IX, Genetic and Evolutionary Computation, pp. 129–151. Springer, New York (2011)
15. Koza, J.: Genetic Programming: on the programming of computers by means of natural selection. MIT Press, Cambridge, MA (1992)
16. Krogh, A., Hertz, J.A.: A simple weight decay can improve generalization. In: Advances in neural information processing systems, pp. 950–957 (1992)
17. Langdon, W.B., Poli, R.: Fitness causes bloat. In: P.K. Chawdhry, R. Roy, R.K. Pant (eds.) Soft Computing in Engineering Design and Manufacturing, pp. 13–22. Springer London (1998). https://doi.org/10.1007/978-1-4471-0427-8_2
18. Lichman, M.: UCI machine learning repository. http://archive.ics.uci.edu/ml (2013)

19. Luke, S., Panait, L.: A comparison of bloat control methods for genetic programming. Evolutionary Computation **14**(3), 309–344 (2006)
20. Mauceri, S., Sweeney, J., McDermott, J.: One-class subject authentication using feature extraction by grammatical evolution on accelerometer data. In: Proceedings of META 2018, 7th International Conference on Metaheuristics and Nature Inspired computing. Marrakesh, Morocco (2018)
21. McDermott, J.: Measuring mutation operators' exploration-exploitation behaviour and long-term biases. In: M. Nicolau, K. Krawiec, M.I. Heywood, M. Castelli, P. García-Sánchez, J.J. Merelo, V.M.R. Santos, K. Sim (eds.) 17th European Conference on Genetic Programming, *LNCS*, vol. 8599, pp. 100–111. Springer, Granada, Spain (2014)
22. McDermott, J., Agapitos, A., Brabazon, A., O'Neill, M.: Geometric semantic genetic programming for financial data. In: Applications of Evolutionary Computation, pp. 215–226. Springer (2014)
23. Moraglio, A.: Towards a geometric unification of evolutionary algorithms. Ph.D. thesis, University of Essex (2007)
24. Moraglio, A., Krawiec, K., Johnson, C.: Geometric semantic genetic programming. In: Proc. PPSN XII: Parallel problem solving from nature, pp. 21–31. Springer, Taormina, Italy (2012)
25. Moraglio, A., Mambrini, A.: Runtime analysis of mutation-based geometric semantic genetic programming for basis functions regression. In: Proceedings of the 15th annual conference on Genetic and evolutionary computation, pp. 989–996. ACM (2013)
26. Ni, J., Drieberg, R.H., Rockett, P.I.: The use of an analytic quotient operator in genetic programming. IEEE Transactions on Evolutionary Computation **17**(1), 146–152 (2013)
27. Nicolau, M., Agapitos, A.: On the effect of function set to the generalisation of symbolic regression models. In: Proceedings of the Genetic and Evolutionary Computation Conference Companion, pp. 272–273. ACM (2018)
28. Nicolau, M., Agapitos, A.: Function sets and their generalisation effect in symbolic regression models (2019). In review
29. Poli, R.: A simple but theoretically-motivated method to control bloat in genetic programming. In: C. Ryan, T. Soule, M. Keijzer, E. Tsang, R. Poli, E. Costa (eds.) Genetic Programming, Proceedings of EuroGP'2003, *LNCS*, vol. 2610, pp. 204–217. Springer-Verlag, Essex (2003)
30. Poli, R., Langdon, W.B., Dignum, S.: On the limiting distribution of program sizes in tree-based genetic programming. In: European Conference on Genetic Programming, pp. 193–204. Springer (2007)
31. Poli, R., Langdon, W.B., McPhee, N.F.: A field guide to genetic programming. Published via http://lulu.com and freely available at http://www.gp-field-guide.org.uk (2008)
32. Rosca, J.P., et al.: Analysis of complexity drift in genetic programming. Genetic Programming pp. 286–294 (1997)
33. Silva, S., Costa, E.: Dynamic limits for bloat control in genetic programming and a review of past and current bloat theories. Genetic Programming and Evolvable Machines **10**(2), 141–179 (2009)
34. Silva, S., Dignum, S.: Extending operator equalisation: Fitness based self adaptive length distribution for bloat free GP. In: EuroGP, pp. 159–170. Springer (2009)
35. Silva, S., Vanneschi, L.: The importance of being flat—studying the program length distributions of operator equalisation. In: R. Riolo, K. Vladislavleva, J. Moore (eds.) Genetic Programming Theory and Practice IX, pp. 211–233. Springer (2011)
36. Springer, M.D.: The algebra of random variables. Wiley (1979)
37. Stephens, T.: GPLearn (2015). https://github.com/trevorstephens/gplearn, viewed 1 April 2019
38. Vanneschi, L., Silva, S., Castelli, M., Manzoni, L.: Geometric semantic genetic programming for real life applications. In: Genetic programming theory and practice xi, pp. 191–209. Springer (2014)

39. Vladislavleva, E.J., Smits, G.F., den Hertog, D.: Order of nonlinearity as a complexity measure for models generated by symbolic regression via Pareto genetic programming. IEEE Transactions on Evolutionary Computation **13**(2), 333–349 (2009)
40. Whigham, P.A.: Inductive bias and genetic programming (1995)
41. Whigham, P.A., McKay, R.I.: Genetic approaches to learning recursive relations. In: X. Yao (ed.) Progress in Evolutionary Computation, *Lecture Notes in Artificial Intelligence*, vol. 956, pp. 17–27. Springer-Verlag (1995)

Chapter 12
Hands-on Artificial Evolution Through Brain Programming

Gustavo Olague and Mariana Chan-Ley

12.1 Introduction

Brain programming is a new kind of artificial evolutionary learning based on neuroscience knowledge and the power of genetic programming. In contrast to state-of-the-art deep learning methodologies, the idea is to look for computational programs embedded within artificial models of the brain. To review the importance of the subject, we can say that Holland provided the first sketch that illustrates the problem in his seminal book [14]. A simple pattern recognizer was used to demonstrate a simple artificial adaptive system. Paradigmatically this research area has not received keen interest from the research community in genetic and evolutionary computation probably due to the difficulty of approaching real-world applications and the lack of an appealing theory. In evolutionary computer vision, the visual problem is studied under the framework of goal-oriented vision [22]. Brain programming follows the route outlined in [22] to define the necessary steps towards not only specifying how to solve a visual problem but also to answer the question: What is the visual task for? Brain programming aims to offer a new theory of visual processing where the paradigm of evolutionary programming can be extensively used in practical image understanding and pattern recognition endeavors.

G. Olague (✉)
CICESE, Ensenada, BC, Mexico
e-mail: olague@cicese.mx

M. Chan-Ley
EvoVisión Laboratory, Ensenada, BC, Mexico
e-mail: mchan@cicese.edu.mx;
http://evovision.cicese.mx

© Springer Nature Switzerland AG 2020   227
W. Banzhaf et al. (eds.), *Genetic Programming Theory and Practice XVII*, Genetic
and Evolutionary Computation, https://doi.org/10.1007/978-3-030-39958-0_12

12.2 Evolution of Visual Attention Programs

In [9] artificial visual programs were evolved for the problem of visual attention. Automation of cognitive modeling is designed through a succession of levels of layers that create an artificial dorsal stream. Visual attention tasks are evolved following a purposive goal-oriented behavior. Results obtained on a well-known testbed confirm that the proposal is able to automatically design visual attention programs outperforming previous human-made systems developed by visual attention experts while providing readable results through a set of mathematical and computational structures. Figure 12.1 presents a sketch to illustrate the analogy of our artificial approach with the natural system. The methodology was further tested on different tasks to assess its suitability to discover solutions that can be tagged as general answers to the question of what is visual attention for?

In [23] the system was tested on the problem of evolving head tracking routines. The system evolves visual operators to obtain several visual and conspicuity maps that are fused into a saliency map, which is converted to a binary image, thus defining the proto-object. Artificial brains are synthesized using multiple visual operators embedded within an intricate hierarchical procedure consisting of several key processes such as center-surround mechanisms, normalization, and pyramid-scale processes. The proposed strategy robustly manages many difficulties such as occlusion, distraction, and illumination, and the resulting programs are real-time

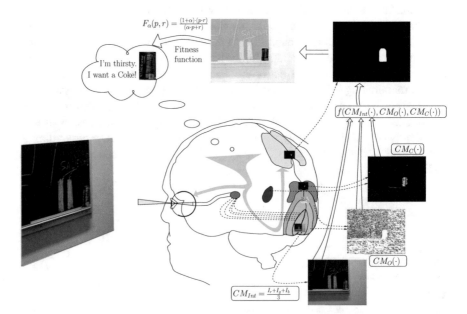

Fig. 12.1 Conceptual model of the artificial artificial dorsal stream. The correspondence between the dorsal stream areas and the stages of the artificial model. The idea is to emulate the transformations that the input image undergoes along the pathway of the natural system

systems that can track a person's head with enough accuracy to automatically control the camera. Extensive experimentation shows that the proposed methodology outperforms several state-of-the-art methods in the challenging problem of head-tracking.

We tested the same system in the automation of video tracking design processes of robotic visual systems [24]. The usual practice of learning a different task is to learn the programs from a database and then test the system with a different set of images representing the learned problem. In that paper, the challenge was to learn to detect a toy dinosaur from a database while testing the evolved programs in a different task considering three distinct robots and visual tracking scenarios. Indeed, the database does not contain information about the visual tracking challenge. When planning an object tracking system detection of moving objects in each frame and correct association of detection to the same object over time need to be approached for the whole sequence. Visual attention is a skill performed by the brain whose functionality is to perceive salient visual features. The automatic design of the acquisition and integration steps of the natural dorsal stream was engineered to emulate its selectivity and goal-driven behavior useful to the task of tracking objects from a video captured with a moving camera. This is a step towards the design of robust and autonomous visual behaviors. The test considers many difficulties due to abrupt object motion, changing appearance patterns of both the object and the scene, nonrigid structures, object-to-object, and object-to-scene occlusions, as well as camera motion, models, and parameters. Tracking relies on the quality of the detection process and automatically designing such a stage could significantly improve tracking methods. Experimental results confirm the validity of the approach, and a comparison with the method of regions with convolutional neural networks (CNN) is provided to illustrate the benefit of the proposed methodology.

The idea of evolving focus of attention studied in an interactive domotic environment produces excellent results [6]. The aim was to self-adapt the focus of attention algorithm to improve the accuracy level of a laser pointer detection system. The proposed technique was compared with previous proposals, and the new method allows to send more accurate orders to home devices. The procedure eradicates false offs, thus preventing orders not signaled by users. Moreover, by adding self-adjusting capabilities with a genetic-fuzzy system, the computer vision algorithm focuses its attention on a narrower area of the image. This work illustrates the application of the evolution of visual attention programs in practical, real-world tasks.

12.2.1 Evolution of Visual Recognition Programs

In [12] the human visual system was used as inspiration for solving object detection and classification tasks. Computational models of an artificial visual cortex (AVC) were evolved to solve challenging classification problems of natural

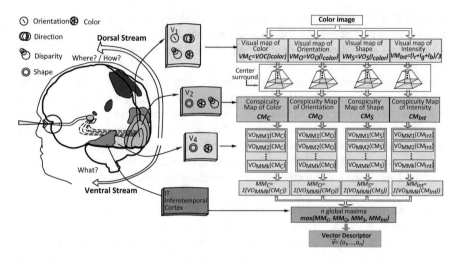

Fig. 12.2 Conceptual model of the artificial visual cortex. The system decomposes the color image into four dimensions (color, orientation, shape, and intensity). Then, a hierarchical structure solves the object classification problem through a function-driven paradigm

images. Figure 12.2 presents a sketch to visualize the general idea of the proposed system for image classification. The idea is to create an image descriptor vector for classification and at the same time finding the object location within the image. That paper presents the paradigm of brain programming from a multi-objective perspective to enhance the performance in the object classification task.

According to brain programming paradigm, each operator within the artificial brain does not represent a solution to the visual task. A single operator might not be of interest since they only hold meaning while working along with other operators within the hierarchical structure. The fitness function is a balance between the artificial dorsal and ventral streams. Each evolved solution is evaluated through two performance measures. The first one measures classification performance with a metric called equal error rate that defines the probability of an algorithm to decide if two instances correspond to the same class. The second objective, based on the F-measure, calculates the correspondence between ground-truth of the object location in the image and the region selected by the algorithm as the possible position. The methodology was tested on the GRAZ-01 and GRAZ-02 databases. The solutions match and in some cases outperform other techniques in the state-of-the-art for classifying such databases.

The artificial visual cortex is a bio-inspired model that is computationally expensive for object classification tasks. This cost is due in part to the high-information content that is necessary to process while working with large databases, with images of high resolution, or in the worst case when they need to operate over video. In [13] the original model was improved through the compute unified device architecture (CUDA) by exploiting computational capabilities of graphics processing units (GPUs). In this work, the artificial visual cortex was coded in the CUDA framework to achieve real-time functionality. As a result, the proposed system can process images in average up to 90 times faster than the original images. Moreover, when the size of the image grows, the proposed AVC-CUDA is faster in comparison with other approaches like CNN-CUDA.

The optimization of the AVC model through genetic and evolutionary programming while tested on challenging object recognition tasks finds reasonable solutions during the initial stage of the search. In [25] a study shows the frequency of discovering optimal visual operators through a simple random search. It exhibits the richness of the paradigm from two complementary viewpoints: (1) the concept of function composition in combination with a hierarchical structure leads to outstanding object recognition programs, and (2) multiple random runs of the search process can discover optimal functions. Experimental results provide evidence that high recognition rates could be achieved in well-known object categorization problems such as Caltech-5, Caltech-101, GRAZ-01, and GRAZ-02.

This paper is organized as follows. After a brief introduction to the problem, Sect. 12.4 reviews the state-of-the-art of classification of digitized art. It explains the reason for selecting this problem to test brain programming. Section 12.5 presents the results that were obtained and a comparison with deep learning. Section 12.5.1 presents an analysis of some runs of the algorithm using an initial population of the best set of solutions discovered in the previous section. Section 12.5.2 presents a sketch about our future research in the domains of visual learning and pattern recognition. Finally, the conclusion is drawn in Sect. 12.6.

12.3   Problem Statement

This paper presents results of a system that automatically designs novel computational models of an artificial ventral stream for the problem of recognizing digitized art. Creating an algorithm for authenticating the type of art from high-resolution digital scans of the original works is considered an open research area. Here, brain programming is tested against a challenging database, and a comparison is made with deep learning to understand the benefits and limits of our approach. A proposal about learning from a set of previously discovered solutions is presented, and results give us some clues for future research. We summarize the concept of object recognition through the following definition.

Definition 12.1 (Object Recognition) Object recognition refers to the task of determining if an object belongs to one or more classes in a collection of images and video sequences.

Here, we are interested in studying the problem of identifying digitized visual arts. While there is no universally accepted definition of art, commonly the word is used to describe something of beauty or a skill which produces an aesthetic result [1, 2, 21, 28]. We chose a database made of photographs of fine art as part of the general category of visual arts. In object categorization, the problem studied is that of identifying the object content of natural images while generalizing across variations inherent to the object class. In all previous studies of the artificial visual cortex, each class describes a kind of specific object. For example, in Caltech-101, the categories correspond to particular objects like cars, airplanes, faces, and so on. ImageNet also includes many other objects in the images that are repetitive for some of the classes, and GRAZ benchmark introduces photographs with several objects within the same image. This paper aims to test brain programming with a different kind of object categorization problem. Here the class is representative of a given visual art, and therefore, the program needs to identify what is common in the set of images that can help to recognize the type of art. This problem is challenging since the methodology needs to build code that is capable of identifying the form of 2D and 3D visual features that render the information confirming the piece of art. which is in itself hard to define.[1]

12.4   Classification of Digitized Art

Koza formulates the paradigm of genetic programming as a way to achieve program induction following the Darwinian principle that structure arises from fitness [16]. He uses as a postulate the central idea raised by Arthur Samuel in the 1950s

> How can computers learn to solve problems without being explicitly programmed? In other words, how can computers be made to do what is needed to be done, without being told exactly how to do it? [16, p. 1]

While this is a final goal for the genetic programming community, for practical reasons, especially when dealing with real-world problems, one needs to make some compromises. In previous research devoted to the question of configuring a photogrammetric network, we avoid the idea of determining the fitness function without knowledge of the derivatives [26]. That research avoids the advice about one of the seven principles pronounced by Koza called "correctness," which at the beginning of evolutionary computation was said that it should not be used in solving every problem. In photogrammetry, modeling-of-data requires that the proposed method not only solves the problem but what is needed is to obtain the solution with the highest accuracy with a minimal amount of resources.

Nowadays, the problem of image classification is on the verge of a digital revolution produced by what is known as deep learning. This method uses the core technology of neural networks as its central paradigm.[2] Since many authors have studied the application of evolutionary computation for neural networks, it is attractive to continue along this line of research. There are advantages:

---

[1] Nowadays, to claim that any computational method (artificial intelligence) is capable of solving similar visual problems needs to be taken carefully since the programs need to be explainable from the artistic viewpoint.

[2] Note that Koza classify neural networks as one of the existing methods that do not seek solutions in the form of computer programs.

- There is a vast experience on the subject.
- Programming packages are available.
- A community ready to work on the subject.

Nevertheless, each has a disadvantage:

- Researchers in evolutionary computation study mostly toy problems.
- Even for today standard, the technology is meager in comparison with the size of the studied problems.
- Not all people interested in the subject has access to the technology necessary to tackle state-of-the-art problems.

To place this in perspective, let us recall three milestones of deep learning. In 2012 researchers from Stanford and Google unveiled to the world the real power of the paradigm, building on previous research into multilayer neural networks [19]. Their study explored unsupervised learning, which does away with the expensive and time-consuming task of manually labeling data before it can be used to train machine learning algorithms. It would accelerate the pace of AI development and open up a new world of possibilities when it came to building machines to do work, which until then could only be done by humans. Specifically, they singled out the fact that their system had become highly competent at recognizing pictures of cats.

The paper described a model which would enable an artificial network to be built containing around one billion connections using a dataset of 10 million 200×200 pixel images. It also conceded that while this was a significant step towards building an "artificial brain," there was still some way to go with neurons in a human brain thought to be joined by a network of around 10 trillion connectors.

In 2012 a convolutional neural network named AlexNet [18] competed in the ImageNet Large Scale Visual Recognition Challenge, where algorithms compete to show their proficiency in recognizing and describing a library of 1000 images. The neural network with 60 million parameters and 500,000 neurons win the image recognition contest. AlexNet is a variant of CNN designs introduced by Yann LeCun [20], who applied the backpropagation algorithm to a variant of Kunihiko Fukushima [11] original CNN architecture called "neocognitron."

AlphaGo is the first computer program to defeat a professional human Go player in 2015, the first program to defeat a Go world champion in 2016, and arguably the strongest Go player in history [27]. The neural networks take a description of the Go board as an input and process it through many different network layers containing millions of neuron-like connections.

Given the size of the neural networks, it is difficult to foresee an improvement of the network architectures for such kind of problems through evolutionary computation [15].

The first papers on brain programming published in 2012 at the EvoStar conference represent the birth of this new methodology based on genetic programming [5, 8]. The studied problems were toy problems for today standard. In particular, for the object recognition problem, the databases were Caltech-5 and Caltech-101. A comparison with HMAX (a kind of hierarchical convolutional network) showed that the proposed system was more straightforward and accurate [5]. The general idea was to apply a function-driven paradigm in contrast to the favorite data-driven model. The system was further improved to solve image classification problems like those posed in the databases of GRAZ-01 and GRAZ-02, which is part of the European network PASCAL's Visual Object Classes Challenge [12]. These datasets are significantly harder to solve in comparison with Caltech-5 and Caltech-101. Since ImageNet is massive researchers prefer to use smaller datasets like MNIST [20] and CIFAR [17] whose images are of reduced size. However, These datasets are simpler to solve in comparison with GRAZ database.

In order to advance in our study, we move to a different problem that represents a compromise between GRAZ and ImageNet that we download from Kaggle web site [3]. The dataset consists of about 9000 art images, including the classes Drawings, Engraving, Painting, Iconography, and Sculpture. We add as the no-class the background class from Caltech-101. This dataset made of different sources and conceived for classifying different styles of art includes high-resolution images of different sizes [4, 7, 10]. The five classes were downloaded from Google images, Yandex images, and the virtual Russian museum. Data is separated on training and validation sets. Examples of the main categories are provided in Figs. 12.3, 12.4, 12.5, 12.6, 12.7, and 12.8 for visualization purposes. Also, we include a description of each database.

Fig. 12.3 This figure presents image examples of the class "Drawings" from the Kaggle database. These images include drawings and watercolors. Note the diversity in color, size, perspective, style, and image content. There are 1108 training images and 122 testing images. Brain programming spent 450 h or 18.75 days for one run of the algorithm

Fig. 12.4 This figure shows image examples of the class "Engraving" from the Kaggle database. This class represents a category of fine art or graphic art usually impress on a flat surface through the practice of incising a design onto a hard, usually flat surface by cutting grooves into it with a burin. The term refers to the arts that rely more on lines or tone than on color. There are 757 training images and 84 testing images. Brain programming spent 544 h or 22.6 days to solve the problem

Fig. 12.5 This figure displays image examples of the class "Painting" from the Kaggle database. Painting is the practice of applying paint, pigment, color, or another medium to a solid surface (support base). It brings elements such as drawings, gesture, composition, narration, or abstraction. Paintings can be naturalistic and representational (as in a still life or landscape), photographic, symbolistic, emotive, or political. There are 2043 training images and 228 testing images. Brain programming finished one run of the algorithm in 597 h or 24.81 days

Fig. 12.6 This figure presents image examples of the class "Iconography" from the Kaggle database. The term refers to the production of religious images, called "icons" in the Byzantine and Orthodox Christian tradition. The dataset contains 2076 training images and 231 testing images. Brain programming finished in 810 h or 33.75 days for this class

Fig. 12.7 This figure includes image examples of the class "Sculpture" from the Kaggle database. The sculpture is a branch of visual arts that operates in three dimensions. It is one of the plastic arts. Sculptural processes include carving (the removal of material) and modeling (the addition of material, as clay), in stone, metal, ceramics, wood, and other materials but, since modernism, there has been almost complete freedom of materials and processes. A wide variety of materials may be worked by removal such as carving, assembled by welding or modeling, or molded or cast. There are 1737 training images and 188 testing images. Brain programming spent 562 h or 23.41 days for the image class

Fig. 12.8 This figure shows image examples of the class "Background" from Caltech-101 database. It includes a wide variety of images of different types, it is considered the Non-class for the problem of binary classification. Note that by mistake this class contains images of the five classes studied in this paper like sculptures, and other similar to engravings and paintings. There are 468 images split in training and validation sets

12.5   Experiments

Experiments were carried out to provide evidence to support the claim that efficient and reliable solutions, for not trivial recognition problems, are discovered through the technique of brain programming. The implementation is programmed in MATLAB running on a Dell Precision T7500 workstation, Intel Xeon eight-core, CPU E5506 at 2.13 GHz, NVIDIA Quadro 4000 with Linux OpenSuse OS. The methodology follows the usual absent/present protocol considering two image sets for learning and testing.

Brain programming is a highly demanding computational paradigm. Instead of applying hundreds or thousands of individuals to create the population, we use a population size of 30 individuals per function. Also, the programs run for 30 generations instead of hundreds or thousands of iterations like in most published research. Note that we use atypical genetic programming parameters, and this is due to the analysis of the visual problem. A common practice in solving challenging problems is to increase the size of the population as a way to enrich the initial population. An important aspect to properly approach any computational problem with genetic programming is the definition of the function and terminal sets. These are the building blocks that genetic programming uses to construct the solutions. In our previous research, we define those sets following our expertise and the literature explaining the inner workings of the visual pathways.

A balance should be found to create programs that can solve non-trivial problems within the state-of-the-art in a reasonable amount of time. Our method spends 450 h or 18.75 days for the class Drawing, 544 h or 22.66 days for the class Engraving, 810 h or 33.75 days for the class Iconography, 597 h or 24.87 days for the class Painting, and 562 h or 23.41 days for the class Sculpture. We use ten workstations to speed the process of optimization. Table 12.1 shows a summary of statistical results after running 15 times our brain programming strategy per class. We include results of two deep-learning methods: from scratch CNN and AlexNet.

Brain programming looks for the optimal strategy within the search space of possible programs. At the moment of making a comparison of the best individual discovered through genetic programming with solutions from the state-of-the-art, it is necessary to use only the best solution. The average and standard deviation give information about the search process, but if we want to know the accuracy

Table 12.1 This table shows a summary of the results of applying brain programming to the classification of digitized art

Run	Drawings	Engraving	Painting	Iconography	Sculpture
1	80.15	80.03	90.27	83.52	81.10
2	84.73	**83.46**	89.71	89.11	80.65
3	78.84	83.30	86.61	87.09	84.21
4	80.65	74.95	92.94	85.60	85.37
5	83.46	78.88	88.43	87.56	84.74
6	79.00	79.70	**98.24**	**89.37**	84.83
7	83.84	76.26	95.93	87.17	84.74
8	74.32	81.99	96.46	86.54	85.64
9	75.94	76.26	90.43	87.01	**89.37**
10	82.57	82.16	90.51	83.71	84.92
11	78.88	74.63	90.74	85.83	85.28
12	**86.01**	79.54	89.47	80.15	86.19
13	79.75	73.48	90.27	84.20	84.83
14	83.33	79.86	92.82	88.72	83.01
15	75.76	81.50	92.50	83.81	84.92
Mean	80.48	79.19	91.40	86.34	84.70
Std. Dev.	3.48	3.19	3.26	2.02	1.93
Min.	74.32	74.14	87.07	83.52	81.10
Max.	86.01	83.46	98.24	89.37	89.37
From scratch CNN	76.18 ± 2.38	79.16 ± 4.88	91.78 ± 2.43	91.49 ± 0.51	81.14 ± 2.51
AlexNet	89.34 ± 0.89	92.50 ± 1.98	96.46 ± 0.36	98.14 ± 0.43	93.81 ± 0.91

Values highlighted with bold font identify the best solutions discovered after applying hands-on artificial evolution. We include results of the deep neural networks for comparison

of a solution, this needs to be computed afterward. The table shows that the best program for the class Painting practically solves the problem and its accuracy is better to AlexNet. It misses only about 40 images from the whole dataset. Regarding the class Drawings, our program reaches close to AlexNet, while for the rest of the classes, the results are better to an underlying convolutional neural network (from scratch CNN).

12.5.1 Beyond Random Search in Genetic Programming

In general, the use of random principles is overused in evolutionary computation. The idea is to adapt the methodology to avoid the unnecessary application of arbitrary or unplanned solutions within the algorithm to advance towards a more goal-oriented methodology. A first idea tested here is to use the best solutions discovered during the previous experiments as the initial population for a new set of experiments of brain programming, see Fig. 12.9. Table 12.2 provides preliminary results achieved for the five classes. We observe that the Class Drawings improved to the point of matching AlexNet's performance and for the Painting class becomes even better. Also, the proposed strategy improves all other classes: Engraving, Iconography, and Sculpture. Tables 12.3, 12.4, 12.5, 12.6, and 12.7 provide the best programs for each studied art problem. It is remarkable the simplicity of solutions in comparison with AlexNet. These programs can be read and are susceptible to improving through analysis. Figure 12.10 shows the results of the five experiments that confirm the benefit of applying the best solutions discovered in previous experiments to run new batches of experiments with excellent results.

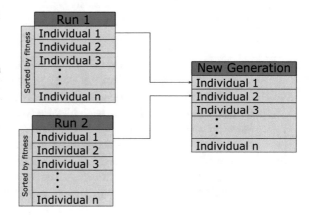

Fig. 12.9 We use the best set of solutions that were discovered in the previous experiments, to create a non-random initial population for a new set of experiments of brain programming. The idea is to continue the evolution from the best local minimum discovered so far

Table 12.2 This table shows a summary of the results after evolving the best set of solutions with brain programming to the classification of digitized art

Run	Drawings	Engraving	Painting	Iconography	Sculpture
1	90.41	84.61	**99.02**	91.11	89.37
2	89.15	84.78	98.24	90.48	89.73
3	**91.32**	84.12	98.24	89.85	89.37
4	90.41	84.94	98.32	90.87	89.37
5	90.78	84.45	98.64	91.11	89.74
6	89.15	84.29	98.32	**91.18**	89.74
7	89.18	83.96	98.32	90.95	89.83
8	89.31	83.96	98.32	90.79	89.37
9	88.78	83.63	98.32	90.72	**89.92**
10	89.15	**86.74**	98.40	89.37	89.83
Original	86.01	83.46	98.24	89.37	89.37
Mean	89.76	84.54	98.41	90.64	89.62
Std. Dev.	0.88	0.87	0.24	0.59	0.23
Best solution	**91.32**	**86.74**	**99.02**	**91.18**	**89.92**
Improvement	5.31	3.28	0.78	1.73	0.25
AlexNet	89.34 ± 0.89	92.50 ± 1.98	96.46 ± 0.36	98.14 ± 0.43	93.81 ± 0.91

Values highlighted with bold font identify the best solutions discovered after applying hands-on artificial evolution. We include results of transfer learning methodology AlexNet for comparison

Table 12.3 This table provides the functions of the best program discovered for the problem Drawings vs. Background

	Best individual
EVO_O[a]	$kSum(imRound(ip\_GaussDy\_1(supremum(kSum(supremum(ip\_GaussDy\_1$ $(supremum(ip\_GaussDx\_1(C)kSum(supremum(ip\_GaussDx\_1(C),$ $imRound(ip\_GaussDx\_1(M))), 0.31))), imRound(ip\_GaussDx\_1(M))), 0.31),$ $kSum(supremum(ip\_GaussDx\_1(C), imRound(ip\_GaussDx\_1(C))), 0.31)))),$ $0.31)$
EVO_C	$kSum(kDiv(kSum(imFloor(imCeil(kDiv(kSum(kSum(kSum(imFloor$ $(imFloor(S)), 0.20), 0.20), 0.20), 0.20), 0.31))), 0.20), 0.31), 0.20)$
EVO_S	$ip\_imdivide(hitmissDmnd(hitmissDsk(hitmissDsk(hitmissDmnd$ $(hitmissDmnd(bottomHat(ip\_imdivide(hitmissDmnd(hitmissDsk(hitmissDsk$ $(hitmissDmnd(bottomHat(dilateDsk(G)))))), R))))))), R)$
MM_1[a]	$ip\_Sqrt(ip\_Sqrt(ip\_Sqrt(ip\_GaussDx\_1(ip\_GaussDy\_1(imagen)))))$
MM_2	$ip\_GaussDy\_1(ip\_imsubtract(ip\_GaussDy\_1(ip\_imsubtract(ip\_GaussDy\_1$ $(ip\_GaussDy\_1(imagen)), ip\_Sqr(ip\_GaussDy\_1(ip\_GaussDy\_1(imagen))))),$ $ip\_GaussDy\_1(ip\_imsubtract(ip\_GaussDy\_1(ip\_GaussDy\_1(imagen)),$ $ip\_GaussDy\_1(ip\_GaussDy\_1(imagen))))))$

[a]EVO stands for evolutionary visual operator and MM for mental maps. Note that image in the MM refers to the conspicuity maps (CM)

Table 12.4 This table provides the functions of the best program discovered for the problem Engraving vs. Background

	Best individual
EVO_O[a]	$ip\_Gauss\_1(imagen)$
EVO_C	$ip\_imcomplement(ip\_Exp(ip\_Exp(ip\_Exp(ip\_Exp(ip$ $\_Exp(RedGreenOpon(image)))))))$
EVO_S	Y
MM_1[a]	$ip\_Logarithm(ip\_Half(ip\_Half(ip\_GaussDx\_1(imagen))))$
MM_2	$ip\_GaussDx\_1(ip\_GaussDx\_1(imagen))$
MM_3	$ip\_Logarithm(ip\_imdivide(ip\_GaussDy\_1(ip\_GaussDy\_1(imagen)),$ $ip\_imadd(ip\_imadd(ip\_GaussDy\_1(ip\_GaussDy\_1(imagen)),$ $ip\_imadd(ip\_GaussDy\_1(ip\_GaussDy\_1(imagen)),$ $ip\_imadd(ip\_GaussDy\_1(ip\_GaussDy\_1(imagen)),$ $ip\_GaussDx\_1(ip\_GaussDy\_1(imagen))))),$ $ip\_imadd(ip\_GaussDy\_1(ip\_GaussDy\_1(imagen)),$ $ip\_GaussDx\_1(ip\_GaussDy\_1(imagen))))))$

[a]EVO stands for evolutionary visual operator and MM for mental maps. Note that these functions are meaningless if the hierarchical structure is not applied to recognize the image

Table 12.5 This table provides the functions of the best program discovered for the problem Painting vs. Background

	Best individual
EVO_O[a]	$ip\_imadd(ip\_GaussDx\_1(ip\_GaussDx\_1(H)), ip\_imabsadd(ip\_GaussDx\_1$ $(ip\_GaussDx\_1(H)), ip\_GaussDx\_1(ip\_GaussDx\_1(H))))$
EVO_C	$RedGreenOpon(image)$
EVO_S	$hitmissSqr(ip\_imadd(erodeDmnd(erodeDmnd(kDiv(K, 0.88))), ip\_imadd$ $(ip\_imsubtract(erodeDsk(B), erodeDsk(B)), ip\_imsubtract(Y, erodeDsk(B)))))$
MM_1[a]	$ip\_imabsadd(ip\_Sqrt(ip\_imabsadd(ip\_imabsadd(ip\_Sqrt(ip\_Logarithm$ $(ip\_GaussDx\_1(ip\_Logarithm(ip\_GaussDx\_1(ip\_GaussDx\_1$ $(ip\_GaussDy\_1(imagen))))))), \quad ip\_GaussDx\_1(ip\_GaussDy\_1(imagen))),$ $ip\_Logarithm(ip\_GaussDx\_1(ip\_GaussDy\_1 \qquad (imagen))))),$ $ip\_Logarithm(ip\_imabs(ip\_imabs(ip\_GaussDx\_1(ip\_Logarithm(ip\_GaussDx\_$ $1(ip\_Logarithm(ip\_GaussDx\_1(ip\_GaussDx\_1(ip\_GaussDy\_1(imagen)))))))))))$

[a]EVO stands for evolutionary visual operator and MM for mental maps

12.5.2   Ideas for a New Kind of Evolutionary Learning

The proposed strategy presents a significant drawback related to the high computational cost. Usually, in machine learning, the goal is to optimize the quality of results through proper management of the set of solutions. In evolutionary computation, this is achieved with a technique that increases diversity by creating sub-populations called "niches", where solutions are only allowed to compete within their sub-groups similar to how species evolve when isolated on islands. Diversity is rewarded through a technique called "fitness sharing", where the difference between members of the population is measured to give an edge in the competition to distinct types of

Table 12.6 This table provides the functions of the best program discovered for the problem Iconography vs. Background

	Best individual
EVO_O[a]	$ip\_GaussDy\_1(kSum(ip\_GaussDy\_1(ip\_GaussDy\_1(ip\_GaussDy\_1$ $(ip\_GaussDy\_1\ (ip\_Gauss\_2(ip\_GaussDx\_1(ip\_GaussDx\_1(C)))))))$, $0.68))$
EVO_C	R
EVO_S	$hitmissSqr(kDiv(skeletonShp(kDiv(skeletonShp(Y), 0.74)), 0.74))$
MM_1[a]	$ip\_Sqrt(ip\_imabsadd(ip\_imabsadd(ip\_imabsadd(ip\_imabsadd$ $(ip\_Sqrt(ip\_imabsadd(ip\_imabsadd(ip\_imabsadd(ip\_imabsadd$ $(ip\_imabsadd(ip\_GaussDx\_1(ip\_GaussDy\_1(imagen)),ip\_Sqrt(ip\_imabsadd$ $(ip\_GaussDx\_1(ip\_GaussDx\_1(imagen)), ip\_GaussDx\_1$ $(ip\_GaussDy\_1(imagen))))),imagen), ip\_Sqrt(ip\_Sqrt(ip\_imabsadd$ $(ip\_imabsadd(ip\_imabsadd(ip\_imabsadd(ip\_imabsadd(ip\_GaussDx\_1$ $(ip\_GaussDy\_1(imagen)), ip\_Sqrt(ip\_imabsadd(ip\_GaussDx\_1$ $(ip\_GaussDx\_1(imagen)), ip\_GaussDx\_1$ $(ip\_GaussDy\_1(imagen))))), imagen),$ $ip\_Sqrt(ip\_imabsadd(imagen, ip\_GaussDx\_1(ip\_GaussDy\_1(imagen))))),$ $imagen),imagen)))), imagen), imagen)), imagen), ip\_Sqrt$ $(ip\_imabsadd(imagen, ip\_GaussDx\_1(ip\_GaussDy\_1(imagen))))),$ $ip\_GaussDy\_1\ (ip\_GaussDy\_1(imagen))), imagen))$
MM_2	$ip\_GaussDy\_1(ip\_GaussDy\_1(imagen))$

[a]EVO stands for evolutionary visual operator and MM for mental maps

Table 12.7 This table provides the functions of the best program discovered for the problem Sculpture vs. Background

	Best individual
EVO_O[a]	$ip\_GaussDx\_1(kSust(ip\_GaussDx\_1(kSust(kSust(ip\_GaussDx\_1$ $(kSust(imFloor(ip\_imabs(ip\_GaussDx\_1(ip\_GaussDy\_1(imagen)))),$ $0.46)), 0.46), 0.46)), 0.46))$
EVO_C	$ip\_imcomplement(imFloor(ip\_Sqrt(imFloor(M))))$
EVO_S	$kSum(M, 0.00)$
MM_1[a]	$ip\_imabs(ip\_GaussDx\_1(ip\_GaussDy\_1(imagen)))$

[a]EVO stands for evolutionary visual operator and MM for mental maps. Note that in this program the function for shape dimension adds zero to the magenta band. In GP it is possible to catch this errors while in neural networks similar problems are harder to identify

solutions. We present some results here to point out some critical issues for future research regarding computational costs.

12.5.3 Running the Algorithm with Fewer Images

In previous research, we remark the facility of brain programming to discover solutions through a simple random search [25]. Table 12.8 shows that it is possible to reduce the computational effort by reducing the size of the subsets for a classification problem. The computational effort could be reduced from several days

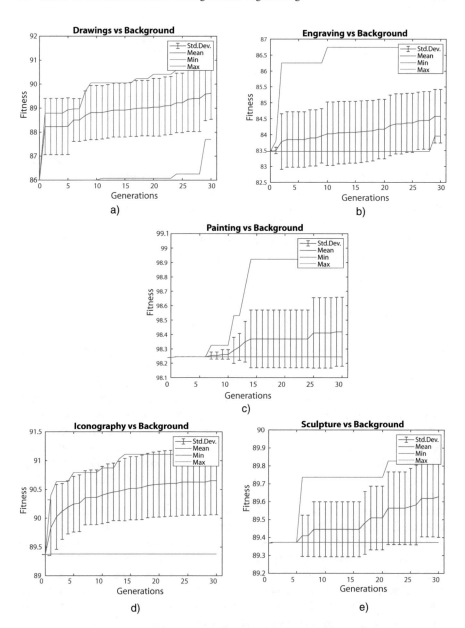

Fig. 12.10 These graphics show the evolution of the maxima discovered for each of the art problems. Note that Painting vs. Background requires more time to ameliorate the solutions with a smooth initial improvement. We observe similar behavior in the Sculpture vs. Background figure. The other graphs present constant progress from the first generations. In general, these figures give evidence that the idea of hands-on evolution works for computationally demanding problems

Table 12.8 Experiments of the computational effort after reducing the number of images of the subsets for the problem Painting vs. Background

No. of images	Time (h)[a]	Fitness	Since Gen	Until Gen
50	18 h	79%	5	50
50	18 h	85%	24	60
100	22 h	88.5%	22	30
200	4 days	84.5%	19	30

[a]All computations were made with matlab parfor

to only a few hours. Nevertheless, fitness can be compromised if care is not taken to balance the problem representation. Indeed, the number of images representing the problem that is being solved needs to be chosen carefully. Here, we observe that fitness is reduced by 10%. Also, we noted that perfect solutions could be achieved randomly for a given subset as already pointed out. This aspect can be misleading since a particular subset with few images could produce a score that brings a solution that is not general. We select a size of 100 images for further experiments.

12.5.4 Running the Algorithm with 100 Images

In a new set of experiments, we attempt to solve the classification problem of Painting vs. Burrito using ten different subsets of 100 images, see Table 12.9. In the first experiments, we chose as the non-class the Background class from Caltech 101. This dataset has images that belong to some of the classes studied in the digitized art problem. In order to avoid this issue, we replace the non-class with a set of images taken from the class Burrito of ImageNet for the rest of the experiments, see Fig. 12.11. We run the algorithm ten times, each run with a different subset, and found three out of four perfect solutions whose scores degrade while being tested with the complete database. We do the same for deep learning to make a comparison with CNN from scratch and transfer learning AlexNet. Note that this latter method gives outstanding results. However, when the best solutions are tested with the Background class, the score degrade significantly. This behavior is one of the main complains of deep learning. The methodology of convolutional neural networks creates models heavily based on the database. Hence, when there are variations to the information, the model reduces its performance. In contrast, brain programming slowly decreases and in some cases maintain the achieved level when the background changes. This novel behavior is due to the function-driven paradigm applied in our approach that attempts to discover a program from the landscape of possible computer programs.

A simple way of understanding this phenomenon is as follows. In normal deep learning, the designer provides the architecture of the deep neural network, like in the case of AlexNet. Then, the network is optimized through conventional learning using gradient descent and the set of given images. The results in Table 12.9 show a small dispersion in comparison with a badly designed network like CNN from scratch. In our GP-based approach this phenomenon of small variance was

Table 12.9 Experiments with subsets of 100 images showing fitness scores for the classification problem of Painting vs. Burrito

Training set	Fitness	Complete database	Painting vs. Background
c1	91.00	81.68	76.24
c2	97.50	95.31	87.32
$c3^{26}$	100	97.27	85.49
c4	99.00	98.03	94.85
$c5^0$	100	66.54	44.10
	97.50	95.91	83.57
c6	100	98.03	85.25
c7	99.00	96.74	83.41
$c8^0$	100	66.43	42.58
	88.5	82.43	77.41
c9	97.00	94.17	82.62
$c10^0$	100	66.62	64.83
	87.00	83.19	76.25
Mean	95.65 ± 4.90	92.90 ± 6.92	83.24 ± 5.70

Some elements of the first column have a superindex indicating that in this run, the program found a perfect solution (score 100%). The number in the superscript indicates the generation when the algorithm discovered the ideal solution. However, these solutions decreased their performance while tested, so we perform another evolution with the same set of images. The second line after superscript shows the result of this evolution

Training set	CNN from scratch			Transfer learning—AlexNet		
	Fitness	Complete	vs. Background	Fitness	Complete	vs. Background
c1	72.00	69.11	54.55	98.5	98.86	80.62
c2	82.50	86.30	73.84	98.00	97.88	76.27
c3	86.00	85.09	70.57	98.5	99.55	81.10
c4	79.00	85.77	75.28	100	99.17	80.78
c5	85.50	87.36	73.76	99.50	98.64	80.06
c6	84.00	80.39	63.64	99.00	99.02	80.54
c7	85.50	86.15	72.41	98.00	98.94	80.54
c8	80.00	83.04	68.10	99.00	99.55	81.26
c9	88.50	90.31	73.21	98.50	98.33	79.74
c10	81.00	76.68	62.28	98.00	99.09	80.86
Mean	82.4 ± 4.7	83.02 ± 6.19	68.76 ± 6.67	98.7 ± 0.67	98.90 ± 0.51	80.84 ± 0.62

Third and fourth columns provide results of the best solutions against the complete database and Caltech 101 background

never reproduced. The designer is attempting to find the best model through an evolutionary and genetic programming-based approach. In this case, the designer is not only looking for architecture but the whole computational model. Hence, there are numerous variations in the final solutions from the several runs being discovered by the algorithm. Note that in the $c3$ training set, the best solution was

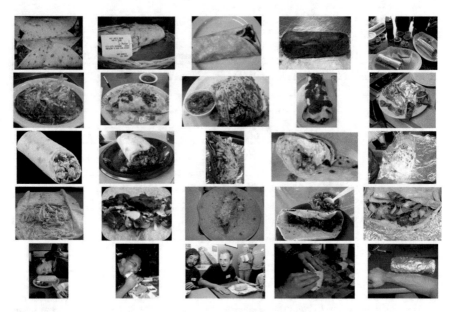

Fig. 12.11 Image examples of the class "Burrito" from the ImageNet database. These images include photographs of a dish called burrito, Mexican and Tex-Mex cuisine that consists of a flour tortilla with various other ingredients. It is wrapped into a closed-ended cylinder that can be picked up, in contrast to a taco, where the tortilla is simply folded around the fillings. The database includes many situations as well as persons with 1300 images. We select this dataset as the non-class for the new set of experiments since this class does not overlap with any of the digitized art classes

discovered in 26 iterations of the algorithm. In this case, the algorithm converged to a local minimum that is good enough to describe the classification problem: the score with the complete database is high. This last experiment was repeated using the Iconography vs. Burrito databases, see Table 12.10. We observe the same pattern for deep learning where minimal dispersion is achieved even for the complete database and Background experiments. Brain programming always had higher-variance independently of the databases. Similar results can be observed in Tables 12.11, 12.12, and 12.13.

12.5.5 Ensemble Techniques and Genetic Programming

Ensemble techniques are meta algorithms that combine several weak classifiers into one predictive model. In general, ensemble techniques use multiple learning algorithms to obtain better predictive performance than could be obtained from any of the constituent learning algorithms alone. The word weak in the jargon of machine learning refers to very poor learners, each performing only just better than

Table 12.10 Experiments with subsets of 100 images showing fitness scores for the classification problem of Iconography vs. Burrito

Training set	Fitness	Complete database	Iconography vs. Background
c1	93.00	64.35	57.04
c2	84.50	73.48	62.79
c3	86.50	77.13	70.42
c4	89.00	84.53	71.13
c5	78.00	70.63	63.81
c6[1]	100	26.83	22.97
	83.50	69.88	64.42
c7	81.00	73.92	68.76
c8	82.5	76.98	69.55
c9	80.5	72.05	67.98
c10	83.5	79.60	72.70
Mean	84.20 ± 4.37	74.26 ± 5.64	66.86 ± 4.78

Some elements of the first column have a superindex indicating that in this run, the program found a perfect solution (score 100%). The number in the superscript indicates the generation when the algorithm discovered the ideal solution. However, these solutions decreased their performance while tested, so we perform another evolution with the same set of images. The second line after superscript shows the result of this evolution

Training set	CNN from scratch			Transfer learning—AlexNet		
	Fitness	Complete	vs. Background	Fitness	Complete	vs. Background
c1	66.50	72.42	58.54	99.50	99.55	81.27
c2	70.00	65.02	56.81	99.50	99.40	81.12
c3	73.50	76.76	65.77	99.00	98.95	80.57
c4	73.00	79.07	65.46	100	99.63	81.35
c5	55.50	64.50	55.78	99.00	99.78	81.51
c6	56.00	77.65	76.08	100	98.88	80.49
c7	73.00	71.97	58.77	100	99.63	81.27
c8	67.00	74.29	65.93	99.50	99.78	81.59
c9	74.50	74.44	60.11	96.50	98.50	81.67
c10	66.50	77.28	72.07	100	99.50	81.27
Mean	67.55 ± 6.91	73.34 ± 5.06	63.53 ± 6.74	99.3 ± 1.06	99.36 ± 0.43	81.21 ± 0.40

Third and fourth columns provide results of the best solutions against the complete database and Caltech 101 background

chance, and which by putting them together it is possible to make an ensemble learner that can perform arbitrarily well. These so-called weak classifiers cannot be taken as such in the case of genetic programming since the programs learned by this evolutionary algorithm follows a process where the solutions are optimized, and their differences are quite significant from genotype and phenotype viewpoints. Next, we present some preliminary results of several techniques called mixture-of-experts.

Following the results of the experiments with the databases Painting vs. Burrito we applied majority voting, see Table 12.9. In this technique, all ensemble members

Table 12.11 Experiments with subsets of 50 images showing fitness scores for the classification problem of Engraving vs. Burrito

Training set	Fitness	Complete database	Engraving vs. Background
c1	96.00	82.89	50.00
c2	87.00	75.37	50.88
c3	85.00	69.17	54.87
c4	87.00	80.97	62.83
c5	83.00	77.29	61.36
c6	79.00	69.91	53.10
$c7^0$	100	65.04	47.35
	84.00	65.04	69.43
$c8^{16}$	100	44.25	44.25
	92.00	74.78	58.23
c9	95.00	80.24	51.33
c10	87.00	76.84	56.34
Mean	87.50 ± 5.38	75.25 ± 5.70	56.87 ± 6.21

Some elements of the first column have a superindex indicating that in this run, the program found a perfect solution (score 100%). The number in the superscript indicates the generation when the algorithm discovered the ideal solution. However, these solutions decreased their performance while tested, so we perform another evolution with the same set of images. The second line after superscript shows the result of this evolution

Training set	CNN from scratch			Transfer learning—AlexNet		
	Fitness	Complete	vs. Background	Fitness	Complete	vs. Background
c1	95.00	96.30	59.57	100	99.71	61.54
c2	85.00	95.77	59.25	99.00	99.71	61.87
c3	90.00	92.06	56.96	99.00	99.56	61.54
c4	91.00	93.65	57.94	100	99.71	61.70
c5	98.00	95.50	59.08	100	99.85	61.87
c6	97.00	90.48	55.97	100	99.85	61.87
c7	85.00	87.04	53.85	99.00	99.85	61.70
c8	91.00	91.01	56.30	99.00	99.41	61.87
c9	76.50	91.80	56.79	100	99.41	61.37
c10	95.00	87.04	83.85	100	99.56	61.87
Mean	90.30 ± 6.75	92.06 ± 3.34	56.96 ± 2.06	99.60 ± 0.52	99.66 ± 0.17	61.72 ± 0.18

Third and fourth columns provide results of the best solutions against the complete database and Caltech 101 background

cast a vote about whether or not the image belongs to a class. We took the decision taken by $(n/2) + 1$ programs. As a result of applying this method we manage to increase the performance of the paintings class to 98.41% with the complete database.

Another mixture-of-experts technique called weighted ensemble was tested with the results of Iconography vs. Burrito shown in Table 12.10. In this method,

Table 12.12 Experiments with subsets of 100 images showing fitness scores for the classification problem of Sculpture vs. Burrito

Training set	Fitness	Complete database	Sculpture vs. Background
c1	80.50	71.66	70.12
$c2^0$	100	35.87	35.87
	83.00	77.48	67.85
c3	81.50	80.74	72.17
c4	80.50	72.52	58.82
c5	90.50	92.47	72.77
$c6^1$	100	38.61	38.61
	89.00	80.91	77.84
c7	93.00	84.76	68.24
c8	81.00	66.18	58.39
c9	82.00	75.68	66.78
c10	90.50	86.22	68.95
Mean	85.15 ± 4.97	78.86 ± 7.79	68.19 ± 5.96

Some elements of the first column have a superindex indicating that in this run, the program found a perfect solution (score 100%). The number in the superscript indicates the generation when the algorithm discovered the ideal solution. However, these solutions decreased their performance while tested, so we perform another evolution with the same set of images. The second line after superscript shows the result of this evolution

Training set	CNN from scratch			Transfer learning—AlexNet		
	Fitness	Complete	vs. Background	Fitness	Complete	vs. Background
c1	73.00	74.65	58.86	97.50	97.98	76.93
c2	64.5	67.05	52.86	99.50	98.89	78.11
c3	72.50	77.65	61.22	99.00	99.40	78.56
c4	73.50	78.57	61.94	100	98.46	77.93
c5	79.50	79.26	62.49	97.00	98.54	77.66
c6	72.00	81.68	64.40	98.50	99.23	77.48
c7	94.50	90.55	71.39	99.00	99.40	78.38
c8	55.50	89.17	70.30	99.00	99.23	78.38
c9	81.00	87.21	68.76	100	98.29	78.20
c10	78.00	73.16	57.67	98.00	99.56	77.20
Mean	74.40 ± 10.29	79.90 ± 7.46	62.99 ± 5.88	98.75 ± 1.00	98.81 ± 0.50	77.88 ± 0.55

Third and fourth columns provide results of the best solutions against the complete database and Caltech 101 background

each classifier has confidence (percentage) assigned as weights for each classifier representing the importance of its decision. Given the top three classifiers, each of them with fitness: 84.53%, 79.60%, and 77.13%. We normalize its weights to 35.04%, 32.99%, and 31.99% respectively. The probability of the ensemble for the class and non-class are computed as follows:

Table 12.13 Experiments with subsets of 50 images showing fitness scores for the classification problem of Drawings vs. Burrito

Training set	Fitness	Complete database	Drawings vs. Background
c1	86.00	78.08	70.22
c2[2]	100	35.17	29.64
	88.00	78.66	72.33
c3	89.00	81.36	65.90
c4	89.00	72.80	63.61
c5[4]	100	35.17	29.64
	92.00	72.92	57.56
c6	96.00	86.87	66.92
c7	95.00	82.42	67.18
c8	89.00	74.09	60.05
c9[0]	100	39.27	22.90
	88.00	91.44	70.93
c10	91.00	83.12	67.68
Mean	90.30 ± 3.20	80.17 ± 6.12	66.23 ± 4.70

Some elements of the first column have a superindex indicating that in this run, the program found a perfect solution (score 100%). The number in the superscript indicates the generation when the algorithm discovered the ideal solution. However, these solutions decreased their performance while tested, so we perform another evolution with the same set of images. The second line after superscript shows the result of this evolution

Training set	CNN from scratch			Transfer learning—AlexNet		
	Fitness	Complete	vs. Background	Fitness	Complete	vs. Background
c1	73.00	85.35	64.50	97.00	98.48	69.85
c2	78.00	72.68	47.58	99.00	98.71	69.47
c3	85.00	82.18	61.45	98.00	97.42	68.58
c4	87.00	82.53	58.27	99.00	98.59	69.85
c5	75.00	79.84	54.58	95.00	97.42	70.10
c6	84.00	86.99	60.56	99.00	99.18	69.97
c7	66.00	77.84	62.34	100	98.48	69.21
c8	79.00	82.06	63.36	100	98.94	69.97
c9	83.00	86.05	64.89	100	98.71	69.85
c10	81.00	85.46	59.03	98.00	98.71	69.34
Mean	79.10 ± 6.38	82.09 ± 4.39	59.66 ± 5.27	98.50 ± 1.58	98.46 ± 0.59	69.62 ± 0.42

Third and fourth columns provide results of the best solutions against the complete database and Caltech 101 background

$$C_1 = (p_n|c_1) * weight_1 + (p_n|c_1) * weight_2 + (p_n|c_1) * weight_3; \qquad (12.1)$$

$$C_2 = (p_n|c_2) * weight_1 + (p_n|c_2) * weight_2 + (p_n|c_2) * weight_3; \qquad (12.2)$$

where C_1 is the probability that the image belongs to the class, and C_2 to the non-class. The value of $weight_k$ is the classifier weight obtained at the moment of computing its fitness with the classifier variance. $(p_n|c_1) = $ is the classifier trust that the n image belongs to class 1, and $(p_n|c_2) = $ is the classifier trust that the n image belongs to class 2. With the above values we can compute the probability of the ensemble for the first image.

$$C_1 = (0.7563 * 0.3504) + (0.9887 * 0.3299) + (0.7893 * 0.3099) = 0.7954;$$

$$C_2 = (0.2437 * 0.3504) + (0.0113 * 0.3299) + (0.2987 * 0.3199) = 0.1748;$$

If we apply this reasoning we obtain a general classifier performance of 84.53%. Of course we give proportional weight to the top three classifiers, but if we change the weights of the ensemble to 40%, 30%, and 30%, the performance of the ensemble scores 85.35% for the complete database.

12.6 Conclusions

This paper presents a summary of the main works that have been developed within the EvoVisión laboratory about a new strategy called brain programming. It also offers the first results of applying the methodology for the problem of classification of digitized art. We report encouraging results and a first proposal to improve the algorithm. The goal of this research is to outline a methodology that can challenge deep learning in current visual perception tasks. Therefore, it is not feasible to leave to the genetic programming methodology the charge of discovering the whole computer program. Moreover, neuroscientific models are the base of brain programming to create visual processing programs. These new ideas represent a source of inspiration for a new kind of evolutionary algorithms that can challenge the state-of-the-art in artificial vision and pattern recognition.

Acknowledgements This research was partially funded by CICESE through the project 634-128, "Programación Cerebral Aplicada al Estudio del Pensamiento y la Visión". The second author graciously acknowledges the scholarship paid by the National Council for Science and Technology of Mexico (CONACyT) under grant 25267-340078.

References

1. Agarwal, S., Karnick, H., Pant, N., Patel, U.: Genre and style based painting classification. In: 2015 IEEE Winter Conference on Applications of Computer Vision, pp. 588–594. IEEE (2015)
2. Arora, R. S., Elgammal, A.: Towards automated classification of fine-art painting style: A comparative study. In: Proceedings of the 21st International Conference on Pattern Recognition (ICPR-2012), pp. 3541–3544. IEEE (2012)

3. Art Images: Drawing/Engraving/Iconography/Painting/Sculpture, https://www.kaggle.com/thedownhill/art-images-drawings-painting-sculpture-engraving
4. Bar, Y., Levy, N., Wolf, L.: Classification of artistic styles using binarized features derived from a deep neural network. In: European Conference on Computer Vision, pp. 71–84. Springer, Cham (2014)
5. Clemente, E., Olague, G., Dozal, L., Mancilla, M.: Object Recognition with an Optimized Ventral Stream Model using Genetic Programming. European Conference on the Applications of Evolutionary Computation, pp. 315–325. EvoApplications (2012)
6. Clemente, E., Chavez de la O, F., Fernández, F., Olague, G.: Self-adjusting Focus of Attention in Combination with a Genetic Fuzzy System for Improving a Laser Environment Control Device System. Applied Soft Computing **32**, 250–265 (2015)
7. Condorovici, R. G., Florea, C., Vrânceanu, R., Vertan, C.: Perceptually-inspired artistic genre identification system in digitized painting collections. In: Scandinavian Conference on Image Analysis, pp. 687–696. Springer (2013)
8. Dozal, L., Olague, G., Clemente, E., Sánchez, M.: Evolving Visual Attention Programs through EVO Features. European Conference on the Applications of Evolutionary Computation, pp. 326–335. EvoApplications (2012)
9. Dozal, L., Olague, G., Clemente, E., Hernández, D.E.: Brain Programming for the Evolution of an Artificial Dorsal Stream. Cognitive Computation **6**(3) 528–557 (2014)
10. Florea, C., Toca, C., Gieseke, F.: Artistic movement recognition by boosted fusion of color structure and topographic description. In: 2017 IEEE Winter Conference on Applications of Computer Vision (WACV), pp. 569–577. IEEE (2017)
11. Fukushima, K.: Neocognitron: A Self-organizing Neural Network Model for a Mechanism of Pattern Recognition Unaffected by Shift in Position. Biological Cybernetics **36**(4), 193–202 (1980)
12. Hernández, D.E., Clemente, E., Olague, G., Briseño, J.L.: Evolutionary Multi-objective Visual Cortex for Object Classification in Natural Images. Journal of Computational Science **17**(part 1), 216–233 (2016)
13. Hernández, D.E., Olague, G., Hernández, B., Clemente, E.: CUDA-based Parallelization of a Bio-inspired Model for Fast Object Classification. Neural Computing and Applications **30**(10), 3007–3018 (2018)
14. Holland, J.H.: Adaptation in Natural and Artificial Systems: An Introductory Analysis with Applications to Biology, Control, and Artificial Intelligence. MIT Press, 211 pages—first appeared in 1975 (1992)
15. Kowaliw, T., McCormack, J., Dorin, A.: Evolutionary automated recognition and characterization of an individual's artistic style. In: IEEE Congress on Evolutionary Computation, pp. 1–8. IEEE (2010)
16. Koza, J.R.: Genetic Programming: On the Programming of Computers by Means of Natural Selection. MIT Press (1992)
17. Krizhevsky, A.: Learning Multiple Layers of Features from Tiny Images. April 2009.
18. Krizhevsky, A., Sutskever, I., Hinton, G.E.: ImageNet Classification with Deep Convolutional Neural Networks. Advances in Neural Information Processing Systems, NIPS (2012)
19. Le, Q.V., Ranzato, M.A., Monga, R., Devin, M., Chen, K., Corrado, G.S., Dean, J., Ng, A.Y.: Building High-level Features using Large Sacle Unsupervised Learning. International Conference on Machine Learning, pp. 507–514 ICML (2012)
20. LeCun, Y., Bottou, L., Bengio, Y., Haffner, P.: Gradient-based Learning Applied to Document Recognition. Proceedings of the IEEE **86**(11), 2278–2324 (1998)
21. Lyu, S., Rockmore, D., Farid, H.: A digital technique for art authentication. Proceedings of the National Academy of Sciences, **101**(49), 17006–17010 (2004)
22. Olague, G.: Evolutionary Computer Vision: The First Footprints. Springer, Heidelberg (2016)
23. Olague, G., Hernández, D.E., Clemente, E., Chan-Ley, M.: Evolving Head Tracking Routines with Brain Programming. IEEE Access **6**, 26254–26270 (2018)

24. Olague, G., Hernández, D.E., Llamas, P., Clemente, E., Briseño, J.L.: Brain Programming as a New Strategy to Create Visual Routines for Object Tracking. Multimedia Tools and Applications **78**(5), 5881–5918, (2018)
25. Olague, G., Clemente, E., Hernández, D.E., Barrera, A., Chan-Ley, M., Bakshi, S.: Artificial Visual Cortex and Random Search for Object Categorization. IEEE Access (2019)
26. Olague, G.: Automated Photogrammetric Network Design using Genetic Algorithms. Photogrammetric Engineering & Remote Sensing **68**(5), 423–431 (2002) Paper awarded the "2003 First Honorable Mention to the Talbert Abrams Award", by ASPRS.
27. Silver, D., Huang, A., Maddison, C.J., Guez, A., Sifre, L., Van den Driessche, G., Schrittwieser J., Antonoglou, I., Panneershelvam, V., Lanctot, M., Dieleman. S., Grewe, D., Nham, J., Kalchbrenner, N., Sustkever, I., Lillicrap, T., Leach, M., Kavukcuoglu, K., Graepel, T., Hassabis, D.: Mastering the Game of Go with Deep Neural Networks and Tree Search. Nature **529**, 484–489 (2016)
28. Zujovic, J.,Gandy, L., Friedman, S., Pardo, B., Pappas, T.: Classifying Paintings by Artistic Genre: An Analysis of Features & Classifiers. In: 2009 IEEE International Workshop on Multimedia Signal Processing, pp. 1–5. IEEE (2009)

Chapter 13
Comparison of Linear Genome Representations for Software Synthesis

Edward Pantridge, Thomas Helmuth, and Lee Spector

13.1 Introduction

Inductive program synthesis is the field of producing executable programs from a set of input-output examples [7, 14, 16]. General software synthesis refers to the sub-field of inductive program synthesis in which the programs produced are expected to be capable of manipulating a variety of data types, control structures, and data structures. The field of genetic programming has produced some the most capable general software synthesis methods, such as PushGP [5], Grammar Guided Genetic Programming [1], and SignalGP [10]. The experiments and discussion in this paper focuses on PushGP.

PushGP synthesizes programs in the Push programming language. Push is a stack-based programming language designed for genetic programming, in which arguments for instructions are taken from typed stacks and return values are placed on the stacks [19]. A Push program is a sequence that may contain instructions, literals, and code blocks. A code block is also a sequence that may contain

E. Pantridge (✉)
Swoop, Inc., Cambridge, MA, USA
e-mail: ed@swoop.com

T. Helmuth (✉)
Hamilton College, Clinton, NY, USA
e-mail: thelmuth@hamilton.edu

L. Spector (✉)
Department of Computer Science, Amherst College, Amherst, MA, USA

School of Cognitive Science, Hampshire College, Amherst, MA, USA

College of Information and Computer Sciences, University of Massachusetts, Amherst, MA, USA
e-mail: lspector@hampshire.edu

© Springer Nature Switzerland AG 2020   255
W. Banzhaf et al. (eds.), *Genetic Programming Theory and Practice XVII*, Genetic
and Evolutionary Computation, https://doi.org/10.1007/978-3-030-39958-0_13

instructions, literals, and code blocks, allowing for hierarchically nested program structures.

When executed by a Push interpreter, the program itself is pushed onto the exec stack, a special stack that keeps track of the executing program. During execution, items from the exec stack are consumed from the program and evaluated sequentially. Literal values are placed onto stacks corresponding to their data types. Instructions are evaluated as functions that pop their arguments from the stacks and push their return values back onto the stacks. When code blocks are interpreted, their contents are unpacked and inserted at the start of the exec stack [19].

Push implementations typically provide instructions and stacks for common data types such as integers, floating point numbers, Boolean values, and strings. It is also possible for users to provide stacks and instructions for any other data types they choose [12, 19]. Since the executing program itself is stored on a stack, instructions can manipulate the executing code itself *as it runs*; this functionality is used to implement both standard and exotic control flow structures using the exec stack.

Like all genetic programming methods, PushGP manipulates "genome" data structures that correspond to executable programs. In its initial design, the PushGP genome structure was also the program structure of nested code blocks. With this representation, it straightforward to implement tree-based genetic operators like those used in tree-based genetic programming [8], but less straightforward to implement operators that act uniformly on program elements at all levels of nesting [17].

More recent implementations of PushGP use the Plush linear genome representation , which can be translated into the hierarchical code block program structure before execution of the program [6]. This layer of indirection provides flexibility with respect to which genetic operators can be applied to genomes, specifically with uniform mutation and crossover operators, which in turn has produced better search performance [3, 6].

This work compares two linear genome representations, each of which takes a different approach to the problem of specifying a nested structure in a flat, linear form. While both Plush and the new genome structure, Plushy, can represent the same set of Push programs, there are trade-offs between them that may affect evolvability and the dynamics of program size and structure.

We begin by detailing the two genome representations, Plush and Plushy. We then present experimental results that allow us to compare their search performance and to examine the effects of each on program structure over evolutionary time. We discuss factors relevant to choosing among the representations in practice, and conclude by recommending the adoption of Plushy genomes in future PushGP work.

13.2   Linear Genomes: Plush vs. Plushy

Plush [6] and Plushy [13] are two different linear data structures that have been used to represent genomes in PushGP systems. Both genome representations encode the

```
(5 x int_gt exec_if (3 x int_sub) (x 2 int_mult))
```

Fig. 13.1 A simple Push program that takes an integer input x and returns $(3 - x)$ if $x > 5$, or $2x$ otherwise. Note that the program contains two code blocks ((x) and (x 2 int_mult)), exemplifying the nested, non-linear structure of Push programs

nested structure of Push programs, and can be translated into executable programs. The process of translating a genome into a program will determine which individual genes should be placed within nested code blocks to produce the structured Push program.

Every instruction gene has a defined number of code blocks expected to follow the instruction, which is the same number in both representations. For example, the int_add instruction sums the top two integers on the integer stack, and thus opens no code blocks. The exec_if instruction takes two code blocks from the exec stack as arguments: one for holding the body of the "then" clause and one for holding the body of the "else" clause. If the top value on the Boolean stack is true, then the code block for the "else" clause is ignored. If the top value on the Boolean stack is false, then the code block for the "then" clause is ignored. Figure 13.1 shows a simple Push program that utilizes this exec_if instruction.

Just as the exec_if is defined to require two code blocks to follow it, other instructions also require specific numbers of code blocks as arguments. Note that this is a feature of the genome specifications we are discussing, not a requirement of the underlying Push programs. In fact, when using Push programs as genomes, there was nothing guaranteeing the presence of code blocks after instructions that made use of them. Often exec_if and similar instructions would just be followed by single instructions instead of code blocks; this can sometimes be useful, but it is often advantageous to have larger code blocks in these positions. When designing Plush, and afterward when designing Plushy, we chose to force instructions that can use code blocks to be followed by them, to increase the use of code blocks in evolved programs [6].

Given that instruction definitions are used to determine where code blocks are opened, it is left up to the genome representation to determine how to store the information denoting where each code block is closed.

The Plush genome representation is a flat sequence of instructions and literals. Each of these tokens is considered a gene of the genome. Each gene also has epigenetic markers that store information that is used when translating the genome into a program—these are "epigenetic" in the sense that they affect translation into Push programs, but do not appear in programs themselves. The distinction between genetic and epigenetic information raises the possibility that the two kinds of information could be varied in different ways or at different times during evolution. While in most prior work with Plush, including the experiments described in this chapter, epigenetic information was only varied during the production of offspring from parents (like genetic information), previous work has used hill-climbing search over variation of epigenetic markers to "learn" during an individual's lifetime [9].

The two kinds of epigenetic markers that have been used in PushGP systems are "close" and "silent" markers, though others could be created [6].

The "close" marker is an integer denoting how many code blocks should be closed directly following that particular gene. This allows the genome to indicate where code blocks are closed using epigenetic markers attached to specific genes. If there are no code blocks open at that location, the close maker value is ignored; if the number of open code blocks is less than the "close" value, then all open blocks are closed. If some code blocks are left open after the entire program has been translated, it is assumed the code blocks are closed at the end of the program.

The "silent" marker is a Boolean flag denoting if the gene is silenced. If true, the gene is skipped during genome translation. Using these markers a genome can hold genetic material that does not influence the resulting program and potentially pass it on to it children.

Figure 13.2 shows a Plush genome that produced the program from Fig. 13.1 when translated. Due to the separation of genes and their epigenetic markers under this representation, the Plush data structure can be thought of as a tabular structure since every gene has a value for every epigenetic marker.

The Plushy genome representation is also a sequence of instruction and literal genes, however there are additional genes used solely for translation. Plushy genomes do not use epigenetic markers, but are instead simply flat sequences of genes. The two additional kinds of genes introduced thus far in Plushy genomes are CLOSE genes and SKIP genes [13].

The CLOSE gene denotes the end of a code block. If there are no code blocks open at that location, the CLOSE is a no-op. If some code blocks are left open after the entire program has been translated, translation continues as if additional CLOSE genes are present until all code blocks are closed.

The SKIP gene causes genome translation to ignore the subsequent gene. Much like the silent epigenetic markers used in Plush genomes, these SKIP genes can be used to suppress genetic material such that it does not appear in the resulting program, yet potentially can be passed down to children. SKIP genes also cause a following SKIP or CLOSE gene to be ignored.

Gene:	5	x	int_gt	exec_if	x	int_sqr	x	2	int_mult
Closes:	0	0	2	0	1	0	0	0	0
Silent:	false	false	false	false	false	true	false	false	false

Fig. 13.2 One potential Plush genome that produces the program from Fig. 13.1 after translation. The definition of the exec_if instruction specifies the opening of two code blocks; one for the "then" clause and one for the "else" clause. The "close" epigenetic marker on the x gene denotes the end of the "then" clause for the exec_if. There is no gene with a non-zero close marker to denote the end of the "else" clause, and thus it is assumed to be at the end of the sequence. Notice that the int_gt instruction closes 2 code blocks despite no code blocks being opened by the previous genes, and thus these close markers are ignored by translation. The int_sqr instruction is not translated into the program because it has a true silent marker

```
5 x int_gt CLOSE exec_if x SKIP int_sqrt CLOSE x 2 int_mult
```

Fig. 13.3 One potential Plushy genome that produces the program from Fig. 13.1 after translation. The definition of the `exec_if` instruction specifies the opening of two code blocks; one for the "then" clause and one for the "else" clause. The end of the "then" clause is denoted by the final CLOSE gene. There is no CLOSE gene to denote the end of the "else" clause, and thus it is assumed to be at the end of the sequence. There is no gene that opens a code block before the first CLOSE and thus it has no effect on translation. The SKIP gene specifies the following gene should not be included in the translation, which explains why `int_sqrt` does not appear in the translated program

Figure 13.3 shows a Plushy genome that produces the program from Fig. 13.1 when translated.

13.2.1 Random Genome Generation

Random genomes are used to seed the initial population of genetic programming runs. Each genome type is generated differently to ensure the logic, structure, and size of the programs in the initial population is diverse.

The instructions and literals in random Plush genomes are typically chosen with a uniform distribution. The close epigenetic markers in Plush are initialized using a probability distribution. Sampling the distribution will give the value of the "close" marker. In previous PushGP research a binomial distribution with $n = 4$ and $p = 1/16$ is used. This yields the following probabilities for assigning values for the "close" marker.

Closes	Probability
0	0.772
1	0.206
2	0.021
3	0.001

We do not use "silent" markers in this work.

When generating Plushy genomes, a set of Push instructions and literals provided by the user is available. Plushy simply adds additional elements to this set for the CLOSE and SKIP genes. As described above, the definition of each instruction in the set denotes how many code blocks are opened by the instruction, based on the number of arguments it takes from the `exec` stack.

If the set of genes were randomly sampled with uniform probability, the CLOSE gene would occur in genomes at a rate of $\frac{1}{|S|}$, where $|S|$ is the number of available genes. This likely provides too few CLOSE genes compared to the number of code blocks opened. Instead, to generate Plushy genomes with a larger proportion of CLOSE genes, we set the probability of sampling a CLOSE gene proportionally to the sum of all "open" counts across all instructions. For example, if there are 10 instructions that each open 1 code block, the CLOSE gene is given a 10 times the probability of being added to the Plushy genome compared to any other instruction. This results in an average of one CLOSE for every code block opened.

13.2.2 Genetic Operators

When using Plush genomes, the genes and their epigenetic markers are two separate values corresponding to the same location in the genome. Genetic operators can affect either the genes, their epigenetic markers, or both. Uniform crossover and mutation operators that keep genes with their epigenetic markers have typically been used [3, 17]. Additionally, specialized genetic operators that do not effect gene values can be applied to epigenetic markers. For example, a *uniform close mutation* operator changes the "close" epigenetic markers by incrementing or decrementing them by one [6]. This close-marker mutation operator is applied to each gene in the genome with some configurable probability.

Plushy genomes do not contain epigenetic markers. Genetic operators that manipulate the genes in the genome are modifying both logic and structure. The uniform genetic operators that are applied to Plush can be applied to Plushy, with the exception of the epigenetic-marker operations. Genetic operators commonly used in the field of genetic algorithms can also be applied to Plushy genomes.

Genetic operators that add random genes to a Plushy genome, such as a mutation, will utilize the same increased probability of adding a CLOSE gene as seen in random genome generation discussed in Sect. 13.2.1.

13.3   Impact on Search Performance

A genome is the data structure manipulated by genetic operator throughout evolution. Different genome structures yield different landscapes to search over. Some landscapes may be more difficult to search through and thus search performance could be degraded when using certain genome representations.

13.3.1 Benchmarks

To evaluate the impact of Plush vs. Plushy genomes on search performance, we tested each on 25 problems from the general software synthesis benchmark suite [5]. These benchmark problems come from coding assignments traditionally given to human programmers in introductory computer science classes. For our detailed analysis of the affects of each genome on evolved program structure, we selected a representative subset of 10 problems, which are:

- **Compare String Lengths**. Given three strings (s_1, s_2, and s_3) return true if $length(s_1) < length(s_2) < length(s_3)$, and false otherwise.
- **Double Letters**. Given a string, print the string, doubling every letter character, and tripling every exclamation point. All other non-alphabetic and non-exclamation characters should be printed a single time each.
- **Last Index of Zero**. Given a vector of integers of length ≤ 50, each integer in the range $[-50, 50]$, at least one of which is 0, return the index of the last occurrence of 0 in the vector.
- **Mirror Image**. Given two lists of integers of the same length ≤ 50, return true if one list is the reverse of the other, and false otherwise.
- **Negative to Zero**. Given a vector of integers in $[-1000, 1000]$ with length ≤ 50, return the vector where all negative integers have been replaced by 0.
- **Replace Space With Newline**. Given a string input, print the string, replacing spaces with newlines. Also, the program should return the integer count of the non-whitespace characters.
- **String Lengths Backwards**. Given a vector of strings with length ≤ 50, where each string has length ≤ 50, print the length of each string in the vector starting with the last and ending with the first.
- **Sum of Squares**. Given an integer $0 < n \leq 100$, return the sum of squaring each positive integer between 1 and n inclusive.
- **Syllables**. Given a string (max length 20, containing symbols, spaces, digits, and lowercase letters), count the number of occurrences of vowels in the string and print that number as X in "The number of syllables is X."
- **Vector Average**. Given a vector of floats with length in $[1, 50]$, with each float in $[-1000, 1000]$, return the average of those floats. Results are rounded to 4 decimal places.

For each benchmark problem, 100 runs were performed with each genome type. All runs were performed with the same configuration of the PushGP system, with the exception of the genome type used. The hyperparameter values used to configure the PushGP system are presented in Fig. 13.4. We use size-neutral

Parameter	Value
Runs per setting	100
Population size	1000
Max number of generations	300
Genetic operator	UMAD, used to make all children
UMAD addition rate	0.09
max genome size	varies per problem, but same for Plush and Plushy

Fig. 13.4 The configuration of the Clojush PushGP system for the experimental runs performed for this research. We leave the tuning of these configurations for each genome type as future research

uniform mutation with addition and deletion (UMAD) to make all children for both Plush and Plushy [3]. UMAD, with an addition rate of 0.09, adds a new random instruction before or after each instruction with 0.09 probability; it then deletes each instruction in the program with probability $\frac{1}{\frac{1}{0.09}+1} \approx 0.08257$ to remain size-neutral on average. Note that we do not use any crossover here; while crossover may play some role in deciding which representation to use, UMAD by itself has outperformed any crossover technique we have tried, so we used it here. The initial and maximum genome sizes vary per problem, and follow the recommendations from the benchmark suite's technical report [4].

13.3.2   Benchmark Results

Figure 13.5 shows the number of solutions found by the genetic programming runs using Plush or Plushy genomes for all problems. Only one of the differences in success rate is significant using a Chi-squared test at $a = 0.05$: the Syllables problem. All other problems show no significance in the difference in numbers of successes. This shows that, at least for these program synthesis problems and hyperparameter settings, the choice of genome between Plush and Plushy has little to no effect on performance. Thus the choice to use Plush or Plushy should be based not on their effects on performance, but instead on other considerations such as flexibility with respect to genetic operators and the required amount of hyperparameter tuning.

13.4   Genome and Program Structure

Push programs have meaningful structure organized by code blocks, which affect the semantics of programs, particularly with respect to control flow. When evaluating genome representations , we therefore consider program structure in addition to search performance.

13.4.1   Sizes

Figure 13.6 shows a comparison of genome lengths produced during PushGP runs for each genome representation for 10 representative problems. Plushy genomes tend to be slightly longer than Plush genomes. This is to be expected because Plushy genomes require explicit `close` genes, increasing the size of genomes, whereas Plush stores "close" markers as epigenitic markers that do not contribute to genome size.

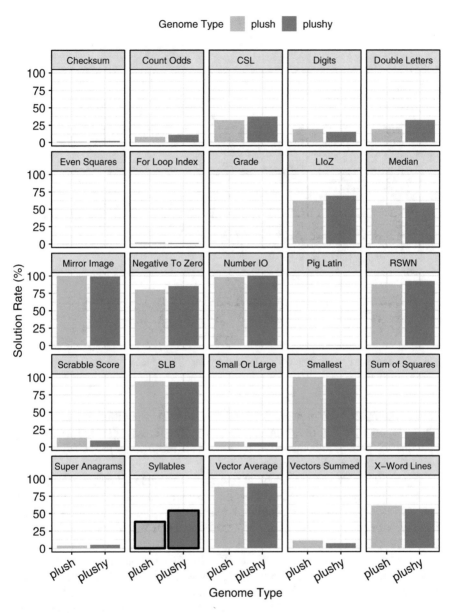

Fig. 13.5 The rate of solutions found using Plush genomes versus Plushy genomes. A genome is a solution if it receives an error of zero on all cases in a previously unseen test set after being simplified by an automatic simplification algorithm [2]. Each genome representation was evaluated across 100 runs for each problem. The difference in solution rates is only significant for one problem, "Syllables", shown with a black outline. On the "Syllables" problem the Plushy genome produces more solutions

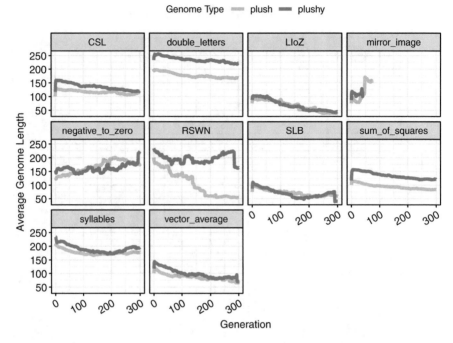

Fig. 13.6 Average Plush and Plushy genome lengths for all benchmark problems, averaged across all benchmark runs

The examples from previous sections also illustrate the difference in size. The Plush genome in Fig. 13.2 contains 9 genes. The Plushy genome in Fig. 13.3 contains 12 genes. Both genomes translate into program in Fig. 13.1, and both genomes only silence/skip one gene.

Figure 13.7 shows a comparison of program sizes produced during PushGP runs for both genome representations. Despite producing longer genomes, the Plushy data structure tends to translate into slightly smaller programs. This further confirms that the difference lengths was due to CLOSE genes, which affect genome lengths but not program sizes.

It seems as though the Replace Space With Newline problem is an outlier for both genome and program sizes. Genome and program sizes tend to be similar for Plush and Plushy experiments in the early generations. In later generations, the Plushy genomes and programs far exceed the size of their Plush counterparts. Since a large number of runs had finished by that point (by finding a solution), this can likely be attributed to a small number of outliers for each drastically changing the average of the remaining runs.

Push programs are nested structures of code blocks. It is possible to measure the maximum depth of a program. We refer the maximum depth of a program as the program's depth. Figure 13.8 shows that Plush genomes and Plushy genomes tend to produce similar program depths. The Sum of Squares problem is a drastic outlier

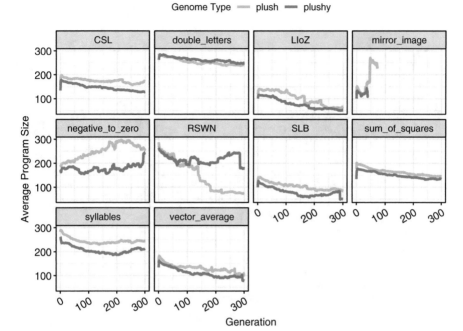

Fig. 13.7 Average Push program sizes produced using Plush and Plushy genomes for all benchmark problems, averaged across all benchmark runs

here, with programs translated from Plush genomes tending to have roughly twice the depth of the programs translated from Plushy genomes.

It is important to note that these genome representations do not have direct effects on program depth, as only the instructions they contain dictate where and how many code blocks are opened. Thus any differences here come about by evolutionary pressures. So, it may be the case that for the Sum of Squares problem the way in which Plushy genomes close code blocks made it evolutionarily advantageous to have more nested instructions than with Plush.

13.4.2   Presence of "Closing" Genes

As PushGP searches for solution programs, it manipulates genomes such that the logic and structure of the resulting programs is varied generation to generation. Figure 13.9 shows the prevalence of "closing" genes in both kinds of genomes as evolution progresses. For Plush genomes, this is the percentage of genes in the genome with a non-zero close epigenetic marker. For Plushy genomes, this is the percentage of CLOSE genes in the genome.

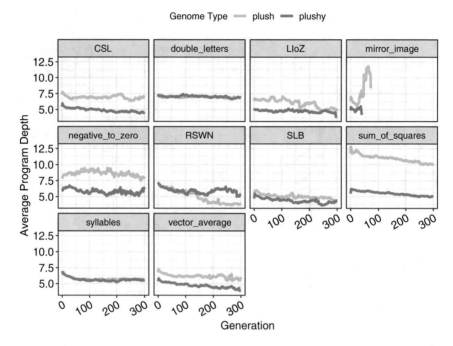

Fig. 13.8 Average program depths produced using Plush and Plushy genomes for all benchmark problems, averaged across all benchmark runs

The levels of "closing" genes stay relatively stable for both Plush and Plushy throughout evolutionary time, often ending with approximately the same percentage of closing genes as in the initial generation. These flat trends indicate that the levels of closing genes are largely dictated by the percentage of close epigenetic markers/CLOSE genes present in random code created during initialization and mutation, and are not reflective of evolutionary pressures toward higher or lower levels. The percentage of close markers with Plush starts around the same level (around 0.25) for every problem, as would be expected with the hard-coded probabilities of close markers as described earlier. The percentage of CLOSE genes in random Plushy genomes depends on the instruction set, and will therefore be different for these different problems, which use differing instruction sets. This explains the high level of CLOSE genes for the Sum of Squares problem, which uses a higher percentage of exec stack instructions (those responsible for opening code blocks) compared to the other problems here. Despite the adaptive prevalence of CLOSE genes offered by Plushy as discussed in Sect. 13.2.1, it is interesting to recall the lack of significantly different solution rates reported in Fig. 13.5. This suggests that the performance of evolution is not particularly sensitive to the prevalence of "closeing" genes.

As demonstrated in Sect. 13.2, when using either Plush or Plushy it is possible to have "closing" genes that have no impact of the structure of the resulting program

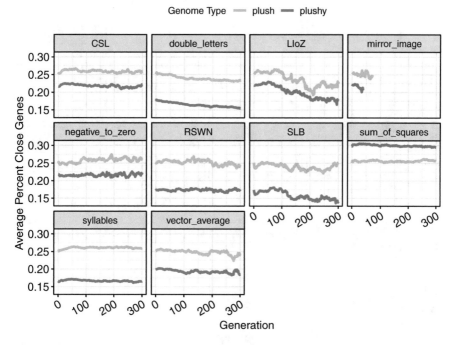

Fig. 13.9 The percentage of "closing" genes observed when using Plush and Plushy genomes for all benchmark problems, averaged across all benchmark runs. For Plush genomes, this is the percentage of genes in the genome with a non-zero close epigenetic marker. For Plushy genomes, this is the percentage of CLOSE genes in the genome

because they occur at locations in the genome where no code blocks are open. In order to compare the number of close genes that are having impact on program structure, we must compare the number of code blocks found in programs translated from each genome representation.

Figure 13.10 shows the average number of code blocks in translated programs divided by the program size.

All problems show that Plush genomes tend to have a slightly higher concentration of code blocks in the translated programs. The range of differences between experiments using Plush genomes and experiments using Plushy genomes is very narrow, suggesting that the genome representation has very little bearing on the concentration of code blocks in a program. The small differences here likely reflect the fact that even though the genome sizes are the same, the Plush genomes will contain more actual instructions compared to Plushy genomes, for which use some genes are CLOSE genes, leading to slightly higher numbers of instructions that open code blocks in Plush genomes.

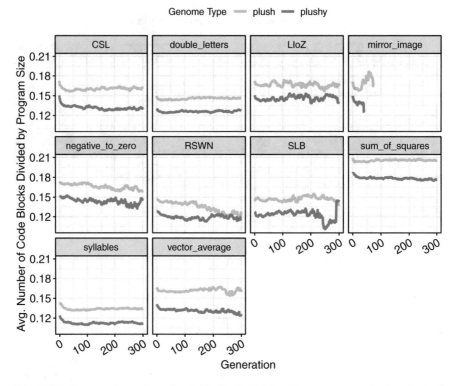

Fig. 13.10 The average number of code blocks divided by the average program size observed when using Plush and Plushy genomes for all benchmark problems, averaged across all benchmark runs. The plot shows a clear similarity between genome representations, especially considering the narrow range of the y-axis

13.5 Other Considerations

Section 13.3 discussed how Plush and Plushy genomes have nearly identical search performance. The various measurements presented in Sect. 13.4 show that programs produced while using the different genome representations are usually similar. This may seem to indicate the choice between Plush and Plushy genomes is inconsequential, but in practice there are important differences regarding their effects on usability and ease of implementation.

13.5.1 Hyperparameter Fitting

Most machine learning systems have a collection of hyperparameters that can be tuned to problem-specific values that improve performance. Typically hyperparameters for genetic programming systems include the population size, mutation rates, and parent selection methods. Grid search is a common method of tuning hyperparameters by exhaustively evaluating sets of values taken from a grid of hyperparameter values.

As mentioned in Sect. 13.2.1, the Plush genome representation requires a probability distribution to generate epigenetic marker values for random code for initialization and mutation. Probability distributions are difficult hyperparameters to tune.

All previous research using Plush genomes assumes a binomial distribution of initial values for epigenetic markers, although this has not been proven to be optimal via theoretical analysis nor empirical experimentation, and in fact has never been tuned. We believe that they are relatively robust to moderate change. However, it is possible that better-tuned values may lead to better performance than has been seen previously. Even if the optimal distribution is a binomial distribution in all cases, there are two hyperparameters to tune (n and p) for initial close marker assignment alone. If the optimal type of probability distribution is problem specific, the number of hyperparameters is unknown. This further complicates the tuning of hyperparameters that is required when using the Plush genome representation. Typically, the computational cost of tuning all hyperparameters drastically increases as the number of hyperparameters increases.

In contrast, when using Plushy genomes the choice of which instructions can appear in the instruction set determines both structure and logic. No additional hyperparameters are required specifically to initialize the CLOSE genes. Furthermore, when using Plushy genomes the proportional rate of CLOSE genes presented in Sect. 13.2.1 agrees with the intuition on how many CLOSE genes should appear in a random program and is suitable for most cases. Using this method of generating random genomes, there are no hyperparameters to tune when using Plushy genomes.

Section 13.2.2 discussed the separate set of genetic operators that can be used to vary the epigenetic markers on genes in Plush genomes. These mutation operations often require their own hyperparameter tuning for values such as mutation rate. When using Plushy genomes, there is no need for genetic operators that vary epigenetic markers, and thus no additional tuning is required.

It is possible that future research will produce genetic operators that specifically target the CLOSE and SKIP genes of Plushy genomes. These operators may expand the space of hyperparameters.

13.5.2 Applicable Search Methods

The field of inductive program synthesis has a large variety of methods undergoing active research across many problem domains. There is no clear superior family of algorithms that dominate the field. It is in the best interest of the field to compare and evaluate as many systems as possible to gain a better understanding of their behaviors.

Nearly every program synthesis method has a different approach to representing programs. This heterogeneity makes comparisons of different search methods on the same problems difficult [11]. The simplicity of the Plushy genome facilitates such comparisons better than the Plush genome because any search method capable of making changes to a sequence of tokens can be used to search over the space of Plushy genomes. Some examples of algorithms that could be used to search for solution Plushy genomes are:

- Evolutionary algorithms such as genetic algorithms and evolution strategies.
- Traditional local search methods such as simulated annealing and hill climbing.
- "Sequence to Sequence" neural architectures that are commonly used to synthesize sentences of natural language.
- Brute force combinatorics.

In contrast, Plush genomes require that each step in the search account for both gene tokens and their epigenetic markers. It is not immediately clear how a given search procedure should coordinate searching through genome space and epigenetic marker space in tandem, illustrating the complexity added by the epigenetic markers compared to CLOSE genes.

13.5.3   Automatic Simplification

Previous research on PushGP has detailed algorithms for automatically simplifying Push programs and Plush genomes [2, 15, 20]. This process uses hill climbing on program or genome size by randomly removing a small set of genes and testing that the program's outputs remain unchanged on the training data. If not, the genes are not removed, and a new random set of genes is removed, and the program is tested again. This process typically reaches a local size optimum within a few thousand iterations [18].

Automatic simplification was originally intended to yield programs that are easier for humans to understand [15, 18, 20], however it has also been shown that applying automatic simplification after evolution often produces programs that generalize better to unseen data [2]. In this sense, automatic simplification can be thought of as a regularization step for evolutionary program synthesis methods.

The solution rates reported in Fig. 13.5 are for simplified programs on a held-out test set (that is, a test set not used during evolution). In this case, we automatically simplified the Push programs, not the genomes, so there should be no difference in simplification for those results. However, previous work described alternative methods for simplifying Plush genomes directly before translation [2], and we could also imagine automatically simplifying Plushy genomes.

When simplifying Plush genomes, we can randomly turn on a small number of "silent" epigenetic markers during a hill-climbing step, effectively removing the genes without losing their information. This allows for backtracking of the hill-climbing by unsilencing those genes at a later time, potentially allowing the process to escape local optima. This leads to smaller Push programs than non-backtracking approaches, though it only produces negligible gains in generalization [2].

When applying the same automatic simplification process to Plushy genomes, it is possible for a set of CLOSE or SKIP genes to be removed without the modification of any genes that encode instructions. We leave it to future research to perform a rigorous study on the impact this has on generalization or interpretability.

13.5.4 Serialization

The main artifact of inductive program synthesis systems is the solution program found by the search. In PushGP, this artifact is typically either an executable Push program or a genome that can be translated into a Push program. In practice these solutions need to be serialized, stored, and recalled for later use.

Serializing Push programs requires the serialization of a nested structure of code blocks, literals, and instructions. One benefit of using a linear genome representation is that the solution's genome is often easier to serialize and de-serialize than the program.

Serializing Plush genomes requires denoting the gene value and the value for all epigenetic markers at every location in the genome. Serializing Plushy genomes only requires serializing the gene values. The simplicity of Plushy cuts down on the size of serialized genomes and improves interpretation and ease of de-serialization.

13.5.5 New Epigenetic Markers for Plush

One of the inspirations for the development of Plush genomes was the ability to add new epigenetic markers to add new data to the genome. We have discussed two such epigenetic markers that have easy translations to the Plushy representation: "close" and "silent" markers. However, we could imagine (and have experimented with) other epigenetic markers that would be more difficult to add to Plushy genomes. For example, we have experimented with the idea of adding "crossover hot-spots", which are locations in the genome where crossover is more likely to occur than in other locations. This is easy to envision as a new epigenetic marker, whereas it is not obvious how this feature would be added to Plushy genomes.

However, we have yet to find a specific use of a new epigenetic marker that actually improves performance of PushGP in practice. We therefore recommend keeping this ability in mind as a possible advantage of Plush—a context in which the complexity of Plush could add to its utility in comparison to Plushy.

13.6   Conclusion

We have compared two genome representations for evolving Push programs, Plush and Plushy. Experiments using the Clojush implementation of PushGP showed that the choice of representation has little effect on the problem-solving power of the genetic programming system, making it impossible to recommend one representation over the other on the basis of problem-solving performance alone. We also explored other qualities of the programs evolved using each representation, and found some minor differences in genome/program sizes, numbers of closing genes, and numbers of code blocks in the translated Push programs. While these differences are interesting and potentially could impact problem-solving performance on other problems, they appear incidental in the problem-solving performance results in this study.

We then discussed at the qualitative aspects of each representation. Plushy requires fewer hyperparameters to be tuned than Plush, since the number of CLOSE genes to include is determined from the instruction set, whereas Plush requires at least one to use (and potentially to tune) hyperparameters that determine the distribution of "close" epigenetic markers in randomly-generated genes. Additionally, the simplicity of Plushy compared to Plush makes it easier to apply non-genetic-programming search methods and to serialize genomes.

After comparing the two representations, we recommend using Plushy genomes for the evolution of Push programs in most settings. Since both representations achieve similar problem-solving performance, Plushy's simplicity makes it more versatile and easier to use.

Acknowledgements Feedback and discussions that improved this work were provided by other members of the Hampshire College Institute for Computational Intelligence, and by participants in the Genetic Programming Theory and Practice workshop.

This material is based upon work supported by the National Science Foundation under Grant No. 1617087. Any opinions, findings, and conclusions or recommendations expressed in this publication are those of the authors and do not necessarily reflect the views of the National Science Foundation.

References

1. Forstenlechner, S., Fagan, D., Nicolau, M., O'Neill, M.: A grammar design pattern for arbitrary program synthesis problems in genetic programming. In: M. Castelli, J. McDermott, L. Sekanina (eds.) EuroGP 2017: Proceedings of the 20th European Conference on Genetic Programming, *LNCS*, vol. 10196, pp. 262–277. Springer Verlag, Amsterdam (2017). https://doi.org/10.1007/978-3-319-55696-3_17
2. Helmuth, T., McPhee, N.F., Pantridge, E., Spector, L.: Improving generalization of evolved programs through automatic simplification. In: Proceedings of the Genetic and Evolutionary Computation Conference, GECCO '17, pp. 937–944. ACM, Berlin, Germany (2017). https://doi.org/10.1145/3071178.3071330. http://doi.acm.org/10.1145/3071178.3071330

3. Helmuth, T., McPhee, N.F., Spector, L.: Program synthesis using uniform mutation by addition and deletion. In: Proceedings of the Genetic and Evolutionary Computation Conference, GECCO '18, pp. 1127–1134. ACM, New York, NY, USA (2018). https://doi.org/10.1145/3205455.3205603. http://doi.acm.org/10.1145/3205455.3205603

4. Helmuth, T., Spector, L.: Detailed problem descriptions for general program synthesis benchmark suite. Technical Report UM-CS-2015-006, Computer Science, University of Massachusetts, Amherst (2015). https://web.cs.umass.edu/publication/details.php?id=2387

5. Helmuth, T., Spector, L.: General program synthesis benchmark suite. In: GECCO '15: Proceedings of the 2015 Annual Conference on Genetic and Evolutionary Computation, pp. 1039–1046. ACM, Madrid, Spain (2015). https://doi.org/10.1145/2739480.2754769. http://doi.acm.org/10.1145/2739480.2754769

6. Helmuth, T., Spector, L., McPhee, N.F., Shanabrook, S.: Linear genomes for structured programs. In: Genetic Programming Theory and Practice XIV. Springer (2017)

7. Kitzelmann, E.: Inductive programming: A survey of program synthesis techniques. In: U. Schmid, E. Kitzelmann, R. Plasmeijer (eds.) Approaches and Applications of Inductive Programming, pp. 50–73. Springer Berlin Heidelberg, Berlin, Heidelberg (2010)

8. Koza, J.R.: Genetic Programming: On the Programming of Computers by Means of Natural Selection. MIT Press, Cambridge, MA, USA (1992). http://mitpress.mit.edu/books/genetic-programming

9. La Cava, W., Helmuth, T., Spector, L., Danai, K.: Genetic programming with epigenetic local search. In: GECCO '15: Proceedings of the 2015 conference on Genetic and Evolutionary Computation Conference, pp. 1055–1062. ACM, Madrid, Spain (2015). https://doi.org/10.1145/2739480.2754763. http://doi.acm.org/10.1145/2739480.2754763

10. Lalejini, A., Ofria, C.: Evolving event-driven programs with signalgp. CoRR **abs/1804.05445** (2018). http://arxiv.org/abs/1804.05445

11. Pantridge, E., Helmuth, T., McPhee, N.F., Spector, L.: On the difficulty of benchmarking inductive program synthesis methods. In: Proceedings of the Genetic and Evolutionary Computation Conference Companion, GECCO '17, pp. 1589–1596. ACM, New York, NY, USA (2017). https://doi.org/10.1145/3067695.3082533. http://doi.acm.org/10.1145/3067695.3082533

12. Pantridge, E., Spector, L.: PyshGP: PushGP in python. In: Proceedings of the Genetic and Evolutionary Computation Conference Companion, GECCO '17, pp. 1255–1262. ACM, Berlin, Germany (2017). https://doi.org/10.1145/3067695.3082468. http://doi.acm.org/10.1145/3067695.3082468

13. Pantridge, E., Spector, L.: Plushi: An embeddable, language agnostic, push interpreter. In: Proceedings of the Genetic and Evolutionary Computation Conference Companion, GECCO '18, pp. 1379–1385. ACM, New York, NY, USA (2018). https://doi.org/10.1145/3205651.3208296. http://doi.acm.org/10.1145/3205651.3208296

14. Perelman, D., Gulwani, S., Grossman, D., Provost, P.: Test-driven synthesis. ACM SIGPLAN Notices **49**(6), 408–418 (2014). https://doi.org/10.1145/2594291.2594297

15. Robinson, A.: Genetic programming: Theory, implementation, and the evolution of unconstrained solutions. Division iii thesis, Hampshire College (2001). http://hampshire.edu/lspector/robinson-div3.pdf

16. Rosin, C.D.: Stepping stones to inductive synthesis of low-level looping programs. CoRR **abs/1811.10665** (2018). http://arxiv.org/abs/1811.10665

17. Spector, L., Helmuth, T.: Uniform linear transformation with repair and alternation in genetic programming. In: Genetic Programming Theory and Practice XI, Genetic and Evolutionary Computation, chap. 8, pp. 137–153. Springer, Ann Arbor, USA (2013). https://doi.org/10.1007/978-1-4939-0375-7_8. http://link.springer.com/chapter/10.1007%2F978-1-4939-0375-7_8

18. Spector, L., Helmuth, T.: Effective simplification of evolved push programs using a simple, stochastic hill-climber. In: GECCO Comp '14: Proceedings of the 2014 conference companion on Genetic and Evolutionary computation companion, pp. 147–148. ACM, Vancouver, BC, Canada (2014). https://doi.org/10.1145/2598394.2598414. http://doi.acm.org/10.1145/2598394.2598414

19. Spector, L., Klein, J., Keijzer, M.: The push3 execution stack and the evolution of control. In: GECCO 2005: Proceedings of the 2005 conference on Genetic and evolutionary computation, vol. 2, pp. 1689–1696. ACM Press, Washington DC, USA (2005). https://doi.org/10.1145/1068009.1068292. http://www.cs.bham.ac.uk/~wbl/biblio/gecco2005/docs/p1689.pdf
20. Spector, L., Robinson, A.: Genetic programming and autoconstructive evolution with the push programming language. Genetic Programming and Evolvable Machines 3(1), 7–40 (2002). https://doi.org/10.1023/A:1014538503543. http://hampshire.edu/lspector/pubs/push-gpem-final.pdf

Chapter 14
Enhanced Optimization with Composite Objectives and Novelty Pulsation

Hormoz Shahrzad, Babak Hodjat, Camille Dollé, Andrei Denissov, Simon Lau, Donn Goodhew, Justin Dyer, and Risto Miikkulainen

14.1 Introduction

Multi-objective optimization is most commonly used for discovering a Pareto front from which solutions that represent useful tradeoffs between objectives can be selected [9, 14–16, 23]. Evolutionary methods are a natural fit for such problems because the Pareto front naturally emerges in the population maintained in these methods. Interestingly, multi-objectivity can also improve evolutionary optimization because it encourages populations with more diversity. Even when the focus of optimization is to find good solutions along a primary performance metric, it is useful to create secondary dimensions that reward solutions that are different in terms of structure, size, cost, consistency, etc. Multi-objective optimization then discovers stepping stones that can be combined to achieve higher fitness along the primary dimension [34]. The stepping stones are useful in particular in problems where the fitness landscape is deceptive, i.e. where the optima are surrounded by inferior solutions [28].

However, not all such diversity is helpful. In particular, candidates that optimize one objective only and ignore the others are less likely to lead to useful tradeoffs,

H. Shahrzad (✉) · B. Hodjat
Cognizant Technology Solutions, Dublin, CA, USA
e-mail: hormoz@cognizant.com; babak@cognizant.com

C. Dollé · A. Denissov · S. Lau · D. Goodhew · J. Dyer
Sentient Investment Management, San Francisco, CA, USA
e-mail: camille.dolle@sentientim.com; andrei.denissov@sentientim.com;
simon.lau@sentientim.com; donn@sentientim.com; justin.dyer@sentientim.com

R. Miikkulainen
Cognizant Technology Solutions, Dublin, CA, USA

The University of Texas at Austin, Austin, TX, USA
e-mail: risto@cognizant.com, risto@cs.utexas.edu

© Springer Nature Switzerland AG 2020   275
W. Banzhaf et al. (eds.), *Genetic Programming Theory and Practice XVII*, Genetic
and Evolutionary Computation, https://doi.org/10.1007/978-3-030-39958-0_14

and they are less likely to escape deception. Prior research demonstrated that it is beneficial to replace the objectives with their linear combinations, thus focusing the search in more useful areas of the search space, and make up for the lost diversity by including a novelty metric in parent selection [39]. This paper improves upon this approach by introducing the concept of novelty pulsation: the novelty selection is turned on and off periodically, thereby allowing exploration and exploitation to leverage each other repeatedly.

This idea is tested in two domains. The first one is the highly deceptive domain of sorting networks [25] used in the original work on composite novelty selection [39]. Such networks consist of comparators that map any set of numbers represented in their input lines to a sorted order in their output lines. These networks have to be correct, i.e. sort all possible cases of input. The goal is to discover networks that are as small as possible, i.e. have as few comparators organized in as few sequential layers as possible. While correctness is the primary objective, it is actually not that difficult to achieve, because it is not deceptive. Minimality, on the other hand, is highly deceptive and makes the sorting network design an interesting benchmark problem. The experiments in this paper show that while the original composite novelty selection and its novelty-pulsation-enhanced version both find state-of-the-art networks up to 20 input lines, novelty pulsation finds them significantly faster. It also beat the state of the art for 20-line network by finding a 91 comparators design, which broke the previous world record of 92 [40].

The second domain is the highly challenging real-world problem of stock trading. The goal is to evolve agents that decide whether to buy, hold, or sell particular stocks over time in order to maximize returns. Compared to original composite novelty method, novelty pulsation finds solutions that generalize significantly better to unseen data. It therefore forms a promising foundation for solving deceptive real-world problems through multi-objective optimization.

14.2   Background and Related Work

Evolutionary methods for optimizing single-objective and multi-objective problems are reviewed, as well as the idea of using novelty to encourage diversity and the concept of exploration versus exploitation in optimization methods. The domains of minimal sorting networks and automated stock trading are introduced and prior work in them reviewed.

14.2.1   Single-Objective Optimization

When the optimization problem has a smooth and non-deceptive search space, evolutionary optimization of a single objective is usually convenient and effective. However, we are increasingly faced with problems of more than one objective and with a rugged and deceptive search space. The first approach often is to combine the objectives into a single composite calculation [14]:

$$Composite\,(O_1,\ O_2,\ldots,O_k) = \sum_{i=1}^{k} \alpha_i O_i^{\beta_i} \qquad (14.1)$$

Where the constant hyper-parameters α_i and β_i determine the relative importance of each objective in the composition. The composite objective can be parameterized in two ways:

1. By folding the objective space, and thereby causing a multitude of solutions to have the same value. Diversity is lost since solutions with different behavior are considered to be equal.
2. By creating a hierarchy in the objective space, and thereby causing some objectives to have more impact than many of the other objectives combined. The search will thus optimize the most important objectives first, which in deceptive domains might result in inefficient search or premature convergence to local optima.

Both of these problems can be avoided by casting the composition explicitly in terms of multi-objective optimization.

14.2.2 Multi-Objective Optimization

Multi-objective optimization methods construct a Pareto set of solutions [16], and therefore eliminate the issues with objective folding and hierarchy noted in Sect. 14.2.1. However, not all diversity in the Pareto space is useful. Candidates that optimize one objective only and ignore the others are less likely to lead to useful tradeoffs, and are less likely to escape deception.

One potential solution is reference-point based multi-objective methods such as NSGA-III [15, 16]. They make it possible to harvest the tradeoffs between many objectives and can therefore be used to select for useful diversity as well, although they are not as clearly suited for escaping deception.

Another problem with purely multi-objective search is crowding. In crowding, objectives that are easier to explore end up with disproportionately dense representation on the Pareto front. NSGA II addresses this problem by using the concept of crowding distance [14], and NSGA III improves upon it using reference points [15, 16]. These methods, while increasing diversity in the fitness space, do not necessarily result in diversity in the behavior space.

An alternative method is to use composite multi-objective axes to focus the search on the area with most useful tradeoffs [39]. Since the axes are not orthogonal, solutions that optimize only one objective will not be on the Pareto front. The focus effect, i.e. the angle between the objectives, can be tuned by varying the coefficients of the composite.

However, focusing the search in this manner has the inevitable side effect of reducing diversity. Therefore, it is important that the search method makes use of whatever diversity exists in the focused space. One way to achieve this goal is to incorporate a preference for novelty into selection.

14.2.3   Novelty Search

Novelty search [31, 33] is an increasingly popular paradigm that overcomes deception by ranking solutions based on how different they are from others. Novelty is computed in the space of behaviors, i.e., vectors containing semantic information about how a solution performs during evaluation. However, with a large space of possible behaviors, novelty search can become increasingly unfocused, spending most of its resources in regions that will never lead to promising solutions.

Recently, several approaches have been proposed to combine novelty with a more traditional fitness objective [17, 19, 20, 37, 38] to reorient search towards fitness as it explores the behavior space. These approaches have helped scale novelty search to more complex environments, including an array of control [3, 13, 37] and content generation [27, 29, 30] domains.

Many of these approaches combine a fitness objective with a novelty objective in some way, for instance as a weighted sum [11], or as different objectives in a multi-objective search [37]. Another approach is to keep the two kinds of search separate, and make them interact through time. For instance, it is possible to first create a diverse pool of solutions using novelty search, presumably overcoming deception that way, and then find solutions through fitness-based search [26]. A third approach is to run fitness-based search with a large number of objective functions that span the space of solutions, and use novelty search to encourage search to utilize all those functions [13, 36, 38]. A fourth category of approaches is to run novelty search as the primary mechanism, and use fitness to select among the solutions. For instance, it is possible to add local competition through fitness to novelty search [30, 31]. Another version is to accept novel solutions only if they satisfy minimal performance criteria [17, 32]. Some of these approaches have been generalized using the idea of behavior domination to discover stepping stones [34, 35].

In the Composite Novelty method [39], a novelty measure is employed to select which individuals to reproduce and which to discard. In this manner, it is integrated into the genetic algorithm itself, and its role is to make sure the focused space that the composite multiple objectives define is searched thoroughly.

14.2.4   Exploration Versus Exploitation

Every search algorithm needs to both explore the search space and exploit the known good solutions in it. Exploration is the process of visiting entirely new regions of a search space, whilst exploitation is the process of visiting regions within the neighborhood of previously visited points. In order to be successful, a search algorithm needs to establish a productive synergy between exploration and exploitation [6].

A common problem in evolutionary search is that it gets stuck in local minima, i.e. in unproductive exploitation. A common solution is to kick-start the search process in such cases by temporarily increasing mutation rates. This solution can be utilized more systematically by making such kick-starts periodic, resulting in methods such as in delta coding and burst mutation [18, 42].

This paper incorporates the kick-start idea into novelty selection. By turning novelty selection on and off periodically allows local search (i.e. exploitation) and novelty search (i.e. exploration) to leverage each other, leading to faster search and better generalization. These effects will be demonstrated in the sorting networks and stock trading domains, respectively.

14.2.5 Sorting Networks

A sorting network of n inputs is a fixed layout of comparison-exchange operations (comparators) that sorts all inputs of size n (Fig. 14.1). Since the same layout can sort any input, it represents an oblivious or data-independent sorting algorithm, that is, the layout of comparisons does not depend on the input data. The resulting fixed communication pattern makes sorting networks desirable in parallel implementations of sorting, such as those in graphics processing units, multi-processor computers, and switching networks [2, 24, 39]. Beyond validity, the main goal in designing sorting networks is to minimize the number of layers, because it determines how many steps are required in a parallel implementation. A tertiary goal is to minimize the total number of comparators in the networks. Designing such minimal sorting networks is a challenging optimization problem that has been the subject of active research since the 1950s [25]. Although the space of possible

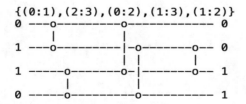

Fig. 14.1 A Four-Input Sorting Network and its representation. This network takes as its input (left) four numbers, and produces output (right) where those number are sorted (small to large, top to bottom). Each comparator (connection between the lines) swaps the numbers on its two lines if they are not in order, otherwise it does nothing. This network has three layers and five comparators, and is the minimal four-input sorting network. Minimal networks are generally not known for large input sizes. Their design space is deceptive which makes network minimization a challenging optimization problem

networks is infinite, it is relatively easy to test whether a particular network is correct: If it sorts all combinations of zeros and ones correctly, it will sort all inputs correctly [25].

Many of the recent advances in sorting network design are due to evolutionary methods [40]. However, it is still a challenging problem, even for the most powerful evolutionary methods, because it is highly deceptive: Improving upon a current design may require temporarily growing the network, or sorting fewer inputs correctly. Sorting networks are therefore a good domain for testing the power of evolutionary algorithms.

14.2.6 Stock Trading

Stock trading is a natural multi-objective domain where return and risk must be balanced [4, 5]. Candidate solutions, i.e. trading agents, can be represented in several ways. Rule-based strategies, sequence modeling with neural networks and LSTMs (Long Short-Term Memory), and symbolic regression using Genetic Programming or Grammatical Evolution are common approaches [1, 12]. Frequency of trade, fundamental versus technical indicators, choice of trading instruments, transaction costs, and vocabulary of order types are crucial design decisions in building such agents.

The goal is to extract patterns from historical time-series data on stock prices and utilize those patterns to make optimal trading decisions, i.e. whether to buy, hold, or sell particular stocks (Fig. 14.2) [10, 41]. The main challenge is to trade in a

Fig. 14.2 Stock Trading Agent. The agent observes the time series of stock prices and makes live decisions about whether to buy, hold, or sell a particular stock. The signal is noisy and prone to overfitting; generalization to unseen data is the main challenge in this domain

manner that generalizes to previously unseen situations in live trading. Some general methods like training data interleaving can be used to increase generalization [21], but their effectiveness might not be enough due to the low signal to noise ratio which is the main source of deception in this domain. The data is extremely noisy and prone to overfitting, and methods that discover more robust decisions are needed.

14.3   Methods

In this section, the genetic representation, the single and multi-objective optimization approaches, the composite objective method, the novelty-based selection method, and the novelty pulsation method are described, using the sorting network domain as an example. These methods were applied to stock trading in an analogous manner.

14.3.1   Representation

In order to apply various evolutionary optimization techniques to the sorting network problem, a general structured representation was developed. Sorting networks of n line can be seen as a sequence of two-leg comparators where each leg is connected to a different input line and the first leg is connected to a higher line than the second:$\{(f_1, s_1), (f_2, s_2), (f_3, s_3), \ldots, (f_c, s_c)\}$.

The number of layers can be determined from such a sequence by grouping successive comparators together into a layer until the next comparator adds a second connection to one of the lines in the same layer. With this representation, mutation and crossover operators amount to adding and removing a comparator, swapping two comparators, and crossing over the comparator sequences of two parents at a single point.

Domain-specific techniques such as mathematically designing the prefix layers [7, 8] or utilizing certain symmetries [40] were not used.

14.3.2   Single-Objective Approach

Correctness is part of the definition of a sorting network: Even if a network mishandles only one sample, it will not be useful. The number of layers can be considered the most important size objective because it determines the efficiency of a parallel implementation. A hierarchical composite objective can therefore be defined as:

$$SingleFitness(m, l, c) = 10{,}000\,m + 100\,l + c \qquad (14.2)$$

Where m, l, and c are the number of mistakes (unsorted samples), number of layers, and number of comparators, respectively.

In the experiments in this paper, the solutions will be limited to less than one hundred layers and comparators, and therefore, the fitness will be completely hierarchical (i.e. there is no folding).

14.3.3 Multi-Objective Approach

In the multi-objective approach, the same dimensions, i.e. the number of mistakes, layers, and comparators m, l, c, are used as three separate objectives. They are optimized by the NSGA-II algorithm [14] with selection percentage set to 10%. Indeed, this approach may discover solutions with just a single layer, or a single comparator, since they qualify for the Pareto front. Therefore, diversity is increased compared to the single-objective method, but this diversity is not necessarily helpful.

14.3.4 Composite Multi-Objective Approach

In order to construct composite axes, each objective is augmented with sensitivity to the other objectives:

$$Composite_1 (m, \ l, \ c) = 10{,}000 \, m + 100 \, l + c \tag{14.3}$$

$$Composite_2 (m, \ l) = \alpha_1 m + \alpha_2 l \tag{14.4}$$

$$Composite_3 (m, \ c) = \alpha_3 m + \alpha_4 c \tag{14.5}$$

The primary composite objective (Eq. 14.3), which will replace the mistake axis, is the same hierarchical fitness used in the single-objective approach. It discourages evolution from constructing correct networks that are extremely large. The second objective (Eq. 14.4), with $\alpha_2 = 10$, primarily encourages evolution to look for solutions with a small number of layers. A much smaller cost of mistakes, with $\alpha_1 = 1$, helps prevent useless single-layer networks from appearing in the Pareto front. Similarly, the third objective (Eq. 14.5), with $\alpha_3 = 1$ and $\alpha_4 = 10$, applies the same principle to the number of comparators.

The values for α_1, α_2, α_3, and α_4 were found to work well in this application, but the approach was found not to be very sensitive to them; A broad range will work as long as they establish a primacy relationship between the objectives.

It might seem like we are adding several hyper-parameters which need to be tuned, but we can estimate them in each domain by picking values that push away trivial or useless solution off the Pareto front.

14.3.5   Novelty Selection Method

In order to measure how novel the solutions are it is first necessary to characterize their behavior. While such a characterization can be done in many ways, a concise and computationally efficient approach is to count how many swaps took place on each line in sorting all possible zero-one combinations during the validity check. Such a characterization is a vector that has the same size as the problem, making the distance calculations fast. It also represents the true behavior of the network: Even if two networks sort the same input cases correctly, they may do it in different ways, and the characterization is likely to capture that difference. Given this behavior characterization, novelty of a solution is measured by the sum of pairwise distances of its behavior vector to those of all the other individuals in the selection pool:

$$NoveltyScore\,(x_i) = \sum_{j=1}^{n} d(b\,(x_i)\,,\ b\,(x_j))\,. \tag{14.6}$$

The selection method also has a parameter called *selection multiplier* (e.g. set to 2 in these experiments), varying between one and the inverse of the elite fraction (e.g. 1/10, i.e. 10%) used in the NSGA-II multi-objective optimization method. The original selection percentage is multiplied by the selection multiplier to form a broader selection pool. That pool is sorted according to novelty, and the top fraction representing the original selection percentage is used for selection. This way, good solutions that are more novel are included in the pool. Figure 14.3 shows an example result of applying Eq. 14.6.

One potential issue is that a cluster of solutions far from the rest may end up having high novelty scores while only one is good enough to keep. Therefore, after the top fraction is selected, the rest of the sorted solutions are added to the selection pool one by one, replacing the solution with the lowest minimum novelty, defined as

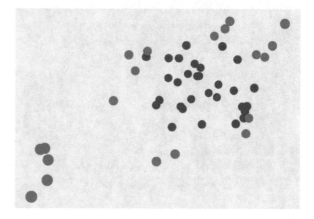

Fig. 14.3 The first phase of novelty selection is to select the solutions (marked green) with the highest Novelty Score (Eq. 14.6)

$$MinimumNovelty\,(x_i) = \min_{1 \leq j \leq n;\ j \neq i} d(b\,(x_i)\,,\ b\,(x_j))\,. \qquad (14.7)$$

Note that this method allows tuning novelty selection continuously between two extremes: By setting *selection multiplier* to one, the method reduces to the original multi-objective method (i.e. only the elite fraction ends up in the final elitist pool), and setting it to the inverse of the elite fraction reduces it to pure novelty search (i.e. the whole population, sorted by novelty, is the selection pool). In practice, low and midrange values for the multiplier work well, including the value 2 used in these experiments. Figure 14.4 shows an example result of applying Eq. 14.7. The entire novelty-selection algorithm is summarized in Fig. 14.5.

To visualize this process, Fig. 14.6 contrasts the difference between diversity that multi-objective method (e.g. NSGA-II) creates (left-side) and diversity that novelty search creates (right-side). In the objective space (top), novelty looks more focused

Fig. 14.4 Phase two of novelty selection eliminates the closest pairs of the green candidates in order to get better overall coverage (blue candidates). The result is a healthy mixture of high-fitness candidates and novel ones (Eq. 14.7)

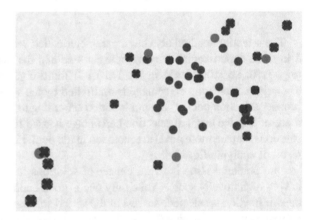

1. *using a selection method (e.g. NSGA-II) pick **selection multiplier** times as many elitist candidates than usual*
2. *sort them in descending order according to their NoveltyScore (Equation 14.6)*
3. *move the usual number of elitist candidates from the top of the list to the result set*
4. *for all remaining candidates in the sorted list:*
   a. *add the candidate to result set*
   b. *remove the candidate with the lowest MinimumNovelty (Equation 14.7)*
   c. *(using a fitness measure as the tie breaker)*
5. *return the resulting set as the elite set*

Fig. 14.5 The novelty selection algorithm

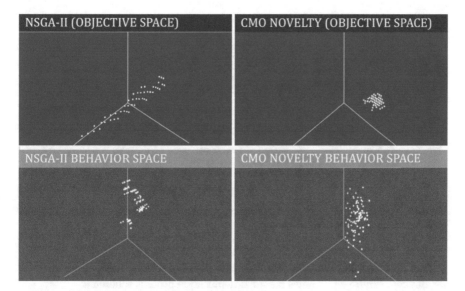

Fig. 14.6 An example demonstrating how novelty selection (right column) creates better coverage of the behavior space (bottom row) than NSGA-II (left column) despite being more focused in the objective space (top row)

and less diverse, but in the behavior space (bottom) it is much more diverse. This type of diversity enables the method to escape deception and find novel solutions, such as the state of the art in sorting networks.

14.3.6 Novelty Pulsation Method

Parent selection is a crucial step in an evolutionary algorithm. In almost all such algorithms, whatever method is used remains unchanged during the evolutionary run. However, when a problem is deceptive or prone to over-fitting, changing the selection method periodically may make the algorithm more robust. It can be used to alternate the search between exploration and exploitation, and thus find a proper balance between them.

In Composite Novelty Pulsation, novelty selection is switched on and off after a certain number of generations. As in delta-coding and burst mutation, once good solutions are found, they are used as a starting point for exploration. Once exploration has generated sufficient diversity, local optimization is performed to find the best possible versions of these diverse points. These two phases leverage each other, which results in faster convergence and more reliable solutions.

Composite Novelty Pulsation adds a new hyper-parameter, P, denoting the number of generations before switching novelty selection on and off. Preliminary experiments showed that $P = 5$ works well in both sorting network and stock

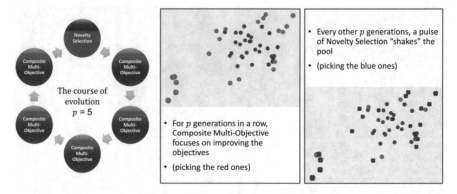

Fig. 14.7 Visulization of how novelty pulsation process alternates between composite multi-objective selection and novelty selection

trading domains; however, in principle it is possible to tune this parameter to fit the domain. Figure 14.7 shows the novelty pulsation process schematic.

14.4 Experiment

Previous work in the sorting networks domain demonstrated that composite novelty can match the minimal known networks up to 18 input line with reasonable computational resources [39, 40]. The goal of the sorting network experiments was to achieve the same result faster, i.e. with fewer resources. The experiments were therefore standardized to a single machine (a multi-core desktop).

In the stock market trading domain, the experiments compared generalization by measuring the correlation between seen and unseen data.

14.4.1 Experimental Setup

Experiments in previous paper [39] already demonstrated that the composite novelty method performs statistically significantly better in the sorting network discovery task than the other methods discussed above. Therefore, this paper focuses on comparing the novelty pulsation method to its predecessor, i.e. the composite novelty method.

In the sorting networks domain, experiments were run with the following parameters:

- Eleven network sizes, 8 through 18;
- Ten runs for each configuration (220 runs in total);
- 10% parent selection rate;

- Population size of 1000 for composite novelty selection and 100 for novelty pulsation. These settings were found to be appropriate for each method experimentally.

In the trading domain, experiments were run with the following parameters:

- Ten runs on five years of historical data;
- Population size of 500;
- 100 generations;
- 10% parent selection rate;
- Performance of the 10 best individuals from each run compared on the subsequent year of historical data, withheld from all runs.

14.4.2   Sorting Networks Results

Convergence time of the two methods to minimal solutions for different network sizes is shown in Fig. 14.8. Novelty pulsation shows an order of magnitude faster convergence across the board. All runs resulted in state-of-the-art sorting networks.

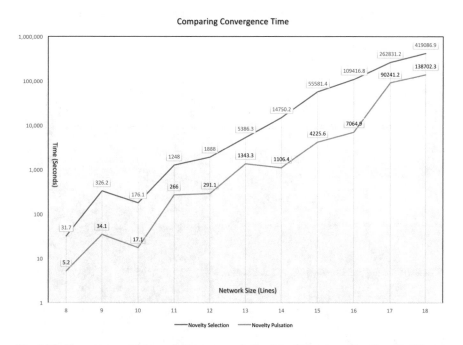

Fig. 14.8 The average runtime needed to converge to the state of the art on networks with different sizes. Novelty pulsation converges significantly faster at all sizes, demonstrating improved balance of exploration and exploitation

An interesting observation is that sorting networks with an even number of lines take proportionately less time to find the state-of-the-art solution than those with odd numbers of lines. This result is likely due to symmetrical characteristics of even-numbered problems. Some methods [40] exploit this very symmetry in order to find state-of-the-art solutions and break previous records, but this domain-specific information was not used in the implementation in this paper. The fact that the method achieves the state-of-the-art results and even breaks one world record (as described in the Appendix) without exploiting domain specific characteristics is itself a significant result.

14.4.3 Stock Trading Results

Figures 14.9 and 14.10 illustrate generalization of the composite novelty selection and novelty pulsation methods, respectively. Points in Fig. 14.10 are noticeably closer to a diagonal line, which means that better training fitness resulted in better testing fitness, i.e. higher correlation and better generalization. Numerically, the seen-to-unseen correlation for the composite novelty method is 0.69, while for

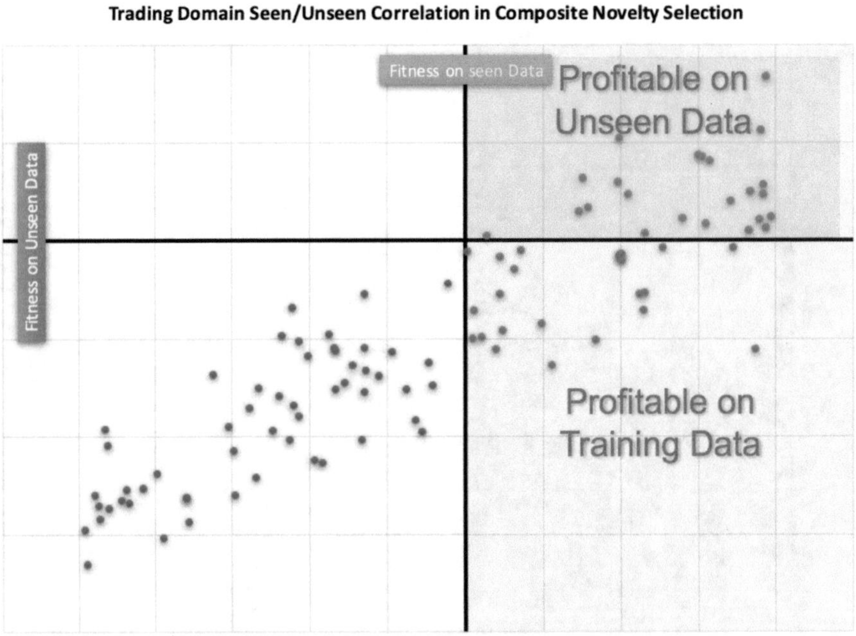

Fig. 14.9 Generalization from seen to unseen data with the Composite Novelty method. The fitness on seen data is on x and unseen is on y. The correlation is 0.69, which is enough to trade but could be improved. Candidates to the right of the vertical line are profitable on seen data, and candidates above the horizontal line are profitable on unseen data

Trading Domain Seen/Unseen Correlation in Composite Novelty Pulsation

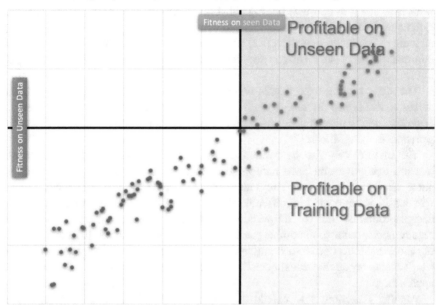

Fig. 14.10 Generalization from seen to unseen data with Composite Novelty Pulsation method. The correlation is 0.89, which results in significantly improved profitability in live trading. It is also notable that the majority of profitable genes on training data are also profitable on unseen data

composite novelty pulsation, it is 0.86. The ratio of the number of profitable candidates on unseen data and training data is also better, suggesting that underfitting is unlikely. In practice, these differences are significant, translating to much improved profitability in live trading.

14.5 Discussion and Future Work

The results in both sorting network and stock trading domains support the anticipated advantages of the composite novelty pulsation approach. The secondary objectives diversify the search, composite objectives focus it on most useful areas, and pulses of novelty selection allow for both accurate optimization and thorough exploration of those areas. These methods are general and robust: they can be readily implemented in standard multi-objective search such as NSGA-II and used in combination with many other techniques already developed to improve evolutionary multi-objective optimization.

The sorting network experiments were designed to demonstrate the improvement provided by novelty pulsation over the previous state of the art. Indeed, it found the best known solutions significantly faster. One compelling direction of future work is to use it to optimize sorting networks systematically, with domain-specific techniques integrated into the search, and with significantly more computing power, including distributed evolution [22]. It is likely that given such power, many new minimal networks can be discovered, for networks with even larger number of input lines.

The stock trading experiments was designed to demonstrate that the approach makes a difference in real-world problems. The main challenge in trading is generalization to unseen data, and indeed in this respect novelty pulsation improved generalization significantly.

The method can also be applied in many other domains, in particular those that are deceptive and have natural secondary objectives. For instance, various game strategies from board to video games can be cast in this form, where winning is accompanied by different dimensions of the score. Solutions for many design problems, such as 3D printed objects, need to satisfy a set of functional requirements, but also maximize strength and minimize material. Effective control of robotic systems need to accomplish a goal while minimize energy and wear and tear. Thus, many applications should be amenable to the composite novelty pulsation approach.

Another direction is to extend the method further into discovering effective collections of solutions. For instance, ensembling is a good approach for increasing the performance of machine learning systems. Usually the ensemble is formed from solutions with different initialization or training, with no mechanism to ensure that their differences are useful. In composite novelty pulsation, the Pareto front consists of a diverse set of solutions that span the area of useful tradeoffs. Such collections should make for a powerful ensemble, extending the applicability of the approach further.

14.6   Conclusion

The composite novelty pulsation method is a promising extension of the composite novelty approach to deceptive problems. Composite objectives focus the search on the most useful tradeoffs (better exploitation), while novelty selection allows escaping deceptive areas (better exploration). Novelty pulsation balances between the exploration and exploitation, finding solutions faster and finding solutions that generalize better. These principles were demonstrated in this paper in the highly deceptive problem of minimizing sorting networks and in the highly noisy domain of stock market trading. Composite novelty pulsation is a general method that can be combined with other advances in population-based search, thus increasing the power and applicability of evolutionary multi-objective optimization.

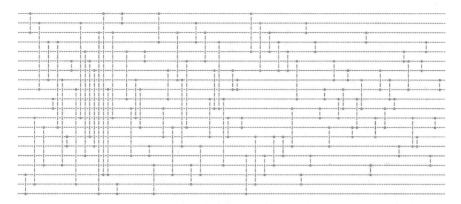

Fig. 14.11   The new 20-line sorting network with 91 comparators, discovered by novelty pulsation

Appendix

The graph of the new world record for 20-line sorting network, which moved the previous record of 92 comparators also discovered by evolution [40] down to 91.

One of the nice properties of Novelty Pulsation Method is the ability to converge with a very small pool size (like only 30 individuals in case of sorting networks). However, it still took almost 2 months to break the world record on the 20-line network running on a single machine (Fig. 14.11). Interestingly, even if it takes the same number of generations for the other methods to get there with a normal pool size of a thousand, those runs will take almost 5 years to converge!

References

1. F. Allen, R. Karjalainen. 1999. Using genetic algorithms to find technical trading rules. *Journal of Financial Economics* 51, 245–271.
2. S. W. A. Baddar. 2009. *Finding Better Sorting Networks*. PhD thesis, Kent State University.
3. J. A. Bowren, J. K. Pugh, and K. O. Stanley. 2016. Fully Autonomous Real-Time Autoencoder Augmented Hebbian Learning through the Collection of Novel Experiences. In Proceedings of ALIFE. 382–389.
4. A. Brabazon, M. O'Neill. 2006. Biologically Inspired Algorithms for Financial Modelling. Springer.
5. R. Bradley, A. Brabazon, M. O'Neill. 2010. Objective function design in a grammatical evolutionary trading system. In: 2010 IEEE World Congress on Computational Intelligence, pp. 3487–3494. IEEE Press.

6. M. Črepinšek, S. Liu, M. Mernik. 2013. Exploration and Exploitation in Evolutionary Algorithms: A Survey. *ACM Computing Surveys* 45, Article 35.
7. M. Codish, L. Cruz-Filipe, and P. Schneider-Kamp. 2014. The quest for optimal sorting-networks: Efficient generation of two-layer prefixes. In *Symbolic and Numeric Algorithms for Scientific Computing (SYNASC), 2014 16th International Symposium on* (pp. 359–366). IEEE.
8. M. Codish, L. Cruz-Filipe, T. Ehlers, M. Muller, and P. Schneider-Kamp. 2016. Sorting networks: To the end and back again. *Journal of Computer and System Sciences.*
9. C. A. C. Coello, G. B. Lamont, and D. A. Van Veldhuizen. 2007. *Evolutionary algorithms for solving multi-objective problems.* Vol. 5. Springer.
10. I. Contreras, J.I. Hidalgo, L. Nunez-Letamendia, J.M. Velasco. 2017. A meta-grammatical evolutionary process for portfolio selection and trading. *Genetic Programming and Evolvable Machines* 18(4), 411–431.
11. G. Cuccu and F Gomez. 2011. When Novelty is Not Enough. In *Evostar.* 234–243.
12. W. Cui, A. Brabazon, M. O'Neill. 2011. Adaptive trade execution using a grammatical evolution approach. *International Journal of Financial Markets and Derivatives* 2(1/2), 4–3.
13. A. Cully, J. Clune, D. Tarapore, and J-B. Mouret. 2015. Robots that can adapt like animals. *Nature* 521, 7553 (2015), 503–507.
14. K. Deb, A. Pratap, S. Agarwal, and T. A. Meyarivan. 2002. A fast and elitist multiobjective genetic algorithm: NSGA-II. *IEEE Trans. on Evolutionary Computation* 6, 2 (2002), 182–197.
15. K. Deb, and H. Jain. 2014. An Evolutionary Many-Objective Optimization Algorithm Using Reference-Point-Based Nondominated Sorting Approach, Part I: Solving Problems With Box Constraints. In *IEEE Transactions on Evolutionary Computation*, vol. 18, no. 4, 577–601.
16. K. Deb, K. Sindhya, and J. Hakanen. 2016. Multi-objective optimization. In *Decision Sciences: Theory and Practice.* 145–184.
17. J. Gomes, P. Mariano, and A. L. Christensen. 2015. Devising effective novelty search algorithms: A comprehensive empirical study. In Proc. of GECCO. 943–950.
18. F. Gomez, and R. Miikkulainen. 1997. Incremental evolution of complex general behavior. *Adaptive Behavior* 5(3–4), pp.317–342.
19. J. Gomes, P. Urbano, and A. L. Christensen. 2013. Evolution of swarm robotics systems with novelty search. *Swarm Intelligence*, 7:115–144.
20. F. J. Gomez. 2009. Sustaining diversity using behavioral information distance. In Proc. of GECCO. 113–120.
21. I. Gonçalves, S. Silva. 2013. Balancing Learning and Overfitting in Genetic Programming with Interleaved Sampling of Training Data. In: Krawiec K., Moraglio A., Hu T., Etaner-Uyar A., Hu B. (eds) Genetic Programming. EuroGP 2013. Lecture Notes in Computer Science, vol 7831. Springer, Berlin, Heidelberg.
22. B. Hodjat, H. Shahrzad, and R. Miikkulainen. 2016. Distributed Age-Layered Novelty Search. In Proc. of ALIFE. 131–138.
23. H. Jain, and K. Deb. 2014. An Evolutionary Many-Objective Optimization Algorithm Using Reference-Point Based Nondominated Sorting Approach, Part II: Handling Constraints and Extending to an Adaptive Approach. In *IEEE Transactions on Evolutionary Computation*, vol. 18, no. 4, 602–622.
24. P. Kipfer, M. Segal, and R. Westermann. 2004. Uberflow: A gpu-based particle engine. In HWWS 2004: Proc. of the ACM SIGGRAPH/EUROGRAPHICS, 115–122.
25. D. E. Knuth. 1998. *Art of Computer Programming: Sorting and Searching*, volume 3. Addison-Wesley Professional, 2 edition.
26. P. Krcah, and D. Toropila. 2010. Combination of novelty search and fitness-based search applied to robot body-brain coevolution. In Proc. of 13th Czech-Japan Seminar on Data Analysis and Decision Making in Service Science.
27. J. Lehman, S. Risi, and J. Clune. 2016. Creative Generation of 3D Objects with Deep Learning and Innovation Engines. In Proc. of ICCC. 180–187.
28. J. Lehman, and R. Miikkulainen. 2014. Overcoming deception in evolution of cognitive behaviors. In Proc. of GECCO.

29. J. Lehman and K. O. Stanley. 2012. Beyond open-endedness: Quantifying impressiveness. In Proc. of ALIFE. 75–82.
30. J. Lehman and K. O. Stanley. 2011. Evolving a diversity of virtual creatures through novelty search and local competition. In Proc. of GECCO. 211–218.
31. J. Lehman and K. O. Stanley. 2011. Abandoning objectives: Evolution through the search for novelty alone. *Evolutionary Computation* 19, 2 (2011), 189–223.
32. J. Lehman and K. O. Stanley. 2010. Efficiently evolving programs through the search for novelty. In Proc. of GECCO. 836–844.
33. J. Lehman and K. O. Stanley. 2008. Exploiting Open-Endedness to Solve Problems Through the Search for Novelty. In Proc. of ALIFE. 329–336.
34. E. Meyerson, and R. Miikkulainen. 2017. Discovering evolutionary stepping stones through behavior domination. In Proc. of GECCO, 139–146. ACM.
35. E. Meyerson, J. Lehman, and R. Miikkulainen. 2016. Learning behavior characterizations for novelty search. In Proc. of GECCO. 149–156.
36. J-B. Mouret and J. Clune. 2015. Illuminating search spaces by mapping elites. CoRR abs/1504.04909 (2015).
37. J-B. Mouret and S. Doncieux. 2012. Encouraging behavioral diversity in evolutionary robotics: An empirical study. *Evolutionary Computation* 20, 1 (2012), 91–133.
38. J. K. Pugh, L. B. Soros, P. A. Szerlip, and K. O. Stanley. 2015. Confronting the Challenge of Quality Diversity. In Proc. of GECCO. 967–974.
39. H. Shahrzad, D. Fink, and R. Miikkulainen. 2018. Enhanced Optimization with Composite Objectives and Novelty Selection. In Proc. of ALIFE. 616–622.
40. V. K. Valsalam, and R. Miikkulainen. 2013. Using symmetry and evolutionary search to minimize sorting networks. *Journal of Machine Learning Research* 14(Feb):303–331.
41. H. White. 2000. A reality check for data snooping. *Econometrica* Sep. 2000; 68(5):1097–126.
42. D. Whitley, K. Mathias, P. Fitzhorn. 1991. Delta coding: An iterative search strategy for genetic algorithms. In ICGA (Vol. 91, pp. 77–84).

Chapter 15
New Pathways in Coevolutionary Computation

Moshe Sipper, Jason H. Moore, and Ryan J. Urbanowicz

15.1 Coevolutionary Computation

In biology, coevolution occurs when two or more species reciprocally affect each other's evolution. Darwin mentioned evolutionary interactions between flowering plants and insects in *Origin of Species*. The term coevolution was coined by Paul R. Ehrlich and Peter H. Raven in 1964.[1]

Coevolutionary algorithms simultaneously evolve two or more populations with coupled fitness [8]. Strongly related to the concept of symbiosis, coevolution can be mutalistic (cooperative), parasitic (competitive), or commensalistic (Fig. 15.1)[2]: (1) In cooperative coevolution, different species exist in a relationship in which each individual (fitness) benefits from the activity of the other; (2) in competitive coevolution, an organism of one species competes with an organism of a different species; and (3) with commensalism, members of one species gain benefits while those of the other species neither benefit nor are harmed.

A cooperative coevolutionary algorithm involves a number of independently evolving species, which come together to obtain problem solutions. The fitness of an individual depends on its ability to collaborate with individuals from other species [2, 8, 9, 15].

---

[1] https://en.wikipedia.org/wiki/Coevolution.
[2] https://en.wikipedia.org/wiki/Symbiosis.

M. Sipper (✉)
Institute for Biomedical Informatics, University of Pennsylvania, Philadelphia, PA, USA

Department of Computer Science, Ben-Gurion University, Beer Sheva, Israel

J. H. Moore · R. J. Urbanowicz
Institute for Biomedical Informatics, University of Pennsylvania, Philadelphia, PA, USA
e-mail: jhmoore@upenn.edu; ryanurb@pennmedicine.upenn.edu

© Springer Nature Switzerland AG 2020
W. Banzhaf et al. (eds.), *Genetic Programming Theory and Practice XVII*, Genetic and Evolutionary Computation, https://doi.org/10.1007/978-3-030-39958-0_15

(a)   (b)   (c)

Fig. 15.1 Coevolution: (**a**) cooperative: Purple-throated carib feeding from and polli-
nating a flower (credit: Charles J Sharp, https://commons.wikimedia.org/wiki/File:Purple-
throated_carib_hummingbird_feeding.jpg); (**b**) competitive: predator and prey—a leopard killing
a bushbuck (credit: NJR ZA, https://commons.wikimedia.org/wiki/File:Leopard_kill_-_KNP_-
_001.jpg); (**c**) commensalistic: Phoretic mites attach themselves to a fly for transport (credit:
Alvesgaspar, https://en.wikipedia.org/wiki/File:Fly_June_2008-2.jpg)

In a competitive coevolutionary algorithm the fitness of an individual is based on
direct competition with individuals of other species, which in turn evolve separately
in their own populations. Increased fitness of one of the species implies a reduction
in the fitness of the other species [5].

We have recently developed two new coevolutionary algorithms, which will be
reviewed herein: OMNIREP and SAFE [10–12].

OMNIREP aims to aid in one of the major tasks faced by an evolutionary
computation (EC) practitioner, namely, deciding how to represent individuals in the
evolving population. This task is actually composed of two subtasks: defining a data
structure that is the representation and defining the encoding that enables to interpret
the representation. OMNIREP discovers *both* a representation and an encoding that
solve a particular problem of interest, by employing two coevolving populations.

SAFE—Solution And Fitness Evolution—stemmed from our recently highlighting a fundamental problem recognized to confound algorithmic optimization:
conflating the objective with the objective function [13]. Even when the former
is well defined, the latter may not be obvious. SAFE is a commensalistic coevolutionary algorithm that maintains two coevolving populations: a population of
candidate solutions and a population of candidate objective functions. To the best
of our knowledge, SAFE is the first coevolutionary algorithm to employ a form of
commensalism.

We first turn to OMNIREP (Sect. 15.2), followed by SAFE (Sect. 15.3), and
ending with concluding remarks (Sect. 15.4). This chapter summarizes our research.
For full details please refer to [10–12]. NB: The code for both OMNIREP and SAFE
is available at https://github.com/EpistasisLab/.

15.2 OMNIREP

One Representation to rule them all, One Encoding to find them,
One Algorithm to bring them all and in the Fitness bind them.
In the Landscape of Search where the Solutions lie.

One of the basic tasks of the EC practitioner is to decide how to represent individuals in the (evolving) population, i.e., precisely specify the genetic makeup of the artificial entity under consideration. As stated by Eiben and Smith [3]: "Technically, a given representation might be preferable over others if it matches the given problem better, that is, it makes the encoding of candidate solutions easier or more natural."

One of the EC practitioner's foremost tasks is thus to identify a representation— a data structure—and its encoding, or *interpretation*. These can be viewed, in fact, as two distinct tasks, though they are usually dealt with simultaneously. To wit, one might define the representation as a bitstring and in the same breath go on to state the encoding, e.g., "the 120-bit bitstring represents 4 numerical values, each encoded by 30 bits, which are treated as signed floating-point values".

OMNIREP uses cooperative coevolution with two coevolving populations, one of representations, the other of encodings. The evolution of each population is identical to a single-population evolutionary algorithm—except where fitness is concerned (Fig. 15.2).

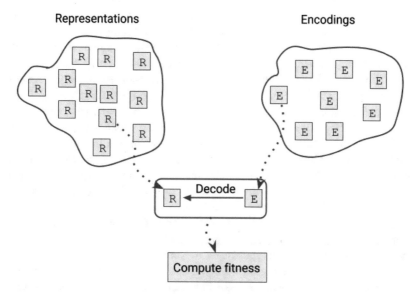

Fig. 15.2 Fitness computation in OMNIREP, where two populations coevolve, one comprising representations, the other encodings. Fitness is computed by combining a representation individual (R) with an encoding individual (E)

Selection, crossover, and mutation are performed as in a standard single-population algorithm. To compute fitness the two coevolving populations cooperate. Specifically, to compute the fitness of a single individual in one population, OMNIREP uses *representatives* from the other population [8]. The representatives are selected via a greedy strategy as the 4 fittest individuals from the previous generation. When evaluating the fitness of a particular representation individual, OMNIREP combines it 4 times with the top 4 encoding individuals, computes 4 fitness values, and uses the average fitness over these 4 evaluations as the final fitness value of the representation individual. In a similar manner OMNIREP uses the average of 4 representatives from the representations population when computing the fitness of an encoding individual.

In [10] we applied OMNIREP successfully to four problems:

- *Bitstring and bit count.* Solve cubic polynomial regression problems, $y = ax^3 + bx^2 + cx + d$, where the objective was to find the coefficients a, b, c, d for a given dataset of x, y values (independent and dependent variables). An individual in the representations population was a bitstring of length 120. An individual in the encodings population was a list of 4 integer values, each of which specified the number of bits allocated to the respective parameter (a, b, c, d) in the representation individual.
- *Floating point and precision.* Solve regression problems, $y = \sum_{j=0}^{49} a_j x^{e_j}$, where $a_j \in \mathbb{R} \cap [0, 1]$, $x \in \mathbb{R} \cap [0, 1]$, $e_j \in \{0, \ldots, 4\}$, $j = 0, \ldots, 49$. An individual in the representations population was a list of 50 real values $\in [0, 1]$ (the coefficients a_j). An individual in the encodings population was a list of 50 integer values, each specifying the precision of the respective coefficient, namely, the number of digits $d \in \{1, \ldots, 8\}$ after the decimal point.
- *Program and instructions.* Find a program that is able to emulate the output of an unknown target program. We considered the evolution of a program composed of 10 lines, each line executing a mathematical, real-valued, univariate function, or instruction. The representation individual was a program comprising 10 lines, each one executing a *generic* instruction of the form x=fi(x), where fi $\in$ {f1, ..., f5}. The program had one variable, x, which was set to a specific value v at the outset, i.e., to each (10-line) program, the instruction x=v was added as the first line. v was thus the program's input. After a program finished execution, its output was taken as the value of x. To run a program one needs to couple it with an encoding individual, which provides the specifics of what each fi performs. Table 15.1 shows an example.
- *Image and blocks.* Herein, we delved into evolutionary art, wherein artwork is generated through an evolutionary algorithm. Our goal was to evolve images that closely matched a given target image, a "standard of beauty" as it were. The representation individual's genome was a list of pixel indexes, with each index considered the start of a same-color block of pixels. The encoding individual was a list equal in length to the representation individual, consisting of tuples (b_i, c_i), where b_i was block i's length, and c_i was block i's color. If a pixel was uncolored by any block it was assigned a default base color. Sample evolved artwork is shown in Fig. 15.3.

Table 15.1 OMNIREP
'program and instructions'
experiment: sample
representation and encoding
individuals, the former being
a 10-line program with
generic instructions, and the
latter being the instruction
meanings

Representation	Encoding
x=v	f1: mul10
x=f1(x)	f2: fabs
x=f2(x)	f3: tan
x=f3(x)	f4: mul10
x=f4(x)	f5: minus2
x=f2(x)	
x=f2(x)	
x=f5(x)	
x=f2(x)	
x=f1(x)	
x=f5(x)	

Fig. 15.3 Sample artwork evolved by OMNIREP

OMNIREP was able to solve all problems successfully. Moreover, it usually found better encodings (e.g., more compact—using less bits or less precision) than fixed-representation schemes, with no degradation in performance. For full details see [10].

15.3 SAFE

We have recently highlighted a fundamental problem recognized to confound algorithmic optimization, namely, *conflating* the objective with the objective function [13]. Even when the former is well defined, the latter may not be obvious. We presented an approach to automate the means by which a good objective function might be discovered, through the introduction of SAFE—Solution And Fitness Evolution—a commensalistic coevolutionary algorithm that maintains two coevolving populations: a population of candidate solutions and a population of candidate objective functions [11, 12].

Consider a robot navigating a maze, wherein the challenge is to evolve a robotic controller such that the robot, when placed in the start position, is able to make its way to the goal. It seems intuitive that the fitness of a given robotic controller be defined as a function of the distance from the robot to the objective, as done, e.g., by Lehman and Stanley [7]. However, reaching the objective may be difficult since the robot is faced with a deceptive landscape, where higher fitness (i.e., being reasonably close to the goal) may not imply that the robot is "almost there". It is quite easy for the robot to attain a fairly good fitness value, yet be stuck behind a wall in a local optimum—quite far from the objective in terms of the path needed to be taken. Indeed, our experiments with such a fitness-based evolutionary algorithm [12] produced the expected failure, demonstrated in Fig. 15.4.

One solution to this conflation problem was offered by Lehman and Stanley [7] in the form of novelty search, which *ignores* the objective and searches for novelty. However, novelty for the sake of novelty alone lacks incentive for solutions that reach and stay at the objective.

Perhaps, though, the issue lies with our ignorance of the *correct* objective function. That is the motivation behind the SAFE algorithm.

SAFE is a coevolutionary algorithm that maintains two coevolving populations: a population of candidate solutions and a population of candidate objective functions. The evolution of each population is identical to a standard, single-population evolutionary algorithm—except where fitness computation is concerned, as shown in Fig. 15.5.

We applied SAFE to two domains: evolving robot controllers to solve mazes [12] and multiobjective optimization [11].

Applying SAFE within the robotic domain, an individual in the solutions population was a list of 16 real values, representing the robot's control vector ("brain"). The controller determined the robot's behavior when wandering the maze, with its phenotype taken to be the final position, or endpoint. The endpoint

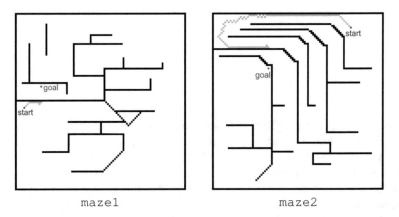

maze1   maze2

Fig. 15.4 In a maze problem a robot begins at the start square and must make its way to the goal square (objective). Shown above are paths (green) of robots evolved by a standard evolutionary algorithm with fitness measured as distance-to-goal, evidencing how conflating the objective with the objective function leads to a non-optimal solution

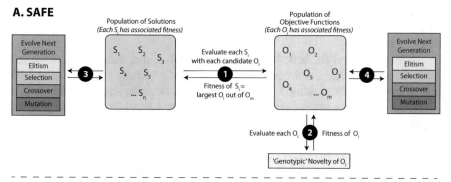

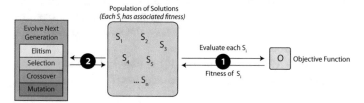

Fig. 15.5 A single generation of SAFE (**a**) vs. a single generation of a standard evolutionary algorithm (**b**). The numbered circles identify sequential steps in the respective algorithms. The objective function can comprise a single or multiple objectives

was used to compute standard distance-to-goal fitness and to compute *phenotypic* novelty: compare the endpoint to all endpoints of current-generation robots *and* to all endpoints in an archive of past individuals whose behaviors were highly novel when they emerged. The final novelty score was then the average of the 15 nearest neighbors.

An individual in the objective-functions population was a list of 2 real values $[a, b]$, each $\in [0, 1]$.

Every solution individual was scored by every candidate objective-function individual in the current population (Fig. 15.5a). Candidate SAFE objective functions incorporated both 'distance to goal' (the evolving a parameter) as well as phenotypic novelty (the evolving b parameter) in order to calculate solution fitness, weighting the two objectives in a simple linear fashion. The best (highest) of these objective-function scores was then assigned to the individual solution as its fitness value.

As for the objective-functions population, determining the quality of an evolving objective function posed a challenge. Eventually we turned to a commensalistic coevolutionary strategy, where the objective functions' fitness did not depend on the population of solutions. Instead, it relied on *genotypic* novelty, based on the objective-function individual's two-valued genome, $[a, b]$. The distance between two objective functions was simply the Euclidean distance of their genomes. Each generation, every candidate objective function was compared to its cohorts in the current population of objective functions and to an archive of past individuals whose behaviors were highly novel when they emerged. The novelty score was the average of the distances to the 15 nearest neighbors, and was used in computing objective-function fitness.

SAFE performed far better than random search and a standard fitness-based evolutionary algorithm, and compared favorably with novelty search. Figure 15.6 shows sample solutions found by SAFE (contrast this with the standard evolutionary algorithm, which always got stuck in a local minimum, as exemplified in Fig. 15.4). For full details see [12].

The second domain we applied SAFE to was multiobjective optmization [11]. A multiobjective optimization problem involves two or more objectives all of which need to be optimized. Applications of multiobjective optimization abound in numerous domains [16].

With a multiobjective optimization problem there is usually no single-best solution, but rather the goal is to identify a set of 'non-dominated' solutions that represent optimal tradeoffs between multiple objectives—the *Pareto front*. Usually, a representative subset will suffice.

Specifically, we applied SAFE to the solution of the classical ZDT problems, which epitomize the basic setup of multiobjective optimization [6, 17]. For example, ZDT1 is defined as:

$$f_1(\mathbf{x}) = x_1 ,$$

$$g(\mathbf{x}) = 1 + 9/(k-1) \sum_{i=2}^{k} x_i ,$$

$$f_2(\mathbf{x}) = 1 - \sqrt{f_1/g} .$$

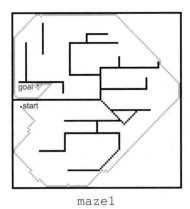
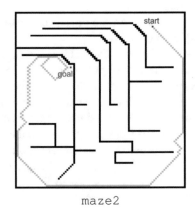

maze1   maze2

Fig. 15.6 Solutions to the maze problems, evolved by SAFE

The two objectives are to minimize both $f_1(\mathbf{x})$ and $f_2(\mathbf{x})$. The dimensionality of the problem is $k = 30$, i.e., solution vector $\mathbf{x} = x_1, \ldots, x_{30}, x_i \in [0, 1]$. The utility of this suite is that the ground-truth optimal Pareto front can be computed and used to determine and compare multiobjective algorithm performance.

SAFE maintained two coevolving populations. An individual in the solutions population was a list of 30 real values. An individual in the objective-functions population was a list of 2 real values $[a, b]$, each in the range $[0, 1]$, defining a candidate set of weights, balancing the two objectives of the ZDT functions: a determined f_1's weighting and b determined f_2's weighting.

Note that, as opposed to many other multiobjective optimizers, SAFE did not rely on measures of the Pareto front (i.e., a Pareto front was not employed to calculate solution fitness, or as a standard for selecting parent solutions to generate offspring solutions).

We tested SAFE on four ZDT problems—ZDT1, ZDT2, ZDT3, ZDT4— recording the evolving Pareto front as evolution progressed. We compared our results with two very recent studies by Cheng et al. [1] and by Han et al. [4], finding that SAFE was able to perform convincingly better on 3 of the 4 problems. For full details see [11].

15.4 Concluding Remarks

The experimentation performed to date is perhaps not definitive yet but we hope to have offered at least proof-of-concept of our two new coevolutionary algorithms. Both have been shown to be successful in a number of domains.

There are several avenues of future research that present themselves, including:

- Study and apply both algorithms to novel domains. We have been looking into applying SAFE to datasets created by the GAMETES system, which models epistasis [14]. We have also created additional art by devising novel encoding-representation couplings for OMNIREP (Fig. 15.7).
- Study the coevolutionary dynamics engendered by OMNIREP and SAFE.
- Cooperative or competitive versions of SAFE (which is currently commensalistic), i.e., finding ways in which the objective-function population depends on the solutions population.
- Examine the incorporation of more sophisticated evolutionary algorithm components (e.g., selection, elitism, genetic operators).

Acknowledgement This work was supported by National Institutes of Health (USA) grants AI116794, LM010098, and LM012601.

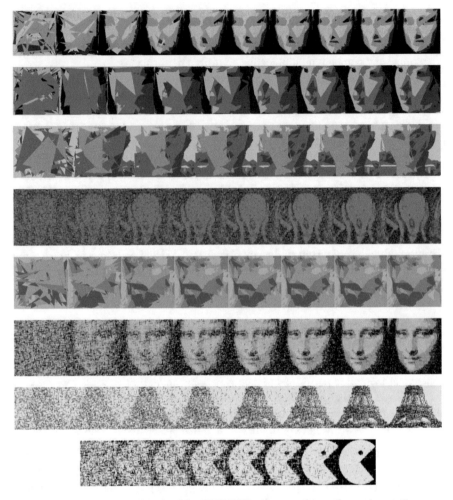

Fig. 15.7 Additional artwork created by OMNIREP using novel encoding-representation couplings (involving polygons, and horizontal and vertical blocks). Each row shows a single evolutionary run, from earlier generations (left) to later generations (right)

References

1. Cheng, T., Chen, M., Fleming, P.J., Yang, Z., Gan, S.: A novel hybrid teaching learning based multi-objective particle swarm optimization. Neurocomputing **222**, 11–25 (2017)
2. Dick, G., Yao, X.: Model representation and cooperative coevolution for finite-state machine evolution. In: 2014 IEEE Congress on Evolutionary Computation (CEC), pp. 2700–2707. IEEE, Piscataway, NJ (2014)
3. Eiben, A.E., Smith, J.E.: Introduction to Evolutionary Computing. Springer-Verlag, Berlin (2003)
4. Han, F., Sun, Y.W.T., Ling, Q.H.: An improved multiobjective quantum-behaved particle swarm optimization based on double search strategy and circular transposon mechanism. Complexity **2018** (2018)

5. Hillis, W.: Co-evolving parasites improve simulated evolution as an optimization procedure. Physica D: Nonlinear Phenomena **42**(1), 228–234 (1990)
6. Huband, S., Hingston, P., Barone, L., While, L.: A review of multiobjective test problems and a scalable test problem toolkit. IEEE Transactions on Evolutionary Computation **10**(5), 477–506 (2006)
7. Lehman, J., Stanley, K.O.: Exploiting open-endedness to solve problems through the search for novelty. In: Proceedings of the Eleventh International Conference on Artificial Life (ALIFE). MIT Press, Cambridge, MA (2008)
8. Pena-Reyes, C.A., Sipper, M.: Fuzzy CoCo: A cooperative-coevolutionary approach to fuzzy modeling. IEEE Transactions on Fuzzy Systems **9**(5), 727–737 (2001)
9. Potter, M.A., De Jong, K.A.: Cooperative coevolution: An architecture for evolving coadapted subcomponents. Evolutionary Computation **8**(1), 1–29 (2000)
10. Sipper, M., Moore, J.H.: OMNIREP: originating meaning by coevolving encodings and representations. Memetic Computing (2019)
11. Sipper, M., Moore, J.H., Urbanowicz, R.J.: Solution and fitness evolution (SAFE): A study of multiobjective problems. In: Proceedings of 2019 IEEE Congress on Evolutionary Computation (CEC). IEEE (2019)
12. Sipper, M., Moore, J.H., Urbanowicz, R.J.: Solution and fitness evolution (SAFE): Coevolving solutions and their objective functions. In: L. Sekanina, T. Hu, N. Lourenço, H. Richter, P. García-Sánchez (eds.) Genetic Programming, pp. 146–161. Springer International Publishing, Cham (2019)
13. Sipper, M., Urbanowicz, R.J., Moore, J.H.: To know the objective is not (necessarily) to know the objective function. BioData Mining **11**(1) (2018)
14. Urbanowicz, R.J., Kiralis, J., Sinnott-Armstrong, N.A., Heberling, T., Fisher, J.M., Moore, J.H.: GAMETES: A fast, direct algorithm for generating pure, strict, epistatic models with random architectures. BioData Mining **5**(1), 16 (2012)
15. Zaritsky, A., Sipper, M.: Coevolving solutions to the shortest common superstring problem. Biosystems **76**(1), 209–216 (2004)
16. Zhou, A., Qu, B.Y., Li, H., Zhao, S.Z., Suganthan, P.N., Zhang, Q.: Multiobjective evolutionary algorithms: A survey of the state of the art. Swarm and Evolutionary Computation **1**(1), 32–49 (2011)
17. Zitzler, E., Deb, K., Thiele, L.: Comparison of multiobjective evolutionary algorithms: Empirical results. Evolutionary Computation **8**(2), 173–195 (2000)

Chapter 16
2019 Evolutionary Algorithms Review

Andrew N. Sloss and Steven Gustafson

16.1 Preface

When attempting to find a perfect combination of chemicals for a specific problem, a chemist will undertake a set of experiments. They know roughly what needs to be achieved but not necessarily how to achieve it. A chemist will create a number of experiments. Each experiment is a combination of different chemicals. Following some theoretical basis for the experiments. The experiments are played out and the promising solutions are identified and gathered together. These new chemical combinations are then used as the basis for the next round of experiments. This procedure is repeated until hopefully a satisfactory chemical combination is discovered.

The reason this discovery method is adopted is because the interactions between the various chemicals is too complicated and potentially unknown. This effectively makes the problem-domain too large to explore. An Evolutionary Algorithm (EA) replaces the decision making by the chemist, using evolutionary principles to explore the problem-space. EAs handle situations that are too complex to be solved with current knowledge or capability using a form of synthetic digital evolution. The exciting part is that the solutions themselves can be original, taking advantage of effects or attributes previously unknown to the problem. EAs provide a framework that can be reused across different domains, they are mostly biologically-inspired algorithms that reside as a subbranch of Artificial Intelligence (AI).

A. N. Sloss (✉)
Arm Inc., Bellevue, WA, USA
e-mail: Andrew.Sloss@arm.com; asloss@arm.com

S. Gustafson
MAANA Inc., Bellevue, WA, USA

© Springer Nature Switzerland AG 2020
W. Banzhaf et al. (eds.), *Genetic Programming Theory and Practice XVII*, Genetic and Evolutionary Computation, https://doi.org/10.1007/978-3-030-39958-0_16

Using Bertrand Russell's method of defining philosophy [63] i.e. *"as soon as definite knowledge concerning any subject becomes possible, this subject ceases to be called philosophy, and becomes a separate science"*. AI research lives in-between philosophy and science. Ideas transition from philosophical thought to applied science and truth. Within Computer Science, the AI field resides at the edge of knowledge and as such includes a distinct part which is more philosophical and another which is more rooted in science. In this review we cover one of the science aspects. AI science incorporates many areas of research e.g. Neural Networks, Bayesian Networks, Evolutionary Algorithms, Correlation, Game Theory, Planning, Vision recognition, Decision making, Natural Language Processing, etc. It is a dynamically changing list as more discoveries are made or developed. Thirdly, *Machine Learning* is the engineering discipline which applies the science to a real world problem.

One of the overriding motivators driving Machine Learning in recent years has been the desire to replace rigid rule-based systems. A strong candidate has been emerging which is both adaptive and outcome-based. This technology relies on data-directed inputs. Jeff Bezos, CEO of Amazon, succinctly described this concept in a letter to shareholders in 2017 [46], *"Over the past decades computers have broadly automated tasks that programmers could describe with clear rules and algorithms. Modern Machine Learning techniques now allow us to do the same for tasks where describing the precise rules is much harder"*. Also Kazuo Yano, Fellow and Corporate Officer of Hitachi Ltd, said in his keynote at the 2018 *Genetic and Evolutionary Computation Conference* (GECCO) [8] that the demand for more flexibility forces us to transition from traditional rule-oriented systems to future outcome-oriented ones.

The adaptability and transition to outcome-oriented systems means, from an end user perspective, there is more uncertainly surrounding the final result. This is construed as being either real or perceived. Rule-based systems are not impervious but tend to be deterministic and understandable e.g. the most notable being the area of safety-critical systems. This uncertainty creates the notion of User Control Attributes (UCA). The UCA include the concepts of *limiters* [74], *explainability* [4], *causality* [58], *fairness* [2, 82] and *correction* [79]. These attributes have seen a lot of scrutiny in recent years due to high-profile public errors, as detailed in the *AI Now 2018 Report* [79]. The report goes into the details of fairness and highlights the various procedures, transparency and accountability required for Machine Learning systems to be safely applied in real social environments. Also worth mentioning is the *International Standard Organization* (ISO), which has formed a study group focusing specifically on *Trustworthiness* [25]. The study will be investigating methods to improve basic trust in Machine Learning systems by exploring transparency, verify-ability, explainability, and control-ability. The study will also look at mitigation techniques. The goal is to improve overall robustness, resiliency, reliability, accuracy, safety, security and privacy; and by doing so hopefully minimize source biases. For this review we will limit the focus to research, while at the same time being cognizant of the dangers of real world deployment.

Control imposes a different level of thinking, where researchers are not just given a problem to solve but the solution requires a model justifying the outcome. That model has to provide the answers to the main questions: Whether the algorithm stays within limits/restrictions? Is the algorithm explainable? Can the algorithm predict beyond historical data? Does the algorithm avoid system biases or even side-step replicating human prejudices [21]? And finally, can the algorithm be corrected? These attributes are not mutually exclusive and in fact intertwine with each other. We can see a trend where modern Machine Learning algorithms will be rated not only on the quality of the results but on how well they cope with the user demanded control attributes. These attributes are to be used as a basis for a new taxonomy.

This is a good point to start discussing the Computer Industry. The industry itself is facing a set of new challenges and simultaneously adding new capabilities, as summarized below:

- **Silicon level**: groups are starting to work on the problem of mass silicon production at the 3-nanometer scale and smaller [22]. This involves designing gates and transistors at a scale unimaginable 10 or even 5-years ago, using enhanced lithographic techniques and grappling with quantum tunneling issues. These unprecedented improvements have allowed other areas higher up the software stack to flourish. Unfortunately these advancements are slowing down, as the current techniques hit both physical and economic limitations. This situation is called the *End of Moore's Law* (EoML) [11, 67].

- **System level**: A number of levels above the silicon lies the system-level which also has seen some impressive advancements with the world-wide web, network infrastructure and data centers that connect everyone with everyone. The scale of the system-level advancements has opened up the possibility of mimicking small parts of the human brain. One such project is called *SpiNNaker* [29]. SpiNNaker was upgraded and switched on in November 2018, it now consists of a million interconnected ARM cores each executing *Spiking Neural Networks* (SNN) [49]. Even with all the hardware capability, estimates suggest that it is only equivalent to about 1% of a human brain.

- **Software design level**: Software design, at the other end of the spectrum, has been constantly pursuing automation. The hope being that automation leads to the ability to handle more complicated problems, which in turn provides more opportunities. New programming languages, new software paradigms and higher level data-driven solutions all contribute to improving software automation.

To handle these new challenges and capabilities requires a continuously changing toolbox of techniques. EAs are one such tool and like other tools they have intrinsic advantages and disadvantages. Recently EAs have seen a resurgence of enthusiasm, which comes at an interesting time since other branches of Machine Learning become mature and crowded. This maturity forces some researchers to explore combinations of techniques. As can be seen in this review a lot of the new focus and vigor is centered upon hybrid solutions, especially important is the area of combining evolutionary techniques with *Artificial Neural Networks* (ANNs).

16.2   Introduction

EAs are not a new subject. In fact as we look back at some of the early computing pioneers we see examples of evolutionary discovery. For example both Alan Turing [76] and John von Neumann [55] formed ideas around Biological Automation, Biological Mathematics and Machine Learning. These forward visionaries focused on the fact that *exploitative* methods could only go so far to solve difficult problems and more *exploratory* methods were required. The main difference between the two techniques is that exploitative focuses on direct local knowledge to obtain a solution whereas exploratory takes effectively a more stochastic approach (leaping into the unknown).

Figure 16.1 shows an idealized view of the changes to hardware capability and algorithmic efficiency over a time period. The figure shows the relationship between improvements in hardware and the types of problems that can be addressed. Before 2010, the computing industry mostly focused on exploitative problems "getting things to work efficiently". Today, due to hardware improvements, we can look at exploratory algorithms that "discover". At the extremes of the *X-Y* axis, top-right and bottom-right lies the respective future potentials of hardware and software i.e. the unknown.

Note: *End of Dennard Scaling* [24] marks the point in time when transistor shrinkage no-longer sustained a constant power density. In other words, static power consumption dominates the power equation for silicon. Forcing the industry to use clever frequency and design duplication techniques to mitigate the problem. *Deep Learning* [34] represents the software resurgence of Neural Nets, due to hardware improvements and the availability of large training data sets.

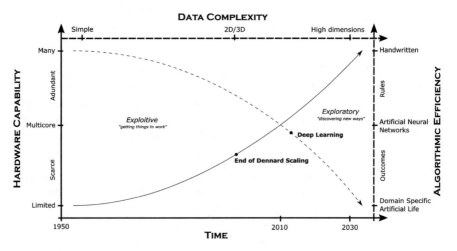

Fig. 16.1 Hardware capability and algorithmic efficiency over an idealized time line

The *Hardware Capability* top-right of the graph represents future hardware concepts, and requires subsequent manufacturing breakthroughs. These future concepts could include alternative computing models, Quantum computing, Neuromorphic architectures, new exotic materials, Asynchronous processing, etc.. By contrast the *Algorithmic Efficiency* bottom-right represents future breakthroughs in subjects like *Artificial Life*, *Artificial General Intelligence* (AGI), etc.; more philosophical goals than either science or engineering. Both require significant advancements beyond what we have today.

With these future developments, the desire is to set a problem-goal and let the "system" find the correct answer. This is extremely simple to state but highly complex to implement. And more importantly, next to impossible to implement without direct insertion of domain specific knowledge into the algorithm in question.

No Free Lunch Theorem [81] states that no algorithm exists which outperforms every other algorithm for every problem. This means to be successful, each problem requires some form of domain specific knowledge to be efficient. The more domain specific knowledge applied to an algorithm the greater the likelihood of beating a stochastic algorithm. A stochastic algorithm can search every problem, without the requirement of domain knowledge. EAs are directed population-based stochastic search algorithms. As hardware capability increases more of these types of problems can be handled. It is the constraints of time and efficiency that forces domain knowledge to be inserted into an algorithm.

This paper provides an up-to-date review of the various EAs, there respective options and how they may be applied to different problem-domains. EAs are a family of biologically-inspired algorithms that take advantage of synthetic methods, namely management of populations, replication, variability and finally selection. All based upon the fundamental theory of *Darwinian* evolution [23]. As a general rule the algorithms are often simple at a high-level but become increasingly complex as more domain knowledge is put into the system.

Another term frequently used to describe these style of algorithms is *metaheuristics*. Metaheauristics is a higher-order concept, it is an algorithm that systematically pursues the identification of the best solution within a problem-space. EAs obviously fall under this class of algorithms and a lot of the academic literature frequently refers to metaheuristics: the main differentiator being biologically inspired algorithms.

For these algorithms to execute, some form of quantitative goal has to be set. This goal is the metric for success. The success metric can also be used as a method to exit the algorithm but most often a function of time is used. The algorithm can also be designed to be continuous i.e. never ending, always evolving. The goal itself can be made up of a single objective or multi-objectives. For multi-objective the search and optimization is towards the *pareto optimal* curve i.e. attempting to satisfy all the objectives in some degree.

As an important side-note, EAs are mostly *derivative-free*, in that majority do not require a derivative function to measure change to determine the optimal result.

Lastly, GECCO 2018 [8], in Kyoto, saw a number of trends. Neuroevolution being one of the more notable ones, Neuroevolution is the method of using EAs to configure ANNs (see Sect. 16.5.2). These new trends will be tracked in more detail in future reviews if they remain popular in the evolutionary research community.

16.2.1 Applications

EAs are applied to problems where traditional exploitative or pure stochastic algorithms fail or find it difficult to reach a conclusion. This is normally due to constraints on resources, high number of dimensions or complex functionality. Solving these problems would require exceeding the available resources. In other words, given infinite resources and infinite compute capability it could be possible for a traditional exploitative or stochastic algorithm to reach a conclusion. By contrast, EAs can be thought of as the algorithms-of-last-resort. The problems in question are inherently complex, size of the problem-domain is extreme or the mere number of objectives make the problem impossibly difficult to explore. In these circumstances the solutions are more likely "good enough" solutions rather than solutions with high precision or accuracy (but this does not preclude precision or accuracy being a goal). EAs tend to be poor candidates for simple problems where standard techniques could easily be used instead.

They can be applied to a broad set of problem types. These types range from variable optimization problems to creating new conceptual designs. In both instances, novelty can occur which may exceed human understanding or ability. These problem-domains can be broken-down into optimization, new design and improvement.

- **Variable optimization** consists of searching a variable space for a "good" solution, where the number of variables being searched is large. Potentially at a magnitude greater than a traditional programming problem. This is where the goal can be strictly defined.
- **New structural design** consists of creating a completely new solution, for example like a program or a mechanical design. A famous example, of a non-anthropomorphized solution, is the evolutionary designed NASA antenna [36]. The antenna design was not necessarily something a human would have created. This is where a particular outcome is desired but it is unknown how that outcome can be constructed.
- **Improvement** is where a known working solution is placed into the system and EAs explore potential better versions. This is where a solution already exists and there is a notion that a potential better solution can be discovered.

Being more specific on the applications side, EAs have been applied to a wide range of problems from leading edge research on self-assembly cellular automata [54] to projecting future city landscapes for town planners (see Sect. 16.6.2.3).

16.3 Fundamentals of Digital Evolution

Before diving directly into the fundamentals we should stress that there are two ways to describe evolution. The first is from a pure biology point-of-view dealing the various interactions of biological systems and the other is from the Computer Science perspective. In this paper we will keep the focus on Computer Science with a hint of biology. Other texts may approach the explanation from a different perspective.

Evolution is a dynamic mechanism that includes a population of entities (potential solutions) where some form of replication, variation and selection occurs on those entities, as shown in Fig. 16.2. This is a stochastic but guided process where the desire is to move towards a fixed goal.

- **Replication**: Is where new entities are formed, either creating a completely new generation of the population or altering specific individuals within the same population (called *steady-state*).
- **Variation**: Making the population diverse. There are in effect two forms of variation, namely *recombination* and *mutation*. Recombination, or more commonly called crossover, creates new entities by combining parts of other entities. By contrast, mutation injects randomness into the population by stochastic-ally changing specific features of the entities.

Fig. 16.2 Idealized Darwinian evolution

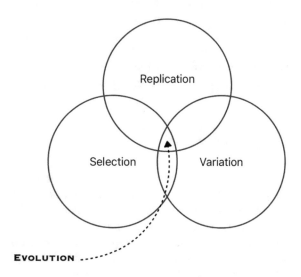

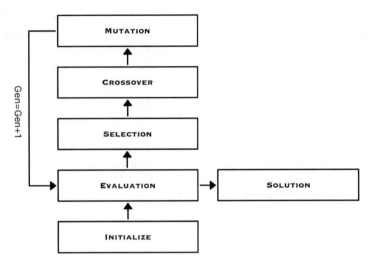

Fig. 16.3 Basic digital process

- **Selection**: Selection is based on Darwin's natural selection, or *Survival Of The Fittest*, where selected entities that show the most promise are carried forward, with variation on the selection being used to create the children of the next generation.

As a recap, digital evolution is one which a population of entities goes through generational changes. Each change starts with a selection from the previous generation. Each entity is evaluated against a known specific goal i.e. the fitness is established and used as input to the selection algorithm. Once a selection is made, replication occurs with different degrees of variation. The variation is either by some form of recombination from the parent selection and/or some stochastic mutation.

Figure 16.3 shows the basic digital process that is adopted by many EAs. The high level synthetic evolutionary process as-shown is a relatively straightforward and simple procedure. Within the process there are many complex nuances and variations, which are influenced by our understanding of evolutionary biology.

16.3.1 Population

A *population* is a set of solution candidates or entities. Each population is created temporally and has the potential to age with each evolutionary cycle i.e. generation. The initial population is either seeded randomly or is sampled from a set of known solutions. There are two ways for the population to be evolved. This first way consists of evolving the entire population to create the next generation. The second way is to evolve individual entities within the population (called Steady-State). Entities can die or thrive between generations.

The size of the population can be fixed or dynamic throughout the evolutionary process. Fixed is when the number of entities is kept constant and dynamic is when the population size changes. If dynamic, a larger initial population may bring some potential advantages. This is especially true when choosing the first strong candidates for further evolution.

A population can also be divided into subgroups, these subgroups are called *demes*. A deme is a biological term that describes a group of evolving organisms with the same taxonomy. They can evolve in isolation within the population [13]. Demes can be made to interact with the other demes. This is often described using an island and canoe metaphor, where the demes are the islands and the interactions occur as canoes move between the islands depositing new entities.

16.3.2 Population Entities

The population entities, or more commonly called *phenotypes*, can be any type provided that type allows recombination and/or mutation to be applied. The main styles are strings and programs. In this context "programs" is a loose term. The strings are fixed length and more likely represent variables or parameters. The programs tend to be more complex. Effectively any structure from a traditional computer programming language to hardware layout, or even biological descriptions, can be a program.

Metaheuristics operate at a high-level making the process generic, allowing idea or concept can be used as the evolved substrate. Including concepts like Deep Learning Neural Network or potentially a causality equation. EAs are neutral on the subject but when it comes to specific problems, the problems themselves, tend to be strictly defined.

There are two important concepts to consider with population entities, namely *niching* and *crowding* [66]. These terms are associated with diversity, which is briefly mentioned in Sect. 16.3.1. Niching means individual entities survive generations in distinct areas of the search space. By contrast, crowding means replacing an individual entity with similar featured individuals.

16.3.3 Generation

A generation is a specific step in an evolutionary population, where individuals in the population undergo change via crossover and/or mutation. For EAs with a fixed-run a restriction is imposed on the number of generations. By contrast, continuous EAs have no upper bounds. EAs can vary from small populations with high number of generations to large populations with significantly lower number of generations. These trade-offs need to be made in terms of computation time. For example, some scenarios will have expensive evaluation functions (so fewer generations with larger populations might be preferred). The number of generations required is both algorithmic and problem-domain specific.

The children in the next generation can also include the parents from the previous generation. This is called *elitism*, where the strongest entities remain in the population.

16.3.4 Representation and the Grammar

Representation, or more commonly called *genotype*, is what EAs manipulate. There are different types of representations which include strings, tree structures, linear structures and direct-graphs. Each representation has advantages and disadvantages, and is dependent on the specific problem-domain. The representation determines what actually gets manipulated when recombination occurs, some lend themselves more to fractal/recursive like procedures and the others are more sequential and linear.

By contrast, the rules are defined by the grammar. EAs grammar provides the expressive boundaries for the representation. For example, in a mathematical domain adding a function like $\sin(x)$ to the grammar provides extra richness and complexity to the representation. Similarly, for string based representations adding or removing a variable to the grammar changes the expressiveness. Other decisions can be made for structural EAs, such-as including more constructors (e.g. *addition*) over destructors (e.g. *subtraction*), vice-versa or more commonly keeping the grammar entirely balanced i.e. same number of constructors as destructors.

Possible grammars include a subset of Python or C, assembly instructions or even pure LLVM intermediate code [47]. These are all potential outputs of EAs.

16.3.5 Fitness

Fitness is the measure of how close a result is to a desired goal. A fitness function is the algorithm used to calculate the fitness value. Calculating the fitness value for an entire population is a time consuming activity. The time taken is related to the complexity of the fitness function and the size of the population being evaluated. The fitness function is used to select individuals for inclusion in future populations. The function can be constant throughout the evolutionary process or it can change depending upon the desired goal or situation. It is the feature of fitness function change that makes EAs highly adaptive.

The fitness can be calculated using different methods e.g. Area Under the Curve (AUC) from a test set, measurement of a robot responding to a set of trials, etc.; each method being problem dependent. For supervised learning it is calculated as the difference between the desired goal and the actual result obtained from the entity. Conversely, for unsupervised learning there are other methods. Once the fitness has been determined the entities can be ranked/sorted by strength. The stronger candidates are more likely to be chosen as parents for the next generation.

16.3.6   Selection

Selection is the method where individuals in the current population are chosen to be the starting parents for the next generations. In digital evolution, parents are not restricted to two; any number can be chosen. A simple method is to use the fitness value to order the population, this method is called *ranked selection*. A population can be organized from the highest to lowest fitness value. The highest entities are then used as the starting parents for the next generation and so forth. As mentioned in Sect. 16.3.3, elitism is where the chosen parents remain in the next population rather than being discarded.

Diversity [66] is an important concept when it comes to a healthy population. Healthy populations are important for discovering "good" solutions. In other words, a diverse population has a higher exploratory capability. This tends to be important especially at the start of the search process. Diversity is directly associated with the amount of variation applied to the entities. It can be argued that local selection schemes [13, 66] (steady-state) are naturally more likely to preserve diversity over global selection scheme. Local selection means evolution is potentially occurring at different rates across the population.

An example of a local selection scheme is called *tournament selection*. As the name implies, selection requires running a tournament between a randomly chosen set of individuals. The "winner" of each tournament is then selected for further evolution.

Note: there are other selection schemes, which are not covered here.

16.3.7   Multi-Objective

As the name implies *multi-objective* is the concept of not having a single objective but multiple objectives. These objectives may act against each other in complex ways i.e. conflict. It is this interaction which makes multi-objective so complex. A typical example in the mobile phone industry is finding the optimum position between performance, power consumption (longevity) and cost. These three objectives can be satisfied to different degrees, effectively giving the end consumer a choice of options. Too much performance may sacrifice longevity + cost, low cost may sacrifice performance + longevity and so-on. EAs are extremely good at exploring multi-objective problems where the fitness is around a compromise along the *Pareto curve*.

16.3.8   Constraints

Constraints are the physical goals, as compared with the objectives which are the logical goals. The physical goals represent the real world limitations. They take a theoretical problem and make it realistic. Constraints are the limitations imposed on the entities. A constraint could be *code size, energy consumption* or *execution time*. EA Researchers have discovered some of the most interesting and potentially best solutions tend to lie somewhere at the edge of the constraint boundaries.

16.3.9   Exploitative-Exploratory Search

EAs use *recombination* and *mutation* for exploitative and exploratory search. The more mutation that occurs the more exploratory the search, and correspondingly the less mutation the more exploitative the search. EAs can be at either end of the spectrum, with only recombination (more exploitative) or only mutation (more exploratory). The ratios of recombination and mutation can be fixed or dynamic. By making the ratios dynamic EAs can adapt to changing circumstances. This shift may occur when the potential "good" solution is perceived to be either near or far. Another way to view this is that mutation is a local search and recombination (or crossover) is a global search. Recombination, despite using only existing genetic material, often takes much larger jumps in the search space than does mutation.

16.3.10   Execution Environment, Modularity and System Scale

EAs can execute as a process within an Operating System, as a self-constructed dynamic program feed into a language interpreter (e.g. *exec(open("ea.py").read())* in Python), or within a simulator, where the simulator can be a physics simulator, biological simulator and/or a processor simulator. EAs are generic and literally any executing model can be used to explore a desired problem-domain.

To handle larger problems some form of modularity has to be adopted. There are many schemes including ones that introduce tree based processing or Byzantine style algorithms i.e. voting systems. Modularity tends to work best when the problem granularity is relatively small and concise, such as a function-call. The normal questions asked are (1) whether all the function-calls use the same input data, (2) whether all information is shared between the functions, or (2) whether a hierarchy of evolution has to be adopted i.e. from function call to full solution or from local to global parameter configuration.

EAs can scale from one process to many processes running on several server clusters [37, 38, 48, 56]. These compute clusters are called *islands*. They operate in either a parallel and/or distributed fashion. A parallel system is a set of evolving islands working together towards a common goal. Genetic material or solutions can be shared. We should highlight that there are a few options when it comes to parallel topologies. By comparison, the distributed approach, which can include parallel islands, is about the physical aspect of running on various hardware systems. With both approaches, scale-out co-ordination becomes an important issue i.e. the management of the islands becomes part of the performance equation.

Lastly, it is important to mention *co-evolution* in the context of scaling. Co-evolution is where two or more evolving populations (effectively *species*) start interfering/cooperating with each other. This is particular important when digital evolution is being used to build much larger systems. Both modularity and system scale add an extra layer of complexity. Scale is a required necessity to answer more complicated problems.

16.3.11   Code Bloat and Clean-Up

The EAs which play with structures can quite easily have to deal with the problem of bloat [60]. Bloat is a byproduct of exploring a problem-domain. Historically this was a major issue with the earlier algorithms due to hardware limitations. Today, modern systems have an abundance of compute and storage. This does not mean the limitation has gone away but it is mitigated to a certain extent. Bloat may be critically important to the evolutionary process. In nature the more that is discovered the more it seems that very little is actually wasted. Non-coding regions are not bloat, they are crucial parts to the process of molecular mechanisms.

There are structures or code sequences that have no value, as-in the result is circumvented by other code or structures. These neutral or noneffective structures are called *introns*. Intron is a biological term referring to the noneffective fragments found in DNA. For software programs, introns are code sequences that are noneffective or neutral; these sequences can be identified and cleaned-up, i.e. eliminated. The elimination can occur either during the evolutionary process itself or at a final stage. Note, introns can be critically important to the process, so early removal can be detrimental.

16.3.12   Non-convergence, or Early Local Optima

There has been a lot of research focusing on the problems of non-convergence and early-local-optima solutions. Non-convergence means that the evolving entities are not making enough progress towards a solution. Early-local-optima means a sub-optimal solution has been found at the beginning of the evolutionary process. This sub-optimal solution has caused the algorithm to limit further exploration, reducing the chance of finding a better solution.

Non-convergence is caused by many factors including the possibility of not having the right data. EAs rely on stochastic processes to move toward, which means that the paths taken are unique, unrepeatable and non-deterministic; unique in the sense that the paths taken are always different, unrepeatable as-in randomness is used to determine the next direction, non-deterministic as-in the length of time to solution is variable. To get to a potential solution may require reruns of the algorithm. Each rerun potentially requiring some form of fine adjustment to help narrow into a good solution. In the end, when everything else fails, more domain specific knowledge may have to be inserted before convergence eventually occurs. It is important to stress that the final solution may very well be deterministic and repeatable, it is the evolutionary process to create the solution which may not be.

Similar to non-convergence is the problem of reaching a local extrema too early, and then EAs iterate persistently around a point not discovering a better solution. Again, the techniques used for non-convergence tend to be used to avoid the local optima scenario. Identification of a local optima can be difficult since the global optima is unknown. After multiple readjustments hopefully the global optima can be discovered.

16.3.13 Other Useful Terms

There are other terms which are worthy of a mention and brief descriptions. The first terms are the *Baldwin Effect* [52] and *Lamarckian Evolution* [19] both important concepts for digital evolution. The Baldwin Effect is about how learned behavior effects evolution. This is important for EAs that improve towards a solution under techniques such as elitism (as briefly mentioned in Sect. 16.3.3). And Lamarckian Evolution which theorizes that children can inherit characteristics from the experiences gained by their parents. Again, an important concept with direct implications for digital evolution.

By contrast, the term *overfitting* [15] describes a situation that should be avoided. It is when the data noise containing irrelevant information and what is actually being discovered combine into a result (configuration parameters). EAs overfit when the discovered parameters satisfies the complete data-set and not the data for the specific exploration, making it effectively useless for any future prediction using other input data sources. The potential concern is the increased risk of a *false-positive* outcome.

Lastly, *Genetic Drift* [1, 9] is a basic evolutionary mechanism found in nature, where some genotype entries between generations leave more descendants or parts than others. These descendants are in the population by random chance. They are neither in the next generation because of a strong attribute nor higher fitness value. In nature this happens to all populations and there is little chance for avoidance. EAs genetic drift can be as a result of a combination of factors, primarily related to selection, fitness function and representation. It happens by unintentional loss of genotypes. For example, random chance that a good genotype solution never gets selected for reproduction. Or, if there is a 'lifespan' to a solution and it dies before it can reproduce. Normally such a genotype only resides in the population for a limited number of generations.

16.4 Traditional Techniques

In this section we briefly cover the traditional and well known EAs. These EAs tend to be older and more mature techniques. The techniques covered are frequently used by industry and research. There are numerous support frameworks available to experiment with [26, 53, 68, 77]. Assume for each technique discussed that there are many more variations available.

Figure 16.4 shows the relationships between the various traditional techniques. For this review we have decided to focus more on Genetic Programming and the various sub-categories. In future reviews this emphasis will likely change.

16.4.1 Evolutionary Strategy, ES

Evolutionary Strategy, ES [48, 50, 66] is one of the oldest EAs, developed in the 1960s at the Technical University of Berlin; it usually only involves mutation and selection. Entities are selected using *truncation selection*. After the entities are evaluated, the entries below the truncation point are systematically removed from the population. The remaining parents are mutated to buildup a new population.

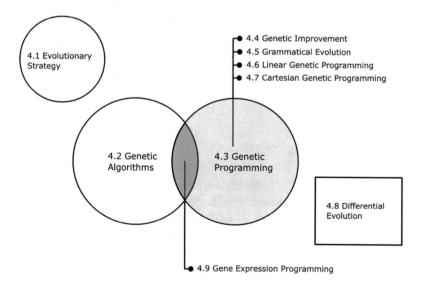

Fig. 16.4 Relationships between traditional EA techniques

Modern implementations include Co-variance Matrix Adaptation—Evolutionary Strategy CMA-ES [32] and an alternative Co-variance Matrix Self Adaptation—Evolutionary Strategy CMSA-ES. CMA-ES is thought to be complicated to implement, and CMSA-ES is a newer alternative and is believed to be easier to implement [12].

What Problems Do ESs Solve? An ES is used to solve continuous parameter optimization. A parameter is defined by its type and interval range (upper and lower bounds). A continuous parameter can take any value within the interval range. The precision determines the minimum change value.

16.4.2 Genetic Algorithms, GA

Genetic Algorithms, GA [31, 48, 66] is the most common and popular among the EAs. GA applies evolution to fixed length strings. The length of the string represents the dimensionality of the problem. These strings represent variables or parameters and are useful when exploring a large number problem-domain space. This space is normally beyond human or traditional methods. As well as being popular, GAs are also the most commonly taught algorithm within the various EAs. Variables or parameters are converted to fixed length strings, the strings are entities in the population and are evolved using crossover windows and mutation. Crossover is ubiquitous but explicit crossover windows are not. By comparison to an ES (Sect. 16.4.1), a GA tends to be more generic.

It should be noted that a recent trend has emerged, in both GAs and ESs, where crossover is dropped and mutation is the sole mechanism for evolution (this differs to the earlier thinking expressed in the classical literature).

What Problems Do GAs Solve? GAs handle optimization and configuration problems where there are too many variables or parameters for a traditional method to succeed. The variables or parameters may interact making a potential solution much harder to identify.

16.4.3 Genetic Programming, GP

Genetic Programming, GP [41–43, 45] in contrast to GAs, manipulate structures and in-particular executable programs or mathematical equations. Early GPs were based on tree representations and used the *LISP* programming language as the grammar. LISP was chosen for its operator richness and was relatively easily to manipulate. More recently there have been other representations introduced and newer languages such as *Python* have become popular as the main target. The recombination carries out the global search, whereas the mutation covers the local search. It is frequently common for mutation to be limited to 5–10% of the population [45]. There are always exceptions, especially if the % mechanisms are dynamically altered between generations [66].

What Problems Do GPs Solve? GPs apply evolutionary techniques to code or functions. GPs handle the manipulation of programs, so that problems that are linear, tree or direct-graph based can be explored. GPs can produce original source code, and in fact find new novel solutions to any structural style problem. In industry, GP are mostly used to discover best fit mathematical equations.

16.4.4   Genetic Improvement, GI

Genetic Improvement, GI [59] is a subclass of GP (Sect. 16.4.3), where instead of a random initial seeded population, a working program is inserted as the starting point to spawn entities of the first population. This is a powerful concept since it does not only search for a better optimized solution but also has the potential to discover and correct faults in the original "working" code.

What Problems Do GIs Solve? Solves an interesting problem, where either the working code is potentially un-optimized and a more optimized version is required or bringing legacy code up to current standards.

16.4.5   Grammatical Evolution, GE

Grammatical Evolution, GE [26] is a powerful technique. It is yet another subclass of GP (Sect. 16.4.3) but instead of using a fixed grammar to evolve-able solutions, the grammar itself is select-able. A good example of GE is the *PonyGE2* [26] tool, written in Python. It takes a standard Backus-Naur Form (BNF) grammar [10] as an input and uses it to evolves solutions. This is a powerful method especially when dealing with more obscure programming languages. GE can also carry out GI (see Sect. 16.4.4). Note the PonyGE2 source code is available on GitHub.

What Problems Does GEs Solve? GE solves the problem of evolving multiple programming languages using the same tool. As long as the language has a BNF-style definition, it can be evolved. This makes GE flexible across a number of problem-domains, and especially ones which require a specific programming language.

16.4.6 Linear Genetic Programming, LGP

Linear Genetic Programming, LGP [13] is a subclass of GP (Sect. 16.4.3) and as the name implies uses a linear structure representation. The linear structure has some advantages over the more complicated tree or directed-graph structures. LGP is particularly useful for problems which are more sequential. For example, optimizing low level assembly output. It also makes the problem of manipulating complex structures easier since it is a linear flow that is being evolved. Constructs like *if-style control flow* or *loops* are superimposed onto the linear structure. The linear aspect of this technique introduces an ordering constraint, which potentially has *Turing Machine* and/or *Turing complete* ramifications.

What Problems Do LPGs Solve? LPG solves problems that are sequential. This is particular useful for optimizing programs and low level assembly style output. Or any problem-domain where the problem being explored is about *sequential ordering*.

16.4.7 Cartesian Genetic Programming, CGP

Rather than linear or tree based, *Cartesian Genetic Programming* CGP [51] is based on Cartesian co-ordinates and directed-graphs. One basic characteristic is that the population is small (e.g. population size around five). The small population goes through a large number of generations. CGP is uniquely qualified to handle specific problems extremely well. EAs themselves can be temporal by default. CGP introduces the concept of spatial awareness into EAs.

What Problems Do CGPs Solve? CGP has been shown to be useful at circuit layout design since the logic components require some form of spatial awareness. Interestingly since CGP is spatial it can also be used to produce artistic designs/patterns. Recent research shows that CGP can achieve competitive results on the Atari benchmark set [80]. CGP can also encode ANNs by adding weights to the links in the graph, allowing them to do neuroevolution (see Sect. 16.5.2).

16.4.8 Differential Evolution, DE

Differential Evolution, DE [48, 66] is an example of a non-biologically inspired algorithm but falls under the metaheuristic category. It is based on iterating a population towards a quality goal. The iteration involves recombination, evaluation and selection. It avoids the need for gradient descent. A new candidate is based on a weighted difference between random candidates to create a third candidate, shifting the population to a higher quality level. Each new population effectively self-organizes.

What Problems Do DEs Solves? Works best on Boolean, Integer spaces and Reals. DE was developed specifically to find the Chebyshev polynomial coefficients and the optimization of digital filter coefficients.

16.4.9   Gene Expression Programming, GEP

Gene Expression Programming, GEP [27, 28] is a subclass of both GA (Sect. 16.4.2) and GP (Sect. 16.4.3). This method borrows from both techniques, as-in it uses fixed length strings which encode expression trees. The expression trees can be of varied size. Evolution occurs on the simple linear, fixed length strings.

What Problems Do GEPs Solve? It offers a powerful linear encoding which is guaranteed to produce valid programs, since GEP follows the syntactic rules of the specific programming language being targeted. This makes it easy to implement powerful genetic operators.

16.5   Specialized Techniques and Concepts

In this section we cover some of the more exotic EAs and extended tools. These EAs are new, hybrids or just miscellaneous concepts. This is not an exhaustive list but a more holistic subset of ideas that do not follow the traditional evolutionary methods. A few of the techniques covered in this section are not technically based on biological or synthetic evolution but play an important role in the process or are placed here due to taxonomy convenience.

Figure 16.5 shows the relationships between the specialized techniques and concepts. These relationships are more tenuous than the relationships found between the various traditional EA techniques.

16.5.1   Auto-Constructive Evolution

Auto-constructive Evolution [69] is where instead of having an overarching algorithm orchestrating the artificial evolution process, entities themselves are given the ability to undergo evolution. This means that children are constructed by their own parents. Parents have an ability to produce children, without the need of a master synthetic algorithm. This is in contrast to the more traditional EAs where the artificial replication occurs at a higher level.

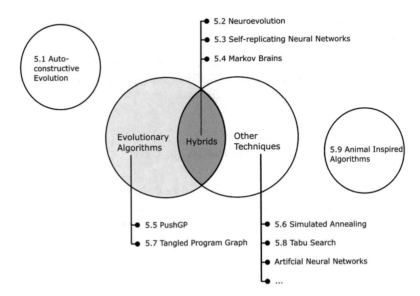

Fig. 16.5 Relationships between specialized techniques and concepts

What Problems Does Auto-Constructive Evolution Solve? Provides a method to carry out micro-evolution. This type of evolution is more inline with the goal of building *Artificial Life*. Potentially has an advantage to scale since it requires no centralized coordination.

16.5.2   *Neuroevolution, or Deep Neuroevolution*

Neuroevolution [7, 61, 62, 73] is classed as a hybrid solution. It has become more popular in recent years. It was a noticeably hot topic in GECCO 2018 [8]. Neuroevolution is the concept of using some form of GA to discover the optimal way to setup a Deep Neural Network (DNN). The entities being optimized are artificial neural networks (ANNs). Or, it can be used in combination with supervised learning and reinforcement learning (RL) techniques. The family of neuroevolution algorithms can be further classified based on how the candidate solutions are encoded and how much they can vary. As mentioned previously, CGP (Sect. 16.4.7) can also be an effective way to evolve ANNs by adding weights to the links.

- **Direct encoding**: parameters of every artificial neuron and connection are part of the solution encoding.
- **Indirect encoding**: is a "recipe" for generating ANNs. The topology can be either fixed or evolving. Fixed means that only the connection weights are optimized, whereas evolving means both the connection weights and the

topology of the ANN are modified. This latter class of algorithms is commonly called Topology and Weight Evolving Artificial Neural Network algorithms (TWEANNs).

Notable examples of TWEANNs are Neuroevolution of Augmenting Topologies (NEAT) [72] and its successor HyperNEAT [71]. The former uses a direct encoding while the latter uses an indirect encoding called "Compositional Pattern Producing Networks" (CPPN) [70]. Another example of a TWEANN approach using an indirect encoding is "Evolutionary Acquisition of Neural Topologies" (EANT2) [65].

What Problems Does Neuroevolution Solve? Neuroevolution combines ideas from Genetic Algorithms and Artificial Neural Networks. It can evolve both ANN weights and topologies making it an attractive alternative to ML hyper-parameter hand-crafting. By contrast with backpropagation, neuroevolution is not limited to differentiable domains [73].

16.5.3   Self-Replicating Neural Networks

Self-replicating Neural Networks [20] is a relatively new idea where the ANNs themselves re-configure. Currently the idea is to have the network learn by producing their own weights as output also to have *regeneration*. Regeneration is the concept of training an ANN by inserting predictions of its own parameters. This technique is still being researched and is at a very early stage. Vigorous exploration is still required but this idea has potential to be become more important. Expect to see forward progress in this area.

What Problems Do Self-Replicating Neural Networks Solve? Self-replication is still a new idea. Early research shows some promise in the area of continual ANN improvement using natural selection.

16.5.4   Markov Brains

Markov Brains [35] is in an early stage. Markov Brains belong to the same hybrid group as neuroevolution. Based on ANNs with some significant differences. Normal ANNs are designed with layers built-up from nodes with the same functional characteristic. Markov Brains are networks built from nodes with different computational characteristics. The components interact with each other, and can connect with external sensors (physical inputs) and actuators (physical outputs).

What Problems Do Markov Brains Solve? This is still relatively early days for Markov Brains but they are showing some early promise especially in unsupervised learning. By being a more flexible substrate than ANNs, they could also lead to a more general understanding of the role recurrence plays in learning. Looking forward to see the next papers on this subject.

16.5.5   PushGP

PushGP [68] is a family of programming languages which have been specifically designed for evolution to be applied-to i.e. an evolution target. It is based on a stack execution model. Each datatype has a separate stack. Code is treated as a manipulated datatype. It has been subjected to continuous research over a number of years and so there are many iterations and implementations. These variations include ones that allow for auto-constructive evolution (see Sect. 16.5.1).

What Problems Does PushGP Solve?   PushGP is designed to be an evolutionary target language rather than forcing a standard programming language to evolve. In other words, instead of using an existing programming language which is not evolutionary friendly, PushGP goes the other way by making it evolutionary friendly.

16.5.6   Simulated Annealing

Simulated Annealing [66] is not an evolutionary algorithm in itself but is frequently used in-conjunction with EAs. It is where a problems starts being exploratory and as it gets nearer to a possible solution moves more to being exploitative. Meaning that more of a stochastic approach is adopted at the beginning and as a good enough solution becomes nearer a more combinational approach is adopted (see Sect. 16.3.9).

What Problems Does Simulated Annealing Solve?   Simulated Annealing is most useful when the problem-domain requires more of an exploratory approach at the beginning but as the solution becomes more in view a more exploitative approach is adopted.

16.5.7   Tangled Program Graph, TPG

Tangled Program Graph (TPG) [40] is another relatively new technique. A method of managing programs to scale. The scale is used to handle more complicated tasks such as game playing. Provides a method to manage continuous evolution of independent and co-evolved populations. Shown to have some promising results when compared with equivalent deep learning methods. And requires significantly lower computation requirement.

What Problems Does TPG Solve? TGP is both efficient and proven to handle complex dynamic problems such as the traditional game playing benchmarks. This is still an early research area and as such should be monitored.

16.5.8   Tabu Search

Tabu search [17] is similar to Simulated Annealing (see Sect. 16.5.6) in that it is often used in conjunction with EAs. Tabu search relaxes the local search method by allowing a not-so-good solution to progress-forward over solutions which have already been visited.

What Problems Does Tabu Search Solve? Tabu search solves the problem of getting off a local maxima by placing solutions which have already been visited onto a tabu list. The list is used as a method to avoid going down known previously explored search paths.

16.5.9   Animal Inspired Algorithms

The *Animal Inspired Algorithms* [66] are biology inspired algorithms. Never-the-less they deserve a reference within the context of EAs. There are a surprisingly large number of animal inspired algorithms. The more famous are swarm, ant and frog algorithms but the actual list is considerably longer. Each animal inspired algorithm provides some unique quality like flying, jumping or walking. They are important algorithms; swarm algorithm, for instance, can be used to control a collection of drones. Ant algorithms can be used to explore an unknown terrain for searching useful resources.

What Problems Do Animal Inspired Algorithms Solve? This is a broad group of specialized algorithms which solve very specific problems.

16.6   Problem-Domain Mapping

The most important part of any algorithm is what can it accomplish. In this section we attempt to map specific problem-domains to potential techniques. We must stress that this is not exhaustive and may change dramatically between reviews. It will act as an important baseline for any future reports.

16.6.1   Specific Problem-Domain Mappings

Here we map the general problem-domains and the specific technique or techniques. These problem-domains are traditional problems found in industry.

16.6.1.1   Variable and Parameter Optimization

Parameter and variable optimization is a process of finding the best set of parameter values for a problem. The problems in themselves are complicated or the number of parameters is extremely large. If neither is true then a more traditional method of optimization may be a better route to a solution.

ES or GA are the normal solutions for this type of problem-domain. It has been shown that a GA can handle up to a million variables, as discussed in the Tutorial on *Next Generation Genetic Algorithms* GECCO 2018 [8, 78]. This makes the search space difficult or impossible for traditional methods and the only remaining true competitor is a pure stochastic method. See sections

- Section 16.4.1: Evolutionary Strategy, ES
- Section 16.4.2: Genetic Algorithms, GA

16.6.1.2   Symbolic and Polynomial Regression

This is the problem when given a data-set finding the equivalent mathematical equation. This is a particularly popular and important activity in many industries. The requirement is to find a matching equation for the data. EAs are adopted when traditional regression methods fail. The automotive industry are actively involved in this area since they frequently need to confirm theoretical equations with practical data. The techniques help find the equation that matches the real data independent of theory.

GP and all subclasses can handle symbolic and polynomial regression.
See sections

- Section 16.4.3: Genetic Programming, GP
- Section 16.4.6: Linear Genetic Programming, LGP
- Section 16.4.7: Cartesian Genetic Programming, CGP
- Section 16.4.5: Grammatical Evolution, GE
- Section 16.5.5: PushGP

16.6.1.3   Automated Code Production

The goal is to produce new code without human involvement. Once the programming language, representation and goal have all been chosen, EAs can explore the problem-domain in search of an optimal solution. If a final code solution is found then it will fall into one of three criteria i.e. precise/accurate, "good enough" or a multi-objective compromise (along the pareto curve).

Historically, EAs have mostly targeted programming languages such as LISP and low-level assembly language. Both have strict input-output formats which can simplify mutation and the joining of code segments. This avoids introducing syntax errors due to interface inconsistencies. Today the popular programming languages are Python (as a replacement for LISP) and Intermediate Representation (as a replacement for assembler instructions). Both offer new opportunities and challenges.

GP, LGP, GE, CGP, and PushGP are all techniques that produce code. Due to the fact that the code is automatically generated it is likely that the end result is difficult or unreadable by humans.

See sections

- Section 16.4.3: Genetic Programming, GP
- Section 16.4.6: Linear Genetic Programming, LGP
- Section 16.4.5: Grammatical Evolution, GE
- Section 16.4.7: Cartesian Genetic Programming, CGP
- Section 16.5.5: PushGP

16.6.1.4   Regular Expression

Automated generation of *regular expression*. This can be achieved using Genetic Programming [30]. Where EAs are used to explore expression strings.

See section

- Section 16.4.3: Genetic Programming

16.6.1.5   Circuit Design

Circuit design is similar to low-level assembly instructions (see Sect. 16.6.1.3) in that it has been successfully explored using EAs. The rules for circuit design is relatively simple and so EAs can explore the problem space relatively easily. This area was explored intensely in the 1990s but has potential to see a revival as system complexity increases.

CGP is particularly good at exploring circuit design since it handles spatial problems using Cartesian co-ordinates.

See sections

- Section 16.4.3: Genetic Programming, GP
- Section 16.4.7: Cartesian Genetic Programming, CGP

16.6.1.6 Code Improvement and Optimization

This is an up and coming area. The evolutionary process starts from a working code base, as compared with an initial random seed population. The existing working code is optimized towards a set of new objectives or is transitioned to fit within a new set of constraints. The new objectives could include specific performance features or any other similar attributes. The constraints could include new code size restrictions or power consumption limitations. The working code is basically used as a seed for the first initial population. Standard evolutionary operators are then applied to search the problem space for a potentially better solution.

As a subtopic, *legacy code improvement* is about taking older existing code and finding a better alternative. This better alternative can either be a metric improvement (e.g. faster, smaller code) or higher quality (hidden or existing anomalies are removed).

EAs in this problem-domain act as extra engineers on the project, where they might or might not produce a better answer from the original. Similar to many parts of engineering, this technique relies heavily on the quality of the goal and the associated test suites.

GP, GI and GE are all referenced to handle code improvement and optimization. See sections

- Section 16.4.3: Genetic Programming, GP
- Section 16.4.4: Genetic Improvement, GI
- Section 16.4.5: Grammatical Evolution, GE

16.6.1.7 Simulator Testing

Simulation testing is an indirect/byproduct of using EAs. It turns out EAs can be extremely good at finding simulator inconsistencies. This is because EAs explore the simulator in a different way than an engineer or scientist would explore a simulator. In fact, there are a number of historical examples where EAs are annoyingly good at discovering faults.

All the EAs are capable of pushing the limits of a simulator.

16.6.1.8 Walking Robot

Robots learning to walk has been a traditionally hard problem. EAs have been involved in learning how to walk for decades. EAs have two attributes which make them particularly useful in handling these types of problems. First, EAs can learn how to improve with each evolutionary cycle (incremental improvements) and second they can adapt to changes in the environment. For example, if a physical component changes or malfunctions EAs can adapt to that change and can continue walking.

GAs are used extensively in Robot Walking Algorithms, for both soft and hard robots.

See section

- Section 16.4.2: Genetic Algorithms, GA

16.6.1.9   Automated Machine Learning

EAs are starting to be used in the new subject of Automated Machine Learning (or more commonly known as AutoML) [61, 62]. AutoML is the automation of machine learning to real problem-domains. The goal is to avoid the labor intensive configuration required to setup a *Deep Neural Network* (DNN). This method also potentially bypasses the requirement for domain experts. Google have shown that AutoML can successfully improve existing ML systems. Google has also shown how Evolutionary AutoML can be used to improve Image Classifiers that were originally designed by humans.

This is becoming an increasingly popular subject, especially as DNNs become more complicated and larger. DNNs are more and more being adopted to solve interesting, real problems but the complexity of setup is causing a slow down in application. The quest today is applying and configuring DNN to more problems quicker and easier. EAs are increasingly being used in this area to discover novel solutions which are more efficient.

In research, GAs are an alternative way to configure a DNN. The GAs provide some form of novelty. For example, in a Google Brain paper [62] *aging* the weaker entities from the population was introduced to DNN configuration. They found that GAs achieved better results, as compared with other methods, when there is limited hardware resources.

One of the more famous public tools in EAs to carry this is out is TPOT [57], source code available on GitHub. It is designed to be a Data Science Assistant, written in Python. The goal is to optimize machine learning pipelines using GP.

See sections

- Section 16.4.2: Genetic Algorithms, GA
- Section 16.4.3: Genetic Programming, GP
- Section 16.5.2: Neuroevolution, or Deep Neuroevolution

16.6.2   *Unusual and Interesting Problem-Domain Mappings*

Here we highlight some of the more unusual problem-domain mappings that have appeared in recent articles. It is expected that these mappings will change the most between reviews.

16.6.2.1   Configuring Neuromorphic Computers

Configuration of a *Neuromorphic Computer* [18] is a more unusual problem. Current techniques map Convolutional Neural Networks (CNNs) to Spiking Neural Networks (SNN). These techniques avoid the dynamic nature and complexity of SNNs. A suggestion is to use an EA to perform these mappings. EAs are used to design simple ANNs to configure the platform. This technique allows the full utilization of a complex SNN to create small networks to solve specific problems. EAs can explore the entire parameter space of the specific hardware.

16.6.2.2   Forecasting Financial Markets

GA are used by institutional quantitative traders and other areas of the financial world [44]. Traders use software packages to set parameters that are optimized using both historical data and a GA. Depending upon the problem, the optimization can vary from which parameters are being used and the associated values to only optimizing the values. Trading comes with some risk but identifying the right parameters that relate to major market turns can be critical.

16.6.2.3   Predicting Future City Landscapes

The Spanish Foundation For Science and Technology [5] have used EAs to predicts the upward growth of cities. They discovered that increases in build height follows similar development as some living systems. A GA takes historical and economic data and uses it to predict the skyline of the future. The GA predicts how the skyscrapers and other buildings increase in height.

16.6.2.4   Designing an Optimized Floor Plan

Using EAs to design internal building floor plans. Floor plan can be complex due to building irregularities. A GA has been applied to optimize complex floor-plans [75]. The GA successfully designed office floor plans that optimized walk times and hallways.

16.6.2.5   Antenna Design

Antennas are complicated and are mostly designed by hand. This is both time-consuming and requires many resources. EAs have been "used to search the design space and automatically find novel antenna designs". In the paper entitled *Automated Antenna Design with Evolutionary Algorithm* [36], a group of NASA scientist successfully achieve designing an antenna using digital evolution. The antenna turned out to be efficient for a variety of applications. It was unique since the final design would not have been created by a human.

16.6.2.6   Defect Identification of Electron Microscopy Images

The US Department of Energy has been using a system called *Multinode Evolutionary Neural Networks for Deep Learning* (MENNDL) [3] to identify defects in electron microscopy images. MENNDL uses NNs to find defects out of changing data. The system runs on the Oakridge Summit supercomputer [6]. Fully utilizing all the available compute-nodes i.e. 18,000 GPUs on 3000 nodes. It analyzes millions of networks using a "scalable, parallel, asynchronous genetic algorithm augmented with a support vector machine to automatically find a superior deep learning network topology and hyper-parameter set" [3]. The scale of this system makes this an impressive hybrid implementation.

16.7   Challenges

Challenges include some personal opinions from the experience we have had navigating the subject of EAs. The points laid out below are opinions so should be debated and discussed. They are not end points.

- It is our observation that the community is relatively small compared with other Machine Learning communities. The size of the community determines the level of vigor that can be applied to validate a new idea or concept. In other words, there is not enough experts in the subject to vigorously prove or disprove a concept. Many ideas, even extremely clever and good ones, go unverified and unchecked by the community. This is a problem since good ideas can go missing due to a lack of support.
- EAs have an inherent difficulty proving they are the best solution to a specific "real-world" problem. There is no automatic methods to compare algorithms that is without any bias. In other words how do we prove, without doubt, that an EA performs better at achieving a time-to-solution than a random search. This has been a consistent issue since it is extremely difficult to prove the results from an EA experiment that is obviously without bias. What compounds the difficulty is that the problem-domains targeted are in themselves inherently complex, that is why an EAs is being used in the first place.
- Recently a lot of work has gone into creating synthetic problem benchmarks but there is concern less work has been applied to "real world" problems with "real world" constraints. Where bench marking and consistency is undoubtedly important, especially when comparing techniques, the most important activity is always applying algorithms to real problems.

- The community is an old community within Machine Learning, with ideas dating back to the early 1950s but many of the original drawbacks of using evolutionary techniques have been removed since modern hardware is both abundant and high performing. This means experiments which were constrained by the hardware resources at the time can now be feasible. Population size and maximum number of generations can be made considerably larger.
- Biology plays important an role but EAs are not organisms. This makes crossing terms between biology and Computer Science difficult. Computer Scientists will use the biological terms loosely for their own purposes whereas the real biological meaning is much more complicated.
- EAs have been proven good at tackling some hard problems but they suffer from a difficulty-of-scale. In complex systems it is important to divide-and-conquer (break the problem down into smaller elements) before attempting to produce a solution. EAs at the bottom level make a lot of sense but they become more problematic as we scale-up the problem. Research work at bigger scale problems is still immature and is also limited, for the most part, with the capabilities of the current hardware available.
- Similar to many other AI disciplines, there is a constant struggle between *make* verses *buy*. There is a tendency for Researchers to re-invent the technology wheel, this is in part due to the required learning curve to attain usefulness. The amount of effort to learn a new framework can be as challenging and time consuming as creating a propriety framework from first principles. This is detrimental to the discipline, as a whole, since the re-invention slows down forward progress. The caveat is that over time frameworks become overly specific to a given problem domain, so applying the same framework to a different problem can be cumbersome.
- Modularity is a method to handle more complex problems by breaking the problem into more manageable components. This falls under the divide-and-conquer strategy or scientific method. EAs solve problems using a bottom-up design and as-such are inherently more difficult to modularize.
- EAs may be deterministic and provable, the process by how the solution was arrived at is non-determinant and if the algorithm is run again there is no guarantee the same result will be found or any result will be found. This is the opposite of some other techniques in Machine Learning.
- EAs are provable. Proof and explainability is becoming increasingly more important. Governments are also stepping in with new regulations, for example the *General Data Protection Regulation* (GDPR) [4]. GDPR is a European regulation which has direct implications on Machine Learning algorithms in general. In Article 22 [2], of the regulation, calls for algorithmic fairness and explainability. In other words, explain how the algorithm is correct, fair and unbiased.
- Being an old subject EAs inherently suffers from the reinvention and rediscovery of already known concepts. Keeping everyone current with what has been published is always a challenge, especially as the amount of scientific information accelerates.

- Even-though EAs could potentially be the next big direction for Machine Learning, the general low funding of the subject may hold back development. This is concerning since EAs are not as well-known as other Machine Learning techniques. This situation makes obtaining core funding for EA related research difficult and very much a secondary focus. The outcome is that only a few full-time researchers worldwide focus solely on these techniques. Also, the broad interdisciplinary knowledge base required is difficult to attain.

16.8   Predictions

In this section we look into the future. What may happen in the next few years with EAs. Again since these are predictions they should be treated with more questions.

- It is likely that in the near future we will see more cross-pollination and collaboration between Machine Learning researchers and molecular biologists, neuroscientists and evolutionary biologists. People are held back by their special-izations and generally dangerous in other areas. The philosopher Paul Feyerabend argued that the most progress is made on the boundaries between subjects [14]. It is at the boundary of biology and computer science where most advancements are likely to occur.
- Pure research on DNNs will slow down and the DNN focus will shift to engi-neering and the application side for the near-term. There is a strong likelihood that Machine Learning research will diversify and be more challenging. Research problems will become hybrid, involving the merging of many techniques. A potential end goal for hybrid systems is Artificial General Intelligence (AGI).
- *Artificial General Intelligence* is a philosophical goal, or concern depending upon whom you ask. This will potentially take decades and many stages before is can be reached, if at all. One stage is the much smaller attainable goal of producing what we call *Domain Specific Artificial Life* (DSAL). Bringing together many disciplines in the desire to create solutions to a specific problem. The term is a fun play on Domain Specific Architectures as advocated by some of the hardware architecture community: small artificial lifeforms to solve specific problems.
- New Artificial Neurons (AN) will be explored and developed. There is potential for the ANs themselves to be re-examined either by vastly increasing the number of interconnects or adding interesting attributes like co-ordinates to the model. Today's ANs have limited interconnects, whereas the biological neurons have substantially more interconnects. Then evolving these models as required. These are ideas that people like Julian Miller have put forward. Whatever future direction is taken, the AN model will most likely change over the next few years. In all likelihood we will end up with many AN models to choose from. This change will occur as our understanding of *Neuroscience* and biological mechanisms increases.

- As James Shapiro points out, there are many genetic mechanisms that could be incorporated into existing and new EAs [64]. One such concept is Horizontal Gene Transfer (HGT) [39]. HGT becomes important to the community, as more advanced complex systems are attempted. HGT is the ability for useful genes (or code segments) to quickly transfer across species (islands). There are Computer Science implications for such ideas. As with Richard Feynman's *"There's plenty of room at the bottom"*, when referring to molecular chemistry, there is plenty of room with evolutionary biology and its application to EAs.
- EAs can potentially be used to explore *causality*. Judea Pearl [58] has given the industry a challenge to explain Machine Learning outcomes and identify the causes of the outcomes. In particular ideas like *counterfactual*, where forward predictions about the future can be made by inserting a change or deciding not to insert a change. This involves not just providing data correlation but creating a model on how the data was created.
- Obfuscation may allow EAs to get involved in security, privacy and data cloaking.
- EAs are already being used in the field of *Quantum Computing* (QC), and we can expect more activity in this area. Either to configure or to handle the complicated data output. 2017 saw the Humies Award [16] go to an Australian Team [33] using EAs applied to Quantum Computing.

16.9   Final Discussion and Conclusion

The 2019 Evolutionary Algorithms Review is a baseline for future reports. We are cognizant that there are many areas which were not covered or were only covered briefly. These areas may become more important as we consider the 2020 review. Traditional EAs continue to provide useful solutions to hard problems. They are mature algorithms. They are becoming particularly useful when placed on modern hardware. The newer trends appear to indicate that hybrid solution may provide the next future capability i.e. combining evolutionary techniques with other techniques. Where EAs are used either to bring knowledge-forwarding or optimizing complex Machine Learning configurations.

We dedicated most of the review to the landscaping of the various techniques. This was planned to act as a baseline for future review development. For problem-domains that affect society, and subsequently industry, we introduced UCA (User Control Attributes) criteria. There are five defined attributes i.e. *limiters*, *explainability*, *causality*, *fairness* and *correction*. They impose an extra level of required thinking, where the algorithms have to be both community friendly and adhere to new government regulations. The UCA will be used as a basis for a new taxonomy for EAs. Future algorithms will not only be rated on their ability to produce an outcome but on the ability to satisfy the UCA criteria in some form or other. More generally this taxonomy could be applied to other forms of Machine Learning.

Current thoughts on applying the UCA to EAs are as follows:

- **Limiters** is the concept of restricting the capability. It is still early days but as EAs handle more problems that either affect the physical environment (i.e. control actuators), or are involved in some privacy aspect, then methods of limiting the capability or cloaking the outcomes become more important. This may require significant external technology and thought.
- **Explainability** revolves around explaining how an algorithm produces an answer. The output solution from EAs are inherently provable since they solve a specific problem but as with all algorithms of this class, explaining how the solution came about is problematic and difficult to repeat.
- **Causality** is about moving to a higher order answer which adds the "Why" component to the answer being searched. The challenge has been set and causality will become more important as we move forward with implementing real world EAs, driven by the desire for the EAs to do more and to understand why a conclusion has been reached.
- **Fairness** is about producing an answer which is balanced and without human prejudice. This may come down to the actual input data selected and the fitness function being used. Algorithmic fairness [79] has to be capable of detecting biases, contesting a decision and finally instigating a potential remedy. A critical method of achieving fairness is making sure that the outcome is vigorously tested. There is potential for a mathematical element of fairness to be incorporated into the evolutionary frameworks. Whereas the society aspects will remain in the human domain for the foreseeable future and may well require careful deliberate biasing to produce a fair outcome. Two features related to fairness, not covered in this review, are the broader subjects of accountability and transparency. A method of algorithmic fairness can be complicated. This is an area that requires future monitoring. Fairness may also involve studying the initial starting states, checking that the biases are not injected right at the start of the process.
- **Correction** is about correcting a problem once it has been identified. When an error is identified/detected/contested, EAs can be assessed on how easily a remedy can be applied. EAs by definition are adaptive, and so correction can occur as part of a continuous evolutionary process (i.e. via environmental-changes) or more manually through a direct change to the fitness function and/or constraints.

Now that we have described the basics of the UCA, we will attempt in the 2020 review to apply them to the various EAs outlined in this review. This provides the new taxonomy.

There is a long way to go with-respect-to EAs. We are just forming a common language by which we can communicate with the biologists. This is an exciting time since the discoveries in biochemistry and synthetic biology are occurring at unprecedented rates and the capabilities of digital hardware are coming to levels that can mimic parts of the biological system. Likewise, biology can also take advantage of some of the newly gained insights from the Computer Science field.

This means that the research runway is long but at the same time we have to realize that the hardware gap in both "effective" processing and interconnects between biology and digital systems is still vast. We may have the vision to achieve biological equivalence but the current state of the hardware is both different and non-optimal for many types of problems. This is particularly interesting as we hit the End of Moore's Law and the possibility of different compute-models being introduced and forced on the industry.

16.10   Feedback

If you have any feedback or suggestions on this review please send them to andrew.sloss@arm.com with the subject line of "2019 Review Feedback" and we will consider the enhancements.

Acknowledgements We would like to acknowledge the following people for their encouragement and feedback during the writing of this review, namely Mbou Eyole, Casey Axe, Paul Gleichauf, Gary Carpenter, Andy Loats, Rene De Jong, Charlotte Christopherson, Leonard Mosescu, Vasileios Laganakos, Julian Miller, David Ha, Bill Worzel, William B. Langdon, Daniel Simon, Emre Ozer, Arthur Kordon, Hannah Peeler and Stuart W. Card.

References

1. Genetic Drift. https://evolution.berkeley.edu/evolibrary/article/evo_24. Last accessed February 4 2019
2. Art. 22 GDPR Automated individual decision-making, including profiling. Intersoft Consulting (2018). https://gdpr-info.eu/art-22-gdpr/. Last accessed November 22, 2018
3. Deep learning for electron microscopy. US Department of Energy (2018). https://m.phys.org/news/2018-12-deep-electron-microscopy.html. Last accessed December 22 2018
4. General Data Protection Regulation GDPR. Intersoft Consulting (2018). https://gdpr-info.eu. Last accessed November 22, 2018
5. A genetic algorithm predicts the vertical growth of cities. Spain Foundation for Science and Technology (2018). https://www.eurekalert.org/pub_releases/2018-05/f-sf-aga052518.php. Last accessed November 17 2018
6. ORNL Launches Summit Supercomputer. US Department of Energy (2018). https://www.ornl.gov/news/ornl-launches-summit-supercomputer. Last accessed April 29 2019
7. Welcoming the Era of Deep Neuroevolution. Uber Engineering (2018). https://eng.uber.com/deep-neuroevolution/. Last accessed December 28 2018
8. Aguirre, H. (ed.): GECCO '18: Proceedings of the Genetic and Evolutionary Computation Conference Companion. ACM, New York, NY, USA (2018)
9. Arnold, J.: Genetic Drift (2001). https://www.sciencedirect.com/topics/neuroscience/genetic-drift. Last accessed April 29 2019
10. Backus, J.W.: The syntax and semantics of the proposed international algebraic language of the Zurich ACM-GAMM Conference. In: IFIP Congress (1959)
11. Bailey, B.: The Impact Of Moore's Law Ending (2018). https://cacm.acm.org/news/232532-the-impact-of-moores-law-ending/fulltext. Last accessed April 29 2019

12. Beyer, H.G., Sendhoff, B.: Covariance matrix adaptation revisited—the CMSA evolution strategy (2008). https://www.researchgate.net/publication/220701715_Covariance_Matrix_Adaptation_Revisited_-_The_CMSA_Evolution_Strategy_-. Last accessed February 6 2019
13. Brameier, M., Banzhaf, W.: Linear Genetic Programming. No. XVI in Genetic and Evolutionary Computation. Springer (2007). URL http://www.springer.com/west/home/default?SGWID=4-40356-22-173660820-0
14. Broad, W.J.: Paul Feyerabend: Science and the Anarchist (2018). https://www.jstor.org/stable/1749231?seq=1#page_scan_tab_contents. Last accessed April 4 2019
15. Brownlee, J.: Overfitting and underfitting with machine learning algorithms (2016). https://machinelearningmastery.com/overfitting-and-underfitting-with-machine-learning-algorithms/. Last accessed April 29 2019
16. Brownlee, J.: Annual "humies" awards for human-competitive results (2018). http://www.human-competitive.org. Last accessed November 12 2018
17. Brownlee, J.: Tabu search (2018). http://www.cleveralgorithms.com/nature-inspired/stochastic/tabu_search.html. Last accessed January 10 2019
18. Buckley, S., McCaughan, A., Chiles, J., P. Mirin, R., Woo Nam, S., Shainline, J., Bruer, G., Plank, J., Schuman, C.: Design of superconducting optoelectronic networks for neuromorphic computing. In: 2018 IEEE International Conference on Rebooting Computing (ICRC), pp. 1–7 (2018)
19. Burkhardt, R.W.: Lamarck, evolution, and the inheritance of acquired characters. Genetics **194** **4**, 793–805 (2013)
20. Chang, O., Lipson, H.: Neural network quine. CoRR **abs/1803.05859** (2018). URL http://arxiv.org/abs/1803.05859
21. Cossins, D.: Discriminating algorithms: 5 times ai showed prejudice (2018). https://www.newscientist.com/article/2166207-discriminating-algorithms-5-times-ai-showed-prejudice/. Last accessed April 26 2019
22. Dahad, N.: Imec, ASML Team on Post-3nm Lithography (2018). https://www.eetimes.com/document.asp?doc_id=1333896. Last accessed January 7 2019
23. Darwin, C.: On the Origin of Species by Means of Natural Selection. Murray, London (1859)
24. Dennard, R.H., Gaensslen, F.H., Rideout, V.L., Bassous, E., LeBlanc, A.R.: Design of ion-implanted mosfet's with very small physical dimensions. IEEE Journal of Solid-State Circuits **9**(5), 256–268 (1974)
25. Diab, W.: About JTC 1/SC 42 Artificial intelligence (2018). https://jtc1info.org/jtc1-press-committee-info-about-jtc-1-sc-42/. Last accessed December 22 2018
26. Fenton, M., McDermott, J., Fagan, D., Forstenlechner, S., Hemberg, E., O'Neill, M.: PonyGE2: Grammatical Evolution in Python. In: Proceedings of the Genetic and Evolutionary Computation Conference Companion, GECCO '17, pp. 1194–1201. ACM, Berlin, Germany (2017)
27. Ferreira, C.: Gene expression programming: a new adaptive algorithm for solving problems. Complex Systems **13**(2), 87–129 (2001)
28. Ferreira, C.: Gene Expression Programming: Mathematical Modeling by an Artificial Intelligence. Springer (2006)
29. Furber, S.: SpiNNaker (2018). http://apt.cs.manchester.ac.uk/projects/SpiNNaker/project/. Last accessed November 11 2018
30. Gibbs, M.: Genetic programming meets regular expressions (2015). https://www.networkworld.com/article/2955126/software/genetic-programming-meets-regular-expressions.html. Last accessed January 2 2019
31. Goldberg, D.E.: Genetic Algorithms in Search, Optimization and Machine Learning, 1st edn. Addison-Wesley Longman Publishing Co., Inc., Boston, MA, USA (1989)
32. Hansen, N.: The cma evolution strategy: A tutorial. CoRR **abs/1604.00772** (2016). URL http://arxiv.org/abs/1604.00772
33. Harper, R., Chapman, R., Ferrie, C., Granade, C., Kueng, R., Naoumenko, D., T. Flammia, S., Peruzzo, A.: Explaining quantum correlations through evolution of causal models. Physical Review A **95** (2016)

34. Hinton, G.E., Osindero, S., Teh, Y.W.: A fast learning algorithm for deep belief nets. Neural Comput. **18**(7), 1527–1554 (2006)
35. Hintze, A., Edlund, J.A., Olson, R.S., Knoester, D.B., Schossau, J., Albantakis, L., Tehrani-Saleh, A., Kvam, P.D., Sheneman, L., Goldsby, H., Bohm, C., Adami, C.: Markov Brains: A Technical Introduction. ArXiv **abs/1709.05601** (2017)
36. Hornby, G., Globus, A., Linden, D., Lohn, J.: Automated antenna design with evolutionary algorithms. Collection of Technical Papers - Space 2006 Conference **1** (2006)
37. Ivan: Parallel and distributed genetic algorithms (2018). https://towardsdatascience.com/parallel-and-distributed-genetic-algorithms-1ed2e76866e3. Last accessed February 7 2019
38. Izzo, D., Ruciński, M., Biscani, F.: The generalized island model. In: Parallel Architectures and Bioinspired Algorithms, pp. 151–169. Springer (2012)
39. Jain, R., Rivera, M.C., Lake, J.A.: Horizontal gene transfer among genomes: The complexity hypothesis. Proceedings of the National Academy of Sciences **96**(7), 3801–3806 (1999)
40. Kelly, S., Heywood, M.I.: Emergent tangled graph representations for Atari game playing agents. In: M. Castelli, J. McDermott, L. Sekanina (eds.) EuroGP 2017: Proceedings of the 20th European Conference on Genetic Programming, *LNCS*, vol. 10196, pp. 64–79. Springer Verlag, Amsterdam (2017). https://doi.org/10.1007/978-3-319-55696-3_5. Best paper
41. Koza, J.R.: Genetic Programming: On the Programming of Computers by Means of Natural Selection. MIT Press, Cambridge, MA, USA (1992)
42. Koza, J.R.: Genetic Programming II: Automatic Discovery of Reusable Programs. MIT Press, Cambridge Massachusetts (1994)
43. Koza, J.R., Andre, D., Bennett III, F.H., Keane, M.: Genetic Programming III: Darwinian Invention and Problem Solving. Morgan Kaufman (1999)
44. Kuepper, J.: Using genetic algorithms to forecast financial market (2018). https://www.investopedia.com/articles/financial-theory/11/using-genetic-algorithms-forecast-financial-markets.asp. Last accessed November 17 2018
45. Langdon, W.B., McPhee, N.F.: A Field Guide to Genetic Programming. LuLu Selfpublishing (2018)
46. Leswing, K.: Jeff Bezos just perfectly summed up what you need to know about artificial intelligence (2018). https://www.businessinsider.com/jeff-bezos-shareholder-letter-on-ai-and-machine-learning-2017-4. Last accessed December 20 2018
47. LLVM.org: The LLVM Compiler Infrastructure Project. https://llvm.org/. Last accessed January 2 2019
48. Luke, S.: Essentials of Metaheuristics, first edn. lulu.com (2009). URL http://cs.gmu.edu/~sean/book/metaheuristics/. Available at http://cs.gmu.edu/~sean/books/metaheuristics/
49. Maass, W.: Networks of Spiking Neurons: The Third Generation of Neural Network Models. Neural Networks **10**, 1659–1671 (1996)
50. Maheswaranathan, N., Metz, L., Tucker, G., Sohl-Dickstein, J.: Guided evolutionary strategies: escaping the curse of dimensionality in random search. CoRR **abs/1806.10230** (2018). URL http://arxiv.org/abs/1806.10230
51. Miller, J.F.: Cartesian genetic programming. In: J.F. Miller (ed.) Cartesian Genetic Programming, Natural Computing Series, chap. 2, pp. 17–34. Springer (2011)
52. Morgan, T.J.H., Griffiths, T.L.: What the Baldwin Effect affects. In: CogSci (2015)
53. Mosescu, L.: Darwin neuroevolution framework (2018). Urlhttps://github.com/tlemo/darwin. Last accessed February 4 2019
54. N. Krasnogor S. Gustafson, D.A.P., Verdegay, J.L.: Systems Self-Assembly, Volume 5: Multidisciplinary Snapshots (Studies in Multidisciplinarity). Elsevier Science (2008)
55. Neumann, J.V.: Theory of Self-Reproducing Automata. University of Illinois Press, Champaign, IL, USA (1966)
56. Ochi, L., Vianna, D., Drummond, L., Victor, A.: A parallel evolutionary algorithm for the vehicle routing problem with heterogeneous fleet. Future Generation Computer Systems **14**, 285–292 (1998)

57. Olson, R.S., Urbanowicz, R.J., Andrews, P.C., Lavender, N.A., Kidd, L.C., Moore, J.H.: Proceedings of Evo Applications 2016, Porto, Portugal, March 30–April 1, 2016, Part I, chap. Automating Biomedical Data Science Through Tree-Based Pipeline Optimization, pp. 123–137. Springer (2016)

58. Pearl, J., Mackenzie, D.: The Book of Why: The New Science of Cause and Effect, 1st edn. Basic Books, Inc., New York, NY, USA (2018)

59. Petke, J., Haraldsson, S.O., Harman, M., Langdon, W.B., White, D.R., Woodward, J.R.: Genetic improvement of software: a comprehensive survey. IEEE Transactions on Evolutionary Computation **22**(3), 415–432 (2018)

60. Purohit, A., Choudhari, N.S.: Code bloat problem in genetic programming (2013). http://www.ijsrp.org/research-paper-0413/ijsrp-p1612.pdf. Last accessed May 26 2019

61. Real, E.: Using evolutionary AutoML to discover neural network architectures (2018). https://ai.googleblog.com/2018/03/using-evolutionary-automl-to-discover.html. Last accessed December 22 2018

62. Real, E., Aggarwal, A., Huang, Y., V Le, Q.: Regularized evolution for image classifier architecture search (2018). https://arxiv.org/abs/1802.01548. Last accessed December 28 2018

63. Russell, B.: The Value of Philosophy. In: S.M. Cahn (ed.) Exploring Philosophy: An Introductory Anthology. Oxford University Press (2009)

64. Shapiro, J.: A 21st century view of evolution: Genome system architecture, repetitive dna, and natural genetic engineering. Gene **345**, 91–100 (2005)

65. Siebel, N.T.: Evolutionary reinforcement learning. http://www.siebel-research.de/evolutionary_learning/. Last accessed January 2 2019

66. Simon, D.: Evolutionary Optimization Algorithms. Wiley (2013). URL https://books.google.com/books?id=gwUwIEPqk30C

67. Simonite, T.: Moore's Law Is Dead. Now What? (2016). https://www.technologyreview.com/s/601441/moores-law-is-dead-now-what/. Last accessed January 7 2019

68. Spector, L.: Push, PushGP and Pushpop (2018). http://faculty.hampshire.edu/lspector/push.html. Last accessed December 28 2018

69. Spector, L., McPhee, N.F., Helmuth, T., Casale, M.M., Oks, J.: Evolution evolves with autoconstruction. In: T. Friedrich, et al. (eds.) GECCO '16 Companion: Proceedings of the Companion Publication of the 2016 Annual Conference on Genetic and Evolutionary Computation, pp. 1349–1356. ACM, Denver, Colorado, USA (2016). https://doi.org/10.1145/2908961.2931727

70. Stanley, K.: Compositional pattern producing networks: A novel abstraction of development. Genetic Programming and Evolvable Machines **8**, 131–162 (2007)

71. Stanley, K.O., D'Ambrosio, D.B., Gauci, J.: A hypercube-based encoding for evolving large-scale neural networks. Artificial Life **15**(2), 185–212 (2009)

72. Stanley, K.O., Miikkulainen, R.: Evolving neural network through augmenting topologies. Evolutionary Computation **10**(2), 99–127 (2002)

73. Such, F.P., Madhavan, V., Conti, E., Lehman, J., Stanley, K.O., Clune, J.: Deep neuroevolution: Genetic algorithms are a competitive alternative for training deep neural networks for reinforcement learning. ArXiv **abs/1712.06567** (2017)

74. Sverdlik, Y.: Google is switching to a self-driving data center management system (2018). https://www.datacenterknowledge.com/google-alphabet/google-switching-self-driving-data-center-management-system. Last accessed January 8 2019

75. Tokmakova, A.: Optimizing floorplans via experimental algorithms (2018). https://archinect.com/news/article/150108746/optimizing-floorplans-via-experimental-algorithms. Last accessed December 22 2018

76. Turing, A.: The chemical basis of morphogenesis. Philosophical Transactions of the Royal Society B **237**, 37–72 (1952)

77. Wall, M.: Galib: Matthew's C++ genetic algorithms library (1996). http://lancet.mit.edu/galib-2.4/. Last accessed February 4 2019

78. Whitley, D., Chicano, F., Ochoa, G., Sutton, A., Tinós, R.: Next generation genetic algorithms (2018). http://gecco-2018.sigevo.org/index.html/tiki-index.php?page=Tutorials Last March 7 2019

79. Whittaker, M., Crawford, K., Dobbe, R., Fried, G., Kaziunas, E., Mathur, V., West, S.M., Richardson, R., Schultz, J.: AI Now 2018 Report (2018). https://ainowinstitute.org/AI_Now_2018_Report.pdf. Last accessed December 20 2018

80. Wilson, D.G., Cussat-Blanc, S., Luga, H., Miller, J.F.: Evolving simple programs for playing Atari games. In: H. Aguirre, et al. (eds.) GECCO '18: Proceedings of the Genetic and Evolutionary Computation Conference, pp. 229–236. ACM, Kyoto, Japan (2018)

81. Wolpert, D.H., Macready, W.G.: No free lunch theorems for optimization. IEEE Transactions on Evolutionary Computation **1**(1), 67–82 (1997)

82. Zaldivar, A.: Introduction to fairness in machine learning (2018). https://developers.googleblog.com/2018/11/introduction-to-fairness-in-machine.html. Last accessed January 8 2019

Chapter 17
Evolving a Dota 2 Hero Bot with a Probabilistic Shared Memory Model

Robert J. Smith and Malcolm I. Heywood

17.1   Introduction

High-dimensional reinforcement learning implies that feature construction as well as policy discovery be performed simultaneously. Recent advances in 'deep rein-forcement learning' approaches have demonstrated that this is possible over a cross-section of task domains defined in terms of computer games, e.g. (Atari) arcade game titles [27]. Previous work has shown that the genetic programming approach of tangled program graphs (TPG) is able to compete with deep reinforcement learning solutions on games of complete information [21]. However, when partial observability plays an increasing role in defining game content (as in first person shooter games), TPG experiences task specific limitations [32]. A recent work proposed an approach for introducing indexed memory into the TPG framework and demonstrated its applicability under VizDoom 'deathmatches' [34]. In this work, we assess the significance of the same approach under a completely different high-dimensional reinforcement learning task, that of evolving a hero to play Defence of the Ancients 2 (Dota 2),[1] a real-time strategy game that has been targeted by OpenAI for developing hero agents using reinforcement learning.[2] The details of the approach adopted by OpenAI are not publicly known, however, it does appear to be based on some form of reinforcement learning.[3]

---

[1] http://blog.dota2.com/?l=english.

[2] https://openai.com/blog/how-to-train-your-openai-five/.

[3] https://openai.com/blog/openai-baselines-ppo/.

R. J. Smith · M. I. Heywood (✉)
Faculty of Computer Science, Dalhousie University, Halifax, NS, Canada
e-mail: rsmith@cs.dal.ca; mheywood@cs.dal.ca

© Springer Nature Switzerland AG 2020
W. Banzhaf et al. (eds.), *Genetic Programming Theory and Practice XVII*, Genetic and Evolutionary Computation, https://doi.org/10.1007/978-3-030-39958-0_17

Our interest is not to compete directly with OpenAI,[4] but to use the challenging environment posed by the Dota 2 environment to validate the previous approach to addressing partial observability in high-dimensional reinforcement learning task. Specifically, Dota 2 state information is limited to line-of-sight, so a hero agent cannot 'see' around objects such as trees or buildings, and descending to the river feature implies that the agent cannot see above the river bank. In short, there are multiple instances of partial observability.

The particular approach taken to designing memory for enabling the TPG framework to operate under partially observable high-dimensional environments are summarized as follows [34]:

- **Memory Synchronization:** TPG incrementally stitches together teams of programs into an increasingly interconnected graph of teams-of-programs. Adding stateful register referencing (Sect. 17.3.2) provides each program with the ability to retain register state, potentially addressing partial observability from the perspective of individual programs, but not collectively. Thus, some form of scalable 'global' indexable memory needs to be assumed, but in such a way that all programs are encouraged to act collaboratively/respectfully, whether they are within the same TPG individual or not. We make this observation because we cannot predict which specific programs will ultimately be combined into each TPG agent. Indexed memory therefore also represents a common communication medium that programs post 'messages' to and receive 'messages' from.
- **What to Write:** The state space of Dota 2 consists of hundreds of attributes when playing a middle lane single opponent configuration, however, this increases to tens of thousands of attributes when playing three lanes with five heroes on each team. In short, even under the minimal setting, a write operation should not write all state information to memory, it will not scale. Instead, we will assume that a program's write operation will use the program's current register state as a surrogate for defining state as 'communicated' through indexed memory. Put another way, we are assuming that each program's register state represents a useful encoding (of state) for communicating to the other programs in the population.
- **Long versus Short term memory:** it is clear that the human mind supports many forms of memory, and supporting both memory that is accessible to all programs versus memory specific to programs represents one component to this. However, explicit support for short and long term memory could also be fruitful; how it is used is down to the evolutionary process to discover. Likewise, how the 'decision' is made regarding a write to/read from long or short term memory should also be an emergent property.

In the following, we develop the topic by first introducing the Dota 2 one-on-one mid-lane task (Sect. 17.2) and survey previous approaches to introducing indexed memory in neural networks and genetic programming (Sect. 17.3). The

---

[4]The computational resources used by OpenAI are in the order of 180 years of gameplay per day.

TPG approach to genetic programming is briefly summarized in Sect. 17.4 and the framework adopted for designing indexed memory to address the above points detailed in Sect. 17.5. Section 17.6 summarizes all the issues specific to interfacing TPG to the Dota 2 game engine (tools, state-space, actions, fitness function). The details of a benchmarking study appear in Sect. 17.7 and conclusions and future work appear in Sect. 17.8.

17.2 The Dota 2 1-on-1 Mid-lane Task

Game play in Dota 2 revolves around developing strategies for defining the behaviour of a 'hero' character. There are three basic categories of hero (strength, agility, intelligence) resulting in a total of 117 specific heroes.[5] We will not be addressing the hero selection issue in our work. Instead, we will assume a specific hero from the agile category—a Shadow Fiend—where the agile category provides heroes with the potential to perform well in a broad range of gaming scenarios [31]. Future research will consider whether we can evolve heroes from the other categories.

The Dota 2 environment is defined in terms of a map in 3-D isometric perspective, however, all the agents playing the game are subject to sensor information constrained by partial observability (line-of-sight/fog-of-war). Thus, from an agent's perspective it is not possible to see 'through' a forest, or above a flight of stairs.

There are two teams, 'The Radiant' represents our team (base at the bottom LHS, Fig. 17.1) and 'The Dire' represent the opponent team (base at the top RHS, Fig. 17.1). The base of each team periodically issue waves of 'creeps' which are simple bots with default behaviours. The creeps advance from each base down each 'lane', where the number of active lanes is a game parameter (the more active lanes, the more difficult the game).

In this work, we are interested in the single mid-lane case, thus creeps will advance from each tower down the leading diagonal (Fig. 17.1). Creeps have predefined behaviours that the hero can potentially manipulate. However, on encountering any form of opponent, they will attack. The hero character for each team attempts to use their powers to develop a strategy to enable them to advance specific aspects of the game to their advantage. Examples might be to support the creep attack or collecting powerup and bounty runes from the environment (increasing team wealth and therefore supports the collection of items which provide additional powers). A game ends when one side either kills the opponent hero twice, or a tower is taken.

The goal we set in this work is to determine whether we can evolve behaviour for the Shadow Fiend hero against the built in Shadow Fiend hero bot under the 'mid-lane' parameterization. We will assume that the opponent hero bot operates

---

[5] https://dota2.gamepedia.com/Heroes.

Fig. 17.1 Initial Dota 2
visual game space. No agents
are shown. This view would
constitute complete
information, which bots do
not possess during game play.
Each team has a tower
(highlighted in red). Should a
team loose their tower the
game is over

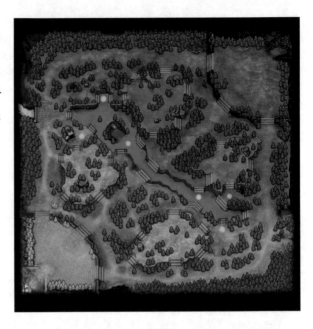

under the hard as opposed to easy or medium level of difficulty. We will then test
the resulting evolved Shadow Fiend hero against all three settings for the opponent
hero bot. This is not a straightforward goal, as we are potentially subject to several
pathologies: (1) fitness disengagement—the opponent already has a challenging
strategy, implying that we are not able to establish a useful gradient for directing
evolution. (2) sparse fitness function—many skills need to be attained in order to
develop effective hero strategies, including navigation, defending, attacking and
collecting bounty. Some of these properties have no direct reward (i.e. navigation)
and others are only indirectly quantified in the fitness function (i.e. defending and
attacking). In short, establishing an effective strategy necessitates the development
of many abilities that are not characterized directly in the fitness function.

17.3   Related Work

17.3.1   Memory in Neural Networks

Letting neurones recycle previous (internal) state information produces a feedback
loop, thus supporting a distributed form of memory, or memory through recurrent
connectivity. However, there are many potential connectivity patterns that could re-
circulate state to provide recurrent properties, e.g. Hopfield network formulations
for content-addressable memory [2], Brain-state-in-a-box [13], Elman recurrent
networks [8]. This diversity of models appears in part due to the impact that different

connectivity schemes have on the properties of the network and the (possible) implications for stability under feedback. One of the most widely employed configurations for recurrent connectivity takes the form of long short-term memory (LSTM) [15].[6] Each LSTM uses a combination of gating elements to define a 'perfect integrator', effectively implementing the expression,

$$y(t + 1) = y(t) + g(t_s) \cdot x(t) \tag{17.1}$$

where t is a discrete time step, y is the LSTM output, x is the input to the LSTM and $g(t_s)$ is the value of a gating neurone used to modulate when the input x appears through the independent temporal index t_s. The problem invariably is that although the LSTM model has had many successes (see [11] for a short survey and on-going developments), the practicalities of getting the LSTM to reliably gate the relevant inputs at the right time has motivated the development of forms of indexed addressable memory for neural networks [9, 10].

In this latter case, the 'Neural Turing Machine' (NTM) represents the most widely assumed starting point [9]. The NTM augments a feedforward neural network with outputs for manipulating a 'bank' of (external) memory using read, write and erase operations. The innovation was to define these operations such that they could be directed using gradient decent. This was achieved by making the operations act on a range of locations relative to the 'position' of the read/write/erase 'head'. Moreover, the range itself could be tuned. Ultimately, explicit mechanisms were designed to encourage read operations that supported associative recall and temporal association and write operations that mimicked a 'least recently used' policy [10].

Initial formulations of the NTM were demonstrated on tasks such as copying arbitrary length sequences to memory and then retrieving them, associative recall of specific sequences of bits, or sorting bit sequences into lists [9]. More realistic tasks were then introduced that included answering graph queries (pertinent to natural language processing and route finding) and puzzle solving on a grid world [10].

Combining a deep learning architecture with a NTM provides the opportunity to learn what specific encoded states (from the deep learning architecture) should be retained in memory for later use. However, the optimal relationship between deep learning network, LSTM or NTM necessary for solving specific tasks remains an unknown. That said, several recent works demonstrate that visual reinforcement learning tasks (the sensor input corresponds to video data) describing environments limited to a first person perspective (i.e., partial observability) can only be addressed using a combination of all three properties [17, 39].

Finally, we note that several NTM results for low dimensional tasks have been demonstrated using neuroevolution [12, 26]. This is significant because neuroevolution is able to assume a simpler interface to indexible memory than

---

[6]LSTM is widely used as it addresses one of the potential pathologies of recurrent connectivity under gradient decent, that of vanishing gradients.

the original NTM (with gradient decent). Neuroevolution is also free to evolve the connectivity for the neural network, thus identifies solutions that are also much simpler than gradient based NTM methods. To date, we are not aware of neuroevolution with external memory as applied to high-dimensional partially observable tasks (e.g. as encountered by bots defining characters in video games), only reactive configurations [30, 36].

17.3.2 Memory in Genetic Programming

Memory mechanisms in GP have a rich history. For the purposes of this review, we recognize two forms of memory as employed with GP: scalar and indexed. Scalar memory is synonymous with register references in linear GP. Thus, linear GP might manipulate code defined in terms of a 'register-to-register' operation: $R[x] := R[x]\langle op \rangle R[y]$ in which $R[\cdot]$ is a register, $x, y \in \{0, \ldots, R_{max} - 1\}$ represent indexes to registers and $\langle op \rangle$ is a two argument operator. Several authors pioneered the use of scalar memory in linear GP [16, 28]. This means that although execution through linearly structured programs is typically sequential,[7] sequential instructions need not use the result of the preceding instruction. In addition, we can also distinguish between stateful and stateless operation in linear GP. *Stateful* operation implies that after execution of the program relative to state $\mathbf{x}(t)$, the values of the registers are *not* reset. Instead, execution at state $\mathbf{s}(t+1)$ commences with the register values as left from the previous execution of the program [1, 14], or

$$R[x]_{t+1} := R[x]_t \langle op \rangle R[y]_t \tag{17.2}$$

where $R[x]_t$ is a reference to register x content relative to external state t. Note, however, that each reference to $R[x]$ at the same update, t, will result in a 'distributed' updating of (recurrent) state. Finally, we also note that the assignment operator $(:=)$ results in the LHS being over-written by the result of the calculation. Poli et al. make the case for a model of scalar memory in which values incrementally change content as opposed to out-right replacing current content [29]. Such a framework was applied to the propagation of values between operations in tree structured GP and the action of operators in register based (linear) GP in regression tasks.

Indexed memory implies that addresses are specified with an explicit read and write operation (i.e. provision of 'load' and 'store' operations in linear GP). Teller investigated various formal properties that might result from supporting indexed memory [38]. Teller also demonstrated the evolution of indexed memory as a basic data structure [37], a topic extensively investigated by Langdon [23]. Other early developments included the evolution of 'mental models', such as recalling the

---

[7]Conditional instructions could change this [6].

content of discrete worlds defined as 4×4 toroidal grids [4, 7]. To do so, a two phase approach to evolving memory content was assumed [4, 7]. In phase 1, the agent can only write to memory, the assumption being that the agent can navigate the entire world noting points of interest *without* reading from memory. In phase 2, the agent is rewarded for systematically revisiting all the points of interest using only the information from memory. Such a formulation would preclude operation in tasks that require memory in order to support navigation. External memory models have also been proposed for use with simple robot controllers e.g. [3]. In this case, a performance measure was needed in order to define what memory content was actually saved and the criteria for replacement. Moreover, each write operation wrote the entire state space to memory. Neither would be feasible in the context of the Dota 2 environment.

17.4 Tangled Program Graphs

The tangled program graph (TPG) framework was previously benchmarked on the Arcade Learning Environment suite of Atari game playing tasks [25], where results competitive with deep learning solutions were demonstrated [19, 21]. In addition, TPG could evolve solutions to multiple ALE titles simultaneously [20, 21]. It was also demonstrated that solutions could be evolved without first down sampling the source video in ALE [22] and sub-tasks described in the VizDoom first person shooter environment [32]. No use is made of task specific instruction sets, instead TPG lets each program perform feature construction from individually indexed pixels [22, 32]. The cooperative nature of this decomposition organizes programs into teams, and ultimately teams into graphs of teams (of programs). Solution complexity then reflects that of the underlying task. Moreover, the computational complexity of the resulting solutions is orders of magnitude lower than that associated with deep learning solutions, resulting in TPG solutions executing in real-time on very modest computing platforms [19, 21, 22, 32].

Figure 17.2 provides an example TPG solution for controlling a bot capable of playing 10 tasks taken from the VizDoom first person shooter environment [40]. Each 'node' of the graph represents a team of programs. The outgoing arcs represent programs with the arrow ending at a specific indexed action, a. Actions can either be atomic (task specific) actions, $a \in \mathscr{A}$, or a pointer to another team, $a \in P$. Each program defines a function indicating the degree of confidence in its corresponding action, i.e. actions are a scalar integer value which is decoded into either an atomic action or a pointer to another team. At initialization, all teams are described in terms of atomic actions. Over time, the action of the variation operators enables surviving teams to index other (surviving) teams, resulting in the emergent discovery of a 'tangled program graph'.

Execution always starts relative to a 'root node', where there is only ever one root node per TPG individual and root nodes are not indexed by other teams. In order to evaluate a node, all the programs at this node are executed on the current state,

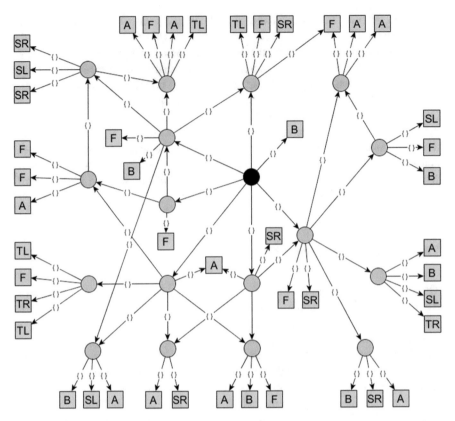

Fig. 17.2 Example TPG solution taken from [32]. Nodes are teams (of programs) with the black node denoting the root team where evaluation always commences. Square nodes are atomic actions. All nodes have at least one atomic node to guarantee that looping behaviour does not occur

$\mathbf{x}(t)$. The program with the maximum output at this node is said to have 'won' the right to suggest its action. If this action is an atomic action, then this is passed to the game engine as the bot's action in this state, resulting in an update to game state, $\mathbf{x}(t) \rightarrow \mathbf{x}(t + 1)$. However, should the action be a pointer to another node, then the process of evaluation repeats at the new node relative to $\mathbf{x}(t)$. Various heuristics ensure that although the graph might be cyclic, any attempt at node revisiting is trapped, and the program with the 'runner up' output used to identify an unvisited graph arc.

This implies that only a fraction of a TPG graph need be evaluated to determine each action. Moreover, no attempt is made to apply a convolution operator with the high-dimensional video input, again contributing to the very low computational footprint of TPG solutions. A tutorial description of the TPG algorithm is available from prior work [21, 22].

17.5   Indexed Memory for TPG

As noted in the introduction, the principle motivation for augmenting TPG with indexed memory is to provide a mechanism for addressing partial observability [34]. However, to do so we also need to bare in mind that TPG solutions consist of multiple programs (Fig. 17.2), so merely having each program address an independent 'bank' of indexed memory will still result in an inability to communicate 'internal state' across an entire TPG individual. Moreover, as TPG solutions are entirely emergent, we cannot a priori identify which programs will ultimately be cooperating to compose a solution. In order to address this issue we adopt an approach first proposed by Spector and Luke in which there is only ever one instance of external memory [35]. As we are interested in reinforcement learning tasks, we carry the concept to the point where state of the indexed memory is *never* reset.[8] Each TPG individual inherits the state of indexed memory as left from the previous evaluation of a TPG individual. Moreover, no reset is performed between generations. From this perspective, memory is as much a communal communication medium as it is memory. Any individual that in some way corrupts the communal state of indexed memory, penalizes the entire population.

Read and write operations will assume different 'views' in order to provide a mechanism to support different temporal durations (long and short term) in indexed memory. Specifically, a read operation will assume that indexed memory, $\mathcal{M}$, is defined as a set of sequential indexes. Such a process can be implemented as a Register-Memory reference with the mode bit setting the operand source to reference indexed memory as opposed to a register or a reference to the application state space. Thus, any read to $\mathcal{M}$ fetches the content from a single indexed memory location and loads it into a register, $R[x]$.

A write operation needs to define both what to write and where to write. We will assume that the content of the set of registers, $\mathcal{R}$, associated with the program performing the write operation represents a suitable definition for what to write, i.e. each program has its own set of R_{max} registers. Writes will define where to write as a probabilistic operation (Algorithm 1), distributing the content of the program's registers, $\mathcal{R}$, across the L columns of external memory $\mathcal{M}$, as per Algorithm 1. This means that columns in the 'middle' of external memory are updated most frequently (short-term memory) whereas columns towards the beginning or end are updated less frequently (long-term memory).

In summary, the read operation treats indexed memory, $\mathcal{M}$, as a set of consecutively addressed locations. Write operations perceive indexed memory as a matrix of $R_{max} \times L$ locations in which the mid region $\frac{L}{2}$ is more likely to be written to and the regions towards the ends are less likely to be written to. This then supports indexed memory with short and long term temporal properties. Section 17.7.1 discuss particular parameter choices for L and β.

---

[8]Indexed memory is initialized *once* at generation zero with NULL content.

Algorithm 1 Write function for External memory $\mathcal{M}$. Function called by a write instruction of the form: Write($\mathcal{R}$) where $\mathcal{R}$ is the vector of register content of the program when the write instruction is called. Step 1 identifies the mid point of memory $\mathcal{M}$, effectively dividing memory into upper and lower memory banks. Step 2 sets up the indexing for each bank such that the likelihood of performing a write decreases as a function of the distribution defined in Step 2a. Step 2(a)i defines the inner loop in terms of the number of registers that can source data for a write. Step 2(a)iA tests for a write to the upper memory bank and Step 2(a)iB repeats the process for the lower bank

Call: Write($\mathcal{R}$)

1. mid = $\frac{L}{2}$
2. for (offset := 0 < mid)

    a. $p_{write} = 0.25 - (\beta \times \text{offset})^2$

       i. for ($j := 0 < R_{max}$)

            A. IF (rnd[0, 1) $\leq p_{write}$)
               THEN ($\mathcal{M}[\text{mid} + \text{offset}][j] = \mathcal{R}[j]$)
            B. IF (rnd[0, 1) $\leq p_{write}$)
               THEN ($\mathcal{M}[\text{mid} - \text{offset}][j] = \mathcal{R}[j]$)

17.6 Dota 2 Game Engine Interface

17.6.1 Developing the Dota 2 Interface

Figure 17.3 summarizes the interaction between the various components comprising our interface to the Dota 2 game engine. The process for delivering data between the Dota 2 game client and an external learning interface (in this case TPG) is created using a combination of the Dota 2 Bot Scripting API (webite) and the Dota 2 Workshop Tools Scripting AI (website).

As seen in Fig. 17.3, the general flow for evaluating the environment and acting on it is loosely defined by a somewhat standard client-server model. After initial configuration is performed (loading the local development script for handling TPG bots, loading the "default" bot behaviour for the opponent, and removing other bots from the game to ensure the game is played as a one-on-one player scenario), the Dota 2 client sends the game state (as defined in Table 17.1) as a JSON structure to the webserver, which runs as a library parallel to the main TPG logic. The webserver extracts the data and stores it in a list of floating point values, which is fed into the main TPG learning algorithm. Once TPG evaluates the game state, it produces an atomic integer action, which is gathered by the webserver and sent back to the game

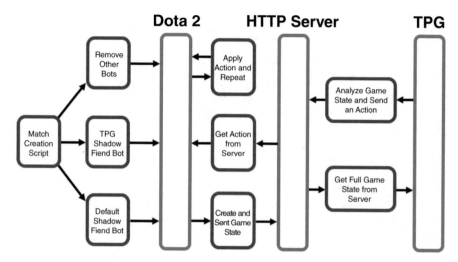

Fig. 17.3 Interaction between components of the Dota 2 bot interface

client. Once the game client receives the integer action, it selects the preset (but minimal) behaviour associated with that action. This process then repeats until the match has concluded.

The Dota 2 client, as a standard feature, does not allow a user to have more than one local development script for bots. This became a problem when selecting heroes. Our goal was to attempt to have TPG play against the built-in bots, however we ran into issues where we could not select both our own hero and the enemy's hero without loading a bot script. However, if we used our local bot script, this would mean TPG would play against itself, which was not a desirable scenario as our opponent should be of known skill in order to facilitate improvement of the TPG agent. However, the Steam client has an online repository for Dota 2 bots which are able to run separately from the local script. This allowed us to create a bot which uses all of the default bot logic, except during the hero selection process, where it always selects Shadow Fiend.

By default, the Dota 2 Bot Scripting API does not have a clean method for handling outside interfaces, likely to prevent forms of cheating. The only viable option for us to enable the exchange of information between the client and TPG was to use the Dota 2 Workshop feature of creating custom HTTP POST requests in order to send data to and from the Dota 2 client. The client and server are able to exchange a single set of 320 floating point and integer values within 7 ms, which can become a technical challenge when running the engine at faster speeds in order to accelerate the learning process and decrease overall run time.

Table 17.1 Attributes used to characterize game state

Attribute type	Attribute count
Self (opponent)	30(26)
Team, level, health (avg, max, regen), Mana (avg, max, regen), movement speed (base, current), base damage (avg., variance), attack (damage, range, speed), Sec. per attack, attack (hit location, time since last, target of current), vision range, strength, agility, intelligence, gold[a], net worth[a], last hits[a], denies[a], location (x-, y-coordinate), Orientation	
Match duration	1
Abilities—6 per self (opponent))	48 (42)
Level, Mana cost, damage, range, Num. of Souls (Necromaster), Cooldown[a], target type, effect flag area	
Creeps (up to 10 nearest)	140
Team, health, max health, movement speed (base, current), base damage (avg., variance), attack (damage, range, speed), Sec. per attack, location (x-, y-coordinate)	
Towers—self (opponent)	6 (6)
Team, health (avg., Max), attack (damage, range, speed)	
Runes	8
Bounty rune location (top, bottom), top regular rune location, top powerup rune (type, status), bottom powerup rune (location, type, status)	
Items—self (opponent)	7 (6)
Inventory (slot 1, . . ., slot 6), healing salve flag[a]	

In the case of attributes estimated for both self and opponent
[a]Indicates a self specific attribute

17.6.2   Defining State Space

A total of eight different categories of feature were assumed from the set of features provided by Valve.[9] The *Self* and *Opponent* categories summarize the basic statistics for each team. Four of these properties represent internal bot state (Current Gold/Net worth, Last hits, Denies) and are therefore unknown in the case of the opponent. *Match time* is merely the duration of the current game. *Abilities* reflect further state information for each hero, and are repeated for all 6 of the Shadow Fiend hero abilities. As such abilities represent skills available to the hero character and change as the character 'level up' during each game (games start with heroes at the lowest level of ability). *Creep* state is characterized in terms of a vector of 14 attributes per creep, for the nearest 10 creeps. One way in which the game can be won/lost is when a tower is taken. Thus, there are six attributes defining state of Radiant and Dire's *mid-lane tower*. *Runes* represent gems that can be collected in the environment, eight attributes summarize type, location and state of such items,

---

[9]Up to 20,000 features are available, however, we are only interested in the case of a 1-on-1 single lane configuration of the game (as opposed to 5 heroes per team over 3 lanes).

and another seven attributes summarize *Items* bought during the course of a game
by each hero. An attribute value of '−1' is assumed for any attributes that are not
measurable due to partial observability or if there is not actually enough of the
property to measure (e.g. there are less than 10 creeps observable/alive).

17.6.3 Defining the Shadow Fiend Action Space

We will assume a minimalist scenario in which we want to provide enough
functionality for our Shadow Fiend hero to support the following five properties,
resulting in a **total of 30 actions**.

Movement: at any point in time heroes have the ability to walk in one of eight
cardinal directions (north, north-east, east, south-east, south, south-west, west,
north-west) or not move, resulting in *9 actions*. When issued, the move command
is applied to a ground location 100 units from the heroes current location.

Attacking opponents: an 'attack action' declares which opponent to attack, and
assumes a default integer value (future work could potentially tune these
defaults). In total TPG will distinguish between 12 possible targets: the opponent
tower, the opponent hero, or the ten nearest opponent creeps, so a total of *12
actions*.

Casting spells: a Shadow Fiend character has a unique set of spell casting
abilities.[10] The 'Shadowraze' is a ranged spell cast with three distances (near,
medium, far), whereas a 'Requiem of Souls' is either cast or not. This results in
another *4 actions*.

Collecting items: a hero can collect 'powerup' and 'bounty' runes while navigating the environment, but to do so it has to explicitly deploy the relevant action.
This results in another *4 actions*.

Hero's bot: the Shadow Fiend hero can control a 'Healing Salve' for which there
is a *single action*, deploy or not.

17.6.4 Fitness Function

A hero has to operate cooperatively with friendly creeps and defensive towers. In
the following we explicitly reward our hero explicitly for successful 'Last Hitting'
and 'Denies', but ignore other tactics such as 'Creep Blocking' and 'Pulling'.[11] In
addition, we include three factors explicitly scored by the game engine: Net worth,
Kills, and Match points.

---

[10]https://dota2.gamepedia.com/Shadow_Fiend.

[11]https://dota2.gamepedia.com/Creep_control_techniques.

Last hits (LH): A 'last hit' refers to the killing blow applied to an opponent or neutral creep. If your hero is the last hitter, then a gold bonus is payed to your hero, increasing your ability to purchase items for improving the capabilities of your hero. Each last hit is awarded **10 points** in the fitness function.

Denies (D): When creep health decreases below 50% you can get a last hit on your own creeps. This is useful because it 'denies' a last hit to the opponent hero, i.e. you prevent the opponent hero from gaining gold or experience points. Each 'deny' is awarded **15 points** in the fitness function, i.e. significantly less common than a last hit.

Net worth (NW): is the overall wealth of your team, so includes all bought items, gold, and any inventory accumulated. We use the net worth as calculated by the game engine.

Kills (K): Each time an opponent hero is killed a reward of **150 points** is given.

Match (M): A match ends with either a hero dying twice, or the successful destruction of the team's tower (Fig. 17.1). If the agent is the winner, the game engine awards **2000 points**.

Not dead (ND): A bonus of **150 points** is awarded if the hero did not die during the match. TPG Shadow Fiend hero fitness, f_i now has the form:

$$f_i = 10 \times LH + 15 \times D + NW + 150 \times K + 150 \times ND + 2000 \times M \qquad (17.3)$$

where LH, D, K reflect counts for the number of times each property is achieved per match, $ND \in [0, 1, 2]$ reflecting the number of times the agent is 'not dead' per match, NW assumes the value estimated by the game engine at the end of the match, and $M \in [0, 1]$.

17.7 Results

17.7.1 TPG Set Up

Parameterization assumes the same set up as an earlier experiment in which bot behaviours were evolved for the death matches in the VizDoom first person shooter environment [34]. External memory is defined by the number of registers a program indexes (R_{max}) and the number of columns, L (Sect. 17.5). Thus, from the perspective of a read operation there are $M = R_{max} \times L$ indexible locations. All our experiments assume $L = 100$ therefore the range of indexes supported during a read operation is $M = 800$. The probability distribution defined in Step 2a of Algorithm 1 is parameterized with $\beta = 0.01$. This means that for 'offset' values ≈ 0 the likelihood of a write operation is 25% (or 4 out of $R_{max} = 8$ registers will

Table 17.2 TPG
Parameterization

Configuration at initialization	
Number of teams (P):	360
Number of programs:	14,400
Instruction distribution:	Equal
Initial Prog. per team (ω):	40
Team population	Program population
Max. instructions: 64	Registers per program (R_{max}): 8
Gap: 50% of root teams	Prob. Delete Instr. (P_{del}): 0.5
P_m: 0.2	Prob. Add Instr. (P_{add}): 0.5
P_d: 0.7	Prob. Mutate Instr. (P_{mut}): 1.0
P_a: 0.7	Prob. Swap Instr. (P_{swp}): 1.0
	P_{nm}, P_{atomic}: 0.2, 0.5

Linear GP representation is assumed in which 10 opcodes
appear: $+, -, \times, \div$, log, exp, Conditional, Write, Read; Two
argument instructions have the form: $R[x] = R[x]\langle op_2 \rangle R[y]$;
Single argument instructions: $R[x] = \langle op_1 \rangle (R[y])$; Conditional
operator: IF $R[x]\langle op_0 \rangle R[y]$ THEN $R[x] = -R[x]$. These definitions are unchanged relative to earlier work on classification
tasks [24]. Read/Write definitions appear in Sect. 17.5

be written to the shortest term memory location as indexed about $\approx \frac{L}{2}$). However,
as 'offset' $\to L$ (and respectively 1), then the likelihood of performing a write tends
to 1%. Overall a write operation will result in 17% of $\mathcal{M}$ changing value under this
parameterization. No claims are made regarding the relative optimality of such a
parameter choice. Table 17.2 summarizes the parameters assumed for TPG.

17.7.2   Training Performance

Figure 17.4 summarizes the development of fitness over the 250 generations
assumed for training. The top curve corresponds to the overall fitness (Eq. (17.3)),
whereas the bottom curves correspond to the contribution to fitness from Last hits
(Fig. 17.4a) and Denies (Fig. 17.4b). Note that the overall fitness is expressed on the
left y-axis on a log scale. The curves are certainly noisy, on account of the stochastic
properties of the game. Given that all training is performed relative to an opponent
built in Shadow Fiend defined as 'hard', we see it as encouraging that the underlying
fitness curve shows a positive trend.

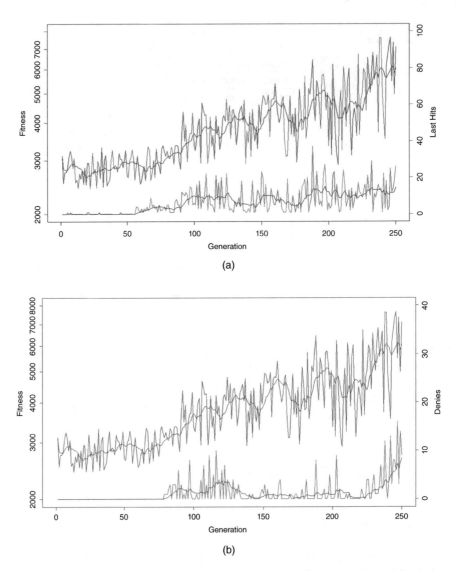

Fig. 17.4 Overall fitness function (left *y*-axis) with development of (**a**) last hits and (**b**) denies (right *y*-axis). Moving average of respective fitness curves shown in black

17.7.3 Assessing Champion TPG Agents Post Training

Post training, the best TPG agent is identified as the individual with highest training fitness. Fifty tournaments are then performed with TPG trained without indexed memory (just scalar stateful registers—hereafter 'reactive TPG'), TPG trained with indexed memory (hereafter 'memory TPG') and three instances of the opponent Shadow Fiend bot from the game engine (corresponds to difficulty levels of 'easy', 'medium' and 'hard'). The distinction between these difficulty levels is summarized in Table 17.3.

Table 17.4 summarizes the number of times that the two forms of TPG were able to win against different difficulty levels of the opponent Shadow Fiend bot, i.e. games are always played until completion. The reactive TPG configuration is not capable of winning games for any level of the opponent. Viewing the resulting game play indicates that the reactive TPG agent is unable to identify the significance of the mid-lane, i.e. where the creeps from its base will march and engage with the opponent creeps. Instead, the reactive TPG hero agent wanders around the world until it either dies (e.g. attacked by creeps specific to forest features) or the opponent team successfully take the TPG agent's tower (marked in red on Fig. 17.1).

Adding the indexed memory to TPG clearly results in competitive performance when playing against 'Easy' and 'Medium' difficulty opponent bots, however, we are not able to compete with the opponent Shadow Fiend at the 'Hard' setting. That said, the fitness curve indicates that additional generations might rase the level of play to the point where we do become competitive at this level. Also shown for the memory TPG agent is the experience points awarded by the game engine, where this is an indication of the ability to 'level up', i.e. gain new abilities/develop the hero character over the course of the game. This is most prominent in the case of the Medium opponent difficulty as games are over too quickly for the Easy and

Table 17.3 Difference between different difficulty settings of the built in Shadow Fiend bot

Difficulty	Reaction time	Last hitting delay	Ability recovery delay
Easy	0.3 s	Yes	6 s
Medium	0.15 s	Yes	3 s
Hard	0.075 s	None	None

Table 17.4 Evaluation of champion TPG agent against opponent Shadow Fiend Bot

Opponent Bot	Reactive TPG	Memory TPG		
Difficulty	# wins	# wins	XP	Gold
Easy	3(6%)	31(62%)	8760	3633
Medium	0(0%)	22(44%)	10,111	3860
Hard	0(0%)	12(24%)	6166	3170

A total of 50 games are played at each difficulty level. XP is the median 'experience points' as awarded to memory TPG play over the 50 games. Median 'gold' reflects the ability of memory TPG to monetize during game play

Hard opponents (i.e. TPG agent wins (looses) to quickly against the Easy (Hard) opponent). We also note that from a behavioural perspective, memory TPG appears to have learnt to navigate and explicitly takes itself to the river feature about the mid-lane where it 'patrols' and engages with the opponents [33].

17.7.4   Characterization of Memory Behaviour

Table 17.5 summarizes the static characteristics of the TPG agent at several levels of granularity. In short, the TPG graph consists of 1369 programs organized into 46 nodes with a median out degree of 19. However, the median number of node visits per decision is 5 (or on average $5 \times 19 \times 51 = 4845$ instructions evaluated per decision).

It is also apparent that all programs perform a read against indexed memory, but only 23% of the programs perform a write to indexed memory (at initialization, all programs consist of all types of instruction). In short, specialization has taken place, with only specific programs being allowed to perform write operations. We note that a similar specialization appeared in VizDoom deathmatches with TPG [34].

Figure 17.5 attempts to capture some of the dynamic properties of indexed memory accesses during the course of games. Figure 17.5a represents a heat map[12] of the memory accesses with darker coloured cells indicating more frequently

Table 17.5 Post training static characterization of champion TPG bot properties

Property	Value
Teams	46
Out-Degree (Med)	19
Learners	1369
Learners/team (Mean)	29.76
Shortest path	Root (1)
Median path	5
Longest path	8
Instructions (Total)	69,664
Instructions/learner (Mean)	50.89
Total read instructions	30,881 (44.33% of total)
Input read instructions	23,818 (34.19% of total)
Memory read instructions	7063 (10.13% of total)
Teams with memory reads	46 (100.0% of total)
Learners with memory reads	1369 (100.0% of total)
Memory write instructions	2786 (3.99% of total)
Teams with memory writes	27 (58.69% of total)
Learners with memory writes	317 (23.16% of total)

---

[12]Heat map produced using [5].

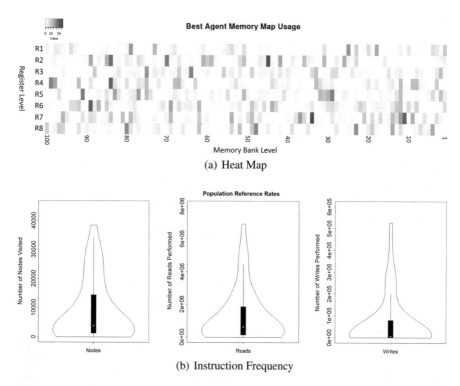

(a) Heat Map

(b) Instruction Frequency

Fig. 17.5 Snapshot of indexed memory utilization over a game (**a**) heat map of memory read/writes (**b**) number of read/write references during a game. Note the difference in scale between reads (10^6) and writes (10^5)

accessed memory locations. Short term memory is associated with the (vertical) middle of the map (indexes around 50) whereas long term memory locations appear towards the ends (indexes around 1 and 100). Certain locations appear to have much higher frequency of memory access than others, and a significant degree of sparseness also appears.

Counting the frequency of indexed memory accesses is also possible (Fig. 17.5b), where the same bias to a lower number of write operations is reported. Note, however, that no attempt is made to remove hitchhiking programs (those programs that never win the right to suggest an action at a node), so further simplification might be possible.

17.8   Conclusion

TPG has previously been demonstrated as providing solutions to visual reinforcement learning tasks from the ALE suite of titles, to a level competitive with deep learning, but without the computational cost [21]. However, game titles from the ALE suite are for the most part, games of complete information. Scaling TPG to games with significant amounts of partial observability requires the addition of indexed memory. A design of indexed memory for this purpose was previously proposed and evaluated under the specific case of 'deathmatches' under the VizDoom first person shooter environment [34]. In this work, we demonstrate that the same appears to be true for the case of evolving a 'Shadow Fiend' hero for the 1-on-1 mid lane configuration of Dota 2.

In summary, for the same design of fitness function, atomic actions and state space, TPG with indexed memory is able to reliably reach the level of performance capable of matching the built in Shadow Fiend hero at the medium setting in 250 generations. Moreover, the fitness curve at this point has not plateaued, implying that further generations would be beneficial. Conversely, TPG without indexed memory is unable to evolve behaviours that even match the opponent bot with an 'easy' difficulty.

In future work, we are interested in investigating whether behaviours can be transferred between different hero characters in order to speed up evolution as well as addressing the issue of multi-lane play and multi-opponent play. Finally, outside of Dota 2, we are interested in knowing whether TPG with indexed memory would also be useful in high-dimensional games of complete information. That is to say, TPG (and GP in general) invariably does not index all the state space, so from the perspective of what GP 'sees' the state space is actually partial, thus indexed memory might be appropriate for high-dimensional reinforcement learning tasks as a whole. We note that research to this end has began to appear, suggesting that TPG with an incremental model for constructing indexed memory is indeed beneficial under high-dimensional games of complete information [18].

Acknowledgements We gratefully acknowledge support from the NSERC CRD program (Canada).

References

1. Agapitos, A., Brabazon, A., O'Neill, M.: Genetic programming with memory for financial trading. In: EvoApplications, *LNCS*, vol. 9597, pp. 19–34 (2016)
2. Aiyer, S.V.B., Niranjan, N., Fallside, F.: A theoretical investigation into the performance of the Hopfield model. IEEE Transactions on Neural Networks **15**, 204–215 (1990)
3. Andersson, B., Nordin, P., Nordahl, M.: Reactive and memory-based genetic programming for robot control. In: European Conference on Genetic Programming, *LNCS*, vol. 1598, pp. 161–172 (1999)
4. Andre, D.: Evolution of mapmaking ability: Strategies for the evolution of learning, planning, and memory using genetic programming. In: IEEE World Congress on Computational Intelligence, pp. 250–255 (1994)
5. Babicki, S., Arndt, D., Marcu, A., Liang, Y., Grant, J.R., Maciejewski, A., Wishart, D.S.: Heatmapper: web-enabled heat mapping for all. Nucleic Acids Research (2016). http://www.heatmapper.ca/

6. Brameier, M., Banzhaf, W.: Linear Genetic Programming. Springer (2007)
7. Brave, S.: The evolution of memory and mental models using genetic programming. In: Proceedings of the Annual Conference on Genetic Programming (1996)
8. Elman, J.L.: Finding structure in time. Cognitive Science **14**, 179–211 (1990)
9. Graves, A., Wayne, G., Danihelka, I.: Neural turing machines. CoRR **abs/1410.5401** (2014)
10. Graves, A., Wayne, G., Reynolds, M., Harley, T., Danihelka, I., Grabska-Barwinska, A., Colmenarejo, S.G., Grefenstette, E., Ramalho, T., Agapiou, J., Badia, A.P., Hermann, K.M., Zwols, Y., Ostrovski, G., Cain, A., King, H., Summerfield, C., Blunsom, P., Kavukcuoglu, K., Hassabis, D.: Hybrid computing using a neural network with dynamic external memory. Nature **538**(7626), 471–476 (2016)
11. Greff, K., Srivastava, R.K., Koutník, J., Steunebrink, B.R., Schmidhuber, J.: LSTM: A search space odyssey. IEEE Transactions on Neural Networks and Learning Systems **28**(10), 2222–2231 (2017)
12. Greve, R.B., Jacobsen, E.J., Risi, S.: Evolving neural turing machines for reward-based learning. In: ACM Genetic and Evolutionary Computation Conference, pp. 117–124 (2016)
13. Grossberg, S.: Content-addressable memory storage by neural networks: A general model and global Liapunov method. In: E.L. Schwartz (ed.) Computational Neuroscience, pp. 56–65. MIT Press (1990)
14. Haddadi, F., Kayacik, H.G., Zincir-Heywood, A.N., Heywood, M.I.: Malicious automatically generated domain name detection using stateful-SBB. In: EvoApplications, *LNCS*, vol. 7835, pp. 529–539 (2013)
15. Hochreiter, S., Schmidhuber, J.: Long short-term memory. Neural Computation **9**(8), 1735–1780 (1997)
16. Huelsbergen, L.: Toward simulated evolution of machine language iteration. In: Proceedings of the Annual Conference on Genetic Programming, pp. 315–320 (1996)
17. Jaderberg, M., Czarnecki, W.M., Dunning, I., Marris, L., Lever, G., Castañeda, A.G., Beattie, C., Rabinowitz, N.C., Morcos, A.S., Ruderman, A., Sonnerat, N., Green, T., Deason, L., Leibo, J.Z., Silver, D., Hassabis, D., Kavukcuoglu, K., Graepel, T.: Human-level performance in 3D multiplayer games with population-based reinforcement learning. Science **364**, 859–865 (2019)
18. Kelly, S., Banzhaf, W.: Temporal memory sharing in visual reinforcement learning. In: W. Banzhaf, E. Goodman, L. Sheneman, L. Trujillo, B. Worzel (eds.) Genetic Programming Theory and Practice, vol. XVII. Springer (2020)
19. Kelly, S., Heywood, M.I.: Emergent tangled graph representations for Atari game playing agents. In: European Conference on Genetic Programming, *LNCS*, vol. 10196, pp. 64–79 (2017)
20. Kelly, S., Heywood, M.I.: Multi-task learning in Atari video games with emergent tangled program graphs. In: ACM Genetic and Evolutionary Computation Conference, pp. 195–202 (2017)
21. Kelly, S., Heywood, M.I.: Emergent solutions to high-dimensional multitask reinforcement learning. Evolutionary Computation **26**(3), 347–380 (2018)
22. Kelly, S., Smith, R.J., Heywood, M.I.: Emergent policy discovery for visual reinforcement learning through tangled program graphs: A tutorial. In: W. Banzhaf, L. Spector, L. Sheneman (eds.) Genetic Programming Theory and Practice, vol. XVI, chap. 3, pp. 37–57. Springer (2019)
23. Langdon, W.B.: Genetic Programming and Data Structures. Kluwer Academic (1998)
24. Lichodzijewski, P., Heywood, M.I.: Symbiosis, complexification and simplicity under GP. In: Proceedings of the ACM Genetic and Evolutionary Computation Conference, pp. 853–860 (2010)
25. Machado, M.C., Bellemare, M.G., Talvitie, E., Veness, J., Hausknecht, M., Bowling, M.: Revisiting the arcade learning environment: evaluation protocols and open problems for general agents. Journal of Artificial Intelligence Research **61**, 523–562 (2018)
26. Merrild, J., Rasmussen, M.A., Risi, S.: Hyperntm: Evolving scalable neural turing machines through hyperneat. In: EvoApplications, pp. 750–766 (2018)

27. Mnih, V., Kavukcuoglu, K., Silver, D., Rusu, A.A., Veness, J., Bellemare, M.G., Graves, A., Riedmiller, M., Fidjeland, A.K., Ostrovski, G., Petersen, S., Beattie, C., Sadik, A., Antonoglou, I., King, H., Kumaran, D., Wierstra, D., Legg, S., Hassabis, D.: Human-level control through deep reinforcement learning. Nature **518**(7540), 529–533 (2015)
28. Nordin, P.: A compiling genetic programming system that directly manipulates the machine code. In: K.E. Kinnear (ed.) Advances in Genetic Programming, pp. 311–332. MIT Press (1994)
29. Poli, R., McPhee, N.F., Citi, L., Crane, E.: Memory with memory in genetic programming. Journal of Artificial Evolution and Applications (2009)
30. Salimans, T., Ho, J., Chen, X., Sutskever, I.: Evolution strategies as a scalable alternative to reinforcement learning. CoRR **abs/1703.03864** (2016)
31. Sapienza, A., Peng, H., Ferrara, E.: Performance dynamics and success in online games. In: IEEE International Conference on Data Mining Workshops, pp. 902–909 (2017)
32. Smith, R.J., Heywood, M.I.: Scaling tangled program graphs to visual reinforcement learning in ViZDoom. In: European Conference on Genetic Programming, *LNCS*, vol. 10781, pp. 135–150 (2018)
33. Smith, R.J., Heywood, M.I.: Evolving Dota 2 Shadow Fiend bots using genetic programming with external memory. In: Proceedings of the ACM Genetic and Evolutionary Computation Conference (2019)
34. Smith, R.J., Heywood, M.I.: A model of external memory for navigation in partially observable visual reinforcement learning tasks. In: European Conference on Genetic Programming, *LNCS*, vol. 11451, pp. 162–177 (2019)
35. Spector, L., Luke, S.: Cultural transmission of information in genetic programming. In: Annual Conference on Genetic Programming, pp. 209–214 (1996)
36. Such, F.P., Madhavan, V., Conti, E., Lehman, J., Stanley, K.O., Clune, J.: Deep neuroevolution: Genetic algorithms are a competitive alternative for training deep neural networks for reinforcement learning. CoRR **abs/1712.06567** (2018)
37. Teller, A.: The evolution of mental models. In: K.E. Kinnear (ed.) Advances in Genetic Programming, pp. 199–220. MIT Press (1994)
38. Teller, A.: Turing completeness in the language of genetic programming with indexed memory. In: IEEE Congress on Evolutionary Computation, pp. 136–141 (1994)
39. Wayne, G., Hung, C.C., Amos, D., Mirza, M., Ahuja, A., Grabska-Barwińska, A., Rae, J., Mirowski, P., Leibo, J.Z., Santoro, A., Gemici, M., Reynolds, M., Harley, T., Abramson, J., Mohamed, S., Rezende, D., Saxton, D., Cain, A., Hillier, C., Silver, D., Kavukcuoglu, K., Botvinick, M., Hasssbis, D., Lillicrap, T.: Unsupervised predictive memory in a goal-directed agent. CoRR **abs/1803.10760** (2018)
40. Wydmuch, M., Kempka, M., Jaśkowski, W.: ViZDoom competitions: Playing doom from pixels. IEEE Transactions on Games **to appear** (2019)

Chapter 18
Modelling Genetic Programming as a Simple Sampling Algorithm

David R. White, Benjamin Fowler, Wolfgang Banzhaf, and Earl T. Barr

18.1 Introduction

Previous attempts to characterise Genetic Programming (GP) have focused on complex schemata, that is 'templates' of full GP trees or subtrees composed of fixed positions that specify the presence of a given function from the function set, and other unspecified 'wildcard' positions that indicate any single function, or an arbitrary subtree; a schema represents a set of expression trees that share some syntactic characteristics. Unfortunately, the complicated definitions of such schemata has made resulting theories difficult to test empirically, and GP schema theory has been criticised for its inability to make significant predictions or inform the development of improved algorithms.

We present a dramatically simplified approach to GP schema theory, which considers only the distribution of functions at a single node in the tree, ignoring second-order effects that describe the context within which the node resides: for example, we do not consider parent-child node relationships. Our schemata are thus

D. R. White
Department of Physics, University of Sheffield, Sheffield, UK
e-mail: d.r.white@sheffield.ac.uk

B. Fowler
Department of Computer Science, Memorial University of Newfoundland, St. John's, NL, Canada
e-mail: fowler@mun.ca

W. Banzhaf (✉)
Department of Computer Science and Engineering, Michigan State University, East Lansing, Okemos, MI, USA
e-mail: banzhafw@msu.edu

E. T. Barr
CREST, University College London, London, UK
e-mail: e.barr@ucl.ac.uk

© Springer Nature Switzerland AG 2020
W. Banzhaf et al. (eds.), *Genetic Programming Theory and Practice XVII*, Genetic and Evolutionary Computation, https://doi.org/10.1007/978-3-030-39958-0_18

simple fixed-position 'point schemata', as illustrated in Fig. 18.1. We show that the behaviour of GP viewed through the sampling of simple schemata is remarkably consistent: GP converges top-down from the root node, with nodes at each level gradually becoming 'frozen' to a given function from the function set, in a recursive fashion; an example of such convergence for the root node on one benchmark is illustrated in Fig. 18.2. We model this behaviour as a competition between simple schemata, where the correlation between rank fitness and schema membership determines the 'winner' and drives convergence. We present empirical evidence to support our model.

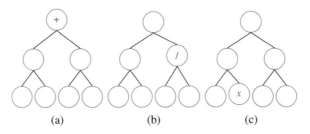

Fig. 18.1 Examples of Simple Single Point Schemata for a function set where the maximum arity is two. A single labelled node indicates a fixed position; all other nodes are unconstrained—provided that they are not on the path from root to the fixed node, a position may not even exist within trees belonging to the schemata. (**a**) + at Root, (**b**) / at Root's second child, (**c**) Terminal x

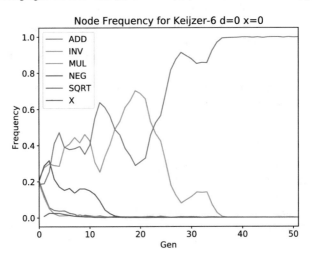

Fig. 18.2 Example convergence of the root node across a population for the Keijzer-6 benchmark. The root node distribution converges to the Sqrt function. Other runs exhibit similar behaviour

18.2   Rationale for Modelling Simple Schemata

Traditional GP, often referred to as tree-based GP, is a simple evolutionary algorithm that explores the space of expression trees. Algorithm 1 provides pseudocode for canonical tree-based GP using tournament selection. Two key observations from this pseudocode are apparent: (1) the variation operators applied *do not consider the semantics of the parents or their children* (see also [7]), and (2) the only point at which fitness is considered by the algorithm is *in selection via relative fitness, i.e. ranking.* We add a third observation from the wider GP literature: (3) GP's strength lies in its versatility: it may easily be applied to a wide variety of problems and demonstrates good performance across a range of problem domains.

Given these observations, any description of how GP searches the space of expression trees is limited to considering (a) syntactic properties of the population and (b) the rank fitness of individuals. Beyond relative fitness within a population, the semantics of individual trees is irrelevant. The building block hypothesis [8] (BBH) conjectures that GP crossover exchanges small subtrees that make above-average contributions to the fitness of an individual, despite GP's disregard for semantic concerns when selecting the location at which that subtree is inserted. That such subtrees should exist seems intuitively unlikely, but far more concerning is the notion that such subtrees should exist across the wide variety of problems GP has been successfully applied to. We are therefore skeptical of the BBH and regard crossover as a constrained macro-mutation operator; this viewpoint is congruent with rigorous empirical studies that have shown crossover to be beneficial on only a limited subset of simple problems [15].

Algorithm 1 Canonical tree-based GP algorithm

Input: $pop\_size, gens, p_{xo}, p_{mut}$
Output: Final population pop

```
 1:  pop ← init_pop(pop_size)
 2:  for 1 to gens do
 3:      for p ∈ pop do
 4:          eval(p)
 5:      end for
 6:      next_gen ← {}
 7:      while size(next_gen) < pop_size do
 8:          op ← select_op(xo, mut, p_xo, p_mut)
 9:          if op == mut then
10:              ind ← tournament_selection(pop, 1)
11:              next_gen ← next_gen ∪ {mutate(ind)}
12:          end if
13:          if op == xo then
14:              ind ← tournament_selection(pop, 2)
15:              next_gen ← next_gen ∪ {xo(ind, ind2)}
16:          end if
17:      end while
18:  end for
19:  return pop
```

We therefore must consider only syntactic features when formulating a model of GP's behaviour, and furthermore consider only those features that are represented in a substantial number of individuals within the population, arriving at the following possible properties that may form part of a behavioural model for GP:

- Tree size
- Tree shape
- Function frequency per individual
- Functions at positions in the tree (simple schemata)
- Higher-order schemata

Tree size and shape depend on the level of bloat in a population; by definition, bloat does not contribute to solving a particular problem, so we disregard tree size and shape as an explanatory factor. Explorative data analysis of *population* function frequency showed some interesting behaviour, but it was too coarse-grained to maintain a strong relationship with semantics.

We are therefore left with simple and higher-order schemata. Given the limited success of higher-order schema theory, we restrict the discussion to models based on function positions within individual trees. Our preliminary investigations into the frequency of different functions at the root node showed promise, and also matched earlier results [12], which considered more complex rooted schemata.

We retain a simpler definition of point schemata; our definition utilises the grid system introduced by Poli and McPhee [10], which is illustrated in Fig. 18.3. Using this coordinate system, we can formally define a simple schema:

Fig. 18.3 Grid system for laying out expression trees, based on Poli and McPhee [10]. The example represents the expression tree $\sqrt{x} + xy$

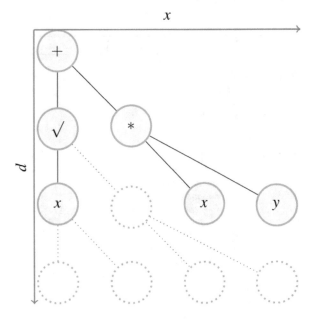

Definition A simple GP schema is a tuple (d, x, f) where d is the depth within a tree (root is defined as depth 0), x is the 'x index' identifying a node at that particular depth as in [10], and f is a function from the function set.

Examples of simple schemata are given in Fig. 18.1.

18.3 Modelling GP

We are interested in modelling the frequency of simple schemata within the population; this is a heritable trait and can therefore be modelled using Price's covariance and selection theorem [11], which gives the expected change in a trait within a population:

$$\Delta z = \frac{1}{\bar{w}} cov(w_i, z_i) + \frac{1}{\bar{w}} E(w_i \Delta z_i) \tag{18.1}$$

Here, z is the prevalence of a trait within the population, Δz is its expected change, $\bar{w}$ is the average number of children per individual, z_i is the amount of the trait within the individual (in our case, 1 if the individual belongs to a schema, and 0 otherwise), and w_i is the number of children of individual i. As our population size is constant, the average fitness (number of children) of an individual is always one and thus the $1/\bar{w}$ may be discarded. In our case, z is the frequency of schema membership, i.e., the proportion of individuals belonging to a given schema.

The expected change in trait z over the population depends on two terms: the change due to selection and the change due to transmission. The change due to selection is determined by the covariance between possession of that trait and fitness that propagates the trait into the next generation, whilst the change due to transmission models the constructive and destructive effects of genetic operators in GP, which may add or remove an individual's schema membership.

18.3.1 Change in Schema Prevalence Due to Selection

We first consider change due to selection. We model only tournament selection, a common selection method, which uses relative (ranked) fitness to select parents. After fitness evaluation, each individual is assigned a rank $r_1 \ldots r_N$ where r_1 is the best individual and r_N is the worst. As given by Blickle and Thiele [2] the probability $p_s(i)$ of selecting an individual i of rank r_i within a population size of n under a tournament of size k is given by:

$$p_s(r_i) = \frac{(n - i + 1)^k - (n - i)^k}{n^k} \tag{18.2}$$

Any selected individual has a single child belonging to the same schema, modulo the opportunity for disruption due to the genetic operators as described below. We define $member(h, r_i)$ to be 1 if the individual ranked r_i is a member of h, and 0 otherwise, We therefore rewrite Eq. (18.1), inserting our function and notation for schema membership:

$$\triangle|h| = cov(p_s(r_i), \ member(h, r_i)) + E(w_i \triangle z_i) \tag{18.3}$$

We can precisely calculate this term in practice, although we require full knowledge of schema membership for each fitness rank. It also represents an expectation, and errors may accumulate if predicting membership several generations ahead: indeed, stochastic variation in the early stages of a run may sometimes push the population away from convergence to the most highly fit schemata to secondary ones. However, such an equation is still useful if it accurately reflects the mechanism that underlies GP, as we may use it to improve on current algorithms.

18.3.2   Change in Schema Prevalence Due to Operators

To model the potential change due to transmission, we must model crossover and mutation. If no crossover or mutation occurs, then the transmission probability is 1 and the case is closed; this is simple reproduction of the parent, and is omitted from Algorithm 1 as we do not use it.

First, consider disruption. A single node schema can only be disrupted if the crossover or mutation point selected is above the node. Given a schema (d, x, f), there are only $d - 1$ such points in the tree i.e. the potential for disruption occurs with probability $\leq d/size(tree)$, otherwise disruption will certainly not occur.

Assuming the selection of a disruptive node for crossover or mutation, there are three possible outcomes: (1) the node is replaced with another node of the same type, preserving the schema; (2) the node is replaced with another node of a different type; (3) the node is removed from the tree due to a change in tree shape. The only possibility that preserves the schema is situation (1), and to do so two conditions are required: first, the inserted subtree must be of a shape such that the node exists in the new tree and, second, the node inserted at position (d, x) must be of type f.

If we assume that the distribution of different functions f out of the function set F is roughly uniform in the population (we can alternatively make a slightly weaker but more complicated assumption with the same conclusion), then the probability of disrupting the schema given the node exists in the donating tree is approximately $(|F|-1)/|F|$. This probability is therefore close to one, and is actually an optimistic lower bound: leaf nodes are disproportionately likely to be inserted, increasing this probability further.

Given that the insertion of a suitable subtree that will preserve a schema is already small, particularly as d grows (because inserted subtrees are disproportionately small), we therefore make a simplifying assumption: that the probability of disruption to the schema given an insertion point between the root and our defining node is approximately one, and therefore the probability of disruption is $\approx d/size(tree)$. Given growth in tree size we can see why convergence close to the root will occur: the probability of disrupting a schema is low and the selection term in Eq. (18.3) will dominate, particularly as average tree size increases.

A similar argument can be used for the constructive case: one of the $(d - 1)/size(tree)$ nodes must be selected for insertion, and in this case some trees may not contain a node at that position, so this is an optimistic upper bound. Given such an insertion point is selected, we must then select for insertion at that point a subtree deep enough with the correct shape to ensure the node exists within the schema, and the inserted node must be of the correct type. Again, as average tree size increases the probability of construction rapidly diminishes and we therefore assume the probability of construction approaches zero as d increases. We conjecture that this is the reason GP systems do not exhibit long-term learning.

18.4 Empirical Data Supporting the Model

How does GP behave with respect to these schemata? We examined a set of three benchmarks, taken from [14] given in Table 18.1, and plotted the convergence behaviour of these schemata. Convergence plots for various schemata are shown in Figs. 18.4, 18.5, 18.6, 18.7, 18.8, and 18.9. Similar results were found across the deeper layers of the tree, with a large proportion converging to no function, i.e. a tree shape that does not involve nodes at position (d, x).

There were two main exceptions to this convergence behaviour: firstly, when the sample size for a particular (d, x) coordinate was small, convergence may not be seen for that position. This was particularly likely to occur deep within the node grid and for problems with large node arities, because a given row in the tree grid has a^k x indices, where k is the greatest arity of functions in the function set. The more significant exception was lower levels in the tree in general, where crossover was more likely to impact convergence: crossover was exponentially more likely to impact a schema at a lower position within the tree than at a higher one, and thus there was a counterbalance to the convergence resulting from rank-based selection. However, the average impact of a node deep in the tree in terms of overall semantics was low, so we did not consider it within our modelling.

Table 18.1 Benchmarks used

Benchmark	Target equation	Paper
Keijzer-6	$\sum_i^x \frac{1}{i}$	Keijzer [3]
Korns-12	$2 - 2.1 \cos(9.8x) \sin(1.3w)$	Korns [4]
Vladislavleva-4	$\frac{10}{5+\sum_{i=1}^{5}(x_i-3)^2}$	Vladislavleva et al. [13]

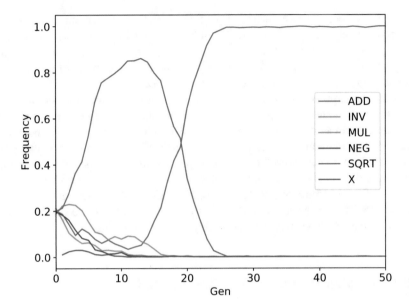

Fig. 18.4 Node frequency for Keijzer6 $d = 1$ $x = 0$

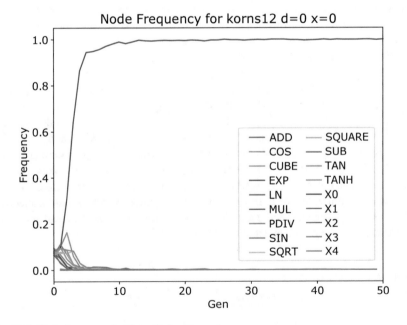

Fig. 18.5 Node frequency for Korns12 $d = 0$ $x = 0$

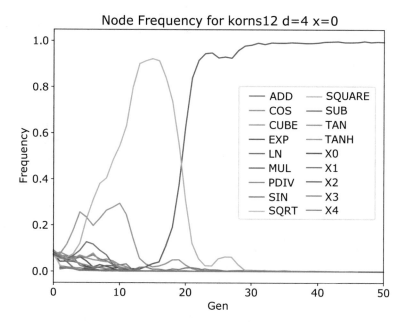

Fig. 18.6 Node frequency for Korns12 $d = 4$ $x = 0$

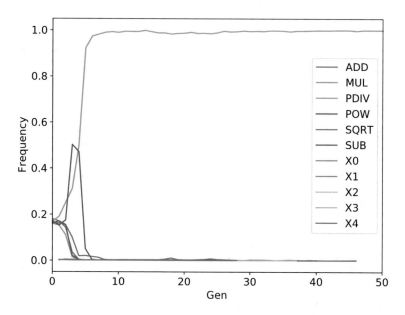

Fig. 18.7 Node frequency for Vladislavleva4 $d = 0$ $x = 0$

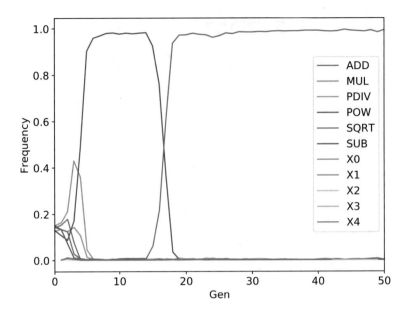

Fig. 18.8 Node frequency for Vladislavleva4 $d = 1 x = 1$

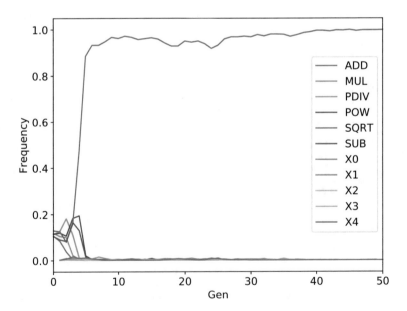

Fig. 18.9 Node frequency for Vladislavleva4 $d = 3 x = 3$

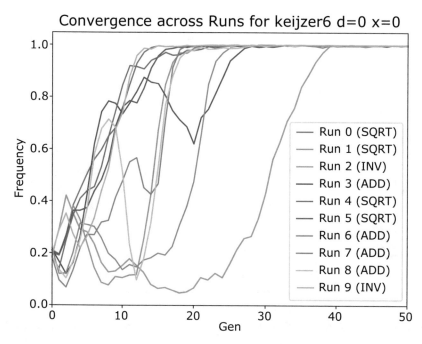

Fig. 18.10 Convergence across many runs for Keijzer6 root node. Runs can be driven towards a particular node based on the sample achieved. A handful of functions dominate each position, in this case ADD and SQRT

Figure 18.10 shows an example of convergence behaviour across runs, where each line is the 'winning' function from a run. It was often the case that two functions demonstrated strong covariance and due to the stochastic nature of GP, convergence to any given function was not certain, but biased towards those with high covariance.

Does covariance between schema membership and rank fitness drive convergence, i.e. is our model accurate? We plot graphs showing both convergence and covariance in Figs. 18.11, 18.12, and 18.13. Schema and frequency are colour-matched, the dashed lines indicating the covariance (right-hand axis). Clear spikes in covariance can be seen prior to population convergence. Note that if we were to anticipate convergence, the best predictor would be covariance, a signal which is clear far in advance of the rising edge of the convergence curve; this suggests it may be possible to exploit covariance measures to accelerate convergence and potentially improve the performance of GP. Note also that over time the clear covariance signal disappears, with covariance tending toward 0. Figure 18.13 shows that after a brief spike in covariance and a subsequent spike in node frequency, all frequencies went to 0 in this example, indicating that node (2,2) is unoccupied.

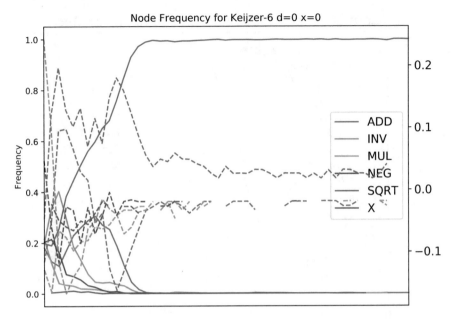

Fig. 18.11 Node frequency (lines, left scale) and covariance (dashed lines, right scale) for Keijzer6 root node

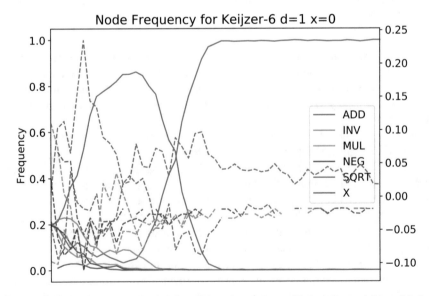

Fig. 18.12 Node frequency (lines, left scale) and covariance (dashed lines, right scale) for Keijzer6 $d = 1$ $x = 0$

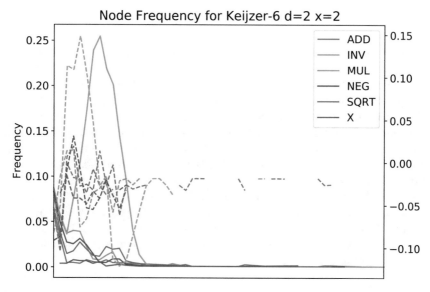

Fig. 18.13 Node frequency (lines, left scale) and covariance (dashed lines, right scale) for Keijzer6 $d = 2$ $x = 2$

18.5   Ways to Improve GP

Based on our model of how GP searches the solution space, we propose a possible improvement to GP: a point schema should be "frozen" if high covariance is found between a given root-node schema and rank fitness. Freezing this node can be achieved by assigning the worst fitness to all individuals in the population that *do not* belong to that schema, forcing convergence. After a new generation has been produced and evaluated, we can check to see if any function dominates the schema: it must dominate both in terms of being substantial in number (e.g. >20% of the population size) and exhibit high covariance. Alternatively, we can simply replace nodes at coordinate (d, x) in all trees with the one we want to freeze (which should have the highest covariance with fitness), and then continue the run.

An even more sophisticated alternative is to replace GP entirely, using a sampling algorithm that descends from the root and systematically samples each schema in a 'racing' algorithm, selecting the node in a particular location that exhibits the highest covariance.

These proposals require more theoretical and empirical work and we anticipate a need for careful comparison with existing algorithms and across more benchmark problems, all of which is beyond the scope of this chapter but will be subject to a future investigation.

18.6   Related Work

There are two areas of work that are related to our modeling: the development of schema theories for GP, and the empirical examination of the role of crossover. One of the motivating results for this work is the rigorous evaluation of crossover in GP [15], which found crossover was not greatly beneficial to the search. That work was itself based on early controversial and influential work by Luke and Spector [5, 6]. These papers cast doubt on the building block hypothesis and the utility of crossover in reassembling solutions which was—at the time—the predominant search operator used by GP researchers. Other early work came to the conclusion that both mutation and crossover operators contribute to the success of GP runs and that it is beneficial to apply both operators with at least 5% probability, at least in Linear GP [1].

GP Schema theory has a long and varied history, with many different approaches proposed by Rosca [12], O'Reilly [9], and Poli and McPhee [10]; the latter included a multi-paper exposition on schema theory for general crossover, and introduced the grid system used in this paper.

18.7   Conclusion

In this chapter, we have demonstrated how GP can be modeled using a simple single-node schema definition, and that empirical results show convergence throughout the population across benchmarks and experimental runs. In our model, crossover and mutation in GP act simply as "sampling" operators, in agreement with previous empirical experimentation. As the sampling occurs disproportionately towards the bottom of the tree, the driving force in GP is selection, making convergence inevitable. We propose modifying GP based on this model.

This work is incomplete: in particular, we need to examine convergence statistically across every node in a tree's grid; more benchmarks, particularly non-symbolic expression benchmarks, should be examined; and a full covariance-based algorithm should be implemented—either as a modification to GP, or else as a re-imagined sampling algorithm.

Future work includes testing this model in other ways: for example, modifying genetic operators to manipulate the disturbance and construction of simple schemata, and confirming the results agree with prediction.

Acknowledgements   WB acknowledges funding from the Koza Endowment provided by MSU.

References

1. Banzhaf, W., Francone, F.D., Nordin, P.: The effect of extensive use of the mutation operator on generalization in genetic programming using sparse data sets. In: H.M. Voigt, W. Ebeling, I. Rechenberg, H.P. Schwefel (eds.) International Conference on Parallel Problem Solving from Nature (PPSN-96), pp. 300–309. Springer (1996)
2. Blickle, T., Thiele, L.: A mathematical analysis of tournament selection. In: L. Eshelman (ed.) International Conference on Genetic Algorithms (ICGA-95), pp. 9–16. Morgan Kaufmann, San Francisco (1995)
3. Keijzer, M.: Improving symbolic regression with interval arithmetic and linear scaling. In: C. Ryan, T. Soule, M. Keijzer, E. Tsang, R. Poli, E. Costa (eds.) European Conference on Genetic Programming, EuroGP 2003, pp. 70–82. Springer, Berlin, Heidelberg (2003)
4. Korns, M.F.: Accuracy in symbolic regression. In: R. Riolo, E. Vladislavleva, J. Moore (eds.) Genetic Programming Theory and Practice IX, pp. 129–151 (2011)
5. Luke, S., Spector, L.: A comparison of crossover and mutation in genetic programming. In: European Conference on Genetic Programming, pp. 240–248. Springer (1997)
6. Luke, S., Spector, L.: A revised comparison of crossover and mutation in genetic programming. In: European Conference on Genetic Programming, pp. 208–213. Springer (1998)
7. Moraglio, A., Krawiec, K., Johnson, C.G.: Geometric Semantic Genetic Programming. In: International Conference on Parallel Problem Solving from Nature, pp. 21–31. Springer (2012)
8. O'Reilly, U.M., Oppacher, F.: Using building block functions to investigate a building block hypothesis for genetic programming. Santa Fe Inst., Santa Fe, NM, Working Paper pp. 94–02 (1994)
9. O'Reilly, U.M., Oppacher, F.: The troubling aspects of a building block hypothesis for genetic programming. In: Foundations of Genetic Algorithms (FOGA-95), vol. 3, pp. 73–88. Elsevier (1995)
10. Poli, R., McPhee, N.F.: General schema theory for genetic programming with subtree-swapping crossover: Part i. Evolutionary Computation 11(1), 53–66 (2003)
11. Price, G.R.: Selection and covariance. Nature 227, 520–521 (1970)
12. Rosca, J.P., Mallard, D.H.: Rooted-tree schemata in genetic programming. In: K.E. Kinnear, W.B. Langdon, L. Spector, P.J. Angeline, U.M. O'Reilly (eds.) Advances in Genetic Programming, Vol 3, pp. 243–271 (1999)
13. Vladislavleva, E.J., Smits, G.F., Den Hertog, D.: Order of nonlinearity as a complexity measure for models generated by symbolic regression via pareto genetic programming. IEEE Transactions on Evolutionary Computation 13, 333–349 (2008)
14. White, D.R., McDermott, J., Castelli, M., Manzoni, L., Goldman, B.W., Kronberger, G., Jaśkowski, W., O'Reilly, U.M., Luke, S.: Better GP benchmarks: community survey results and proposals. Genetic Programming and Evolvable Machines 14(1), 3–29 (2013)
15. White, D.R., Poulding, S.: A rigorous evaluation of crossover and mutation in genetic programming. In: European Conference on Genetic Programming, pp. 220–231. Springer (2009)

Chapter 19
An Evolutionary System for Better Automatic Software Repair

Yuan Yuan and Wolfgang Banzhaf

19.1 Introduction

Automatic software repair [13, 39, 49] aims to fix bugs in software automatically, generally relying on a specification. When a test suite is considered as the specification, the paradigm is called *test-suite based repair* [39]. The test suite should contain at least one negative (i.e., initially failing) test that triggers the bug to be fixed and a number of positive (i.e., initially passing) tests that define the expected program behavior. In terms of test-suite based repair, a bug is regarded to be *fixed* or *repaired*, if a created patch makes the entire test suite pass. Such a patch is referred to as a *test-adequate patch* [33] or a *plausible patch* [44].

Evolutionary repair approaches [49] are a popular category of techniques for test-suite based repair. These approaches determine a search space potentially containing correct patches, then use evolutionary computation (EC) techniques, particularly genetic programming (GP) [2, 4, 21], to explore that search space. A major characteristic of evolutionary repair approaches is that they have high potential to fix multi-location bugs, since GP can manipulate multiple likely faulty locations at a time. However, GenProg [12, 25, 27, 51], the most well-known approach of this kind, does not fulfill the potential in multi-location bug fixing according to large-scale empirical studies [33, 44], partly due to the search ability of its underlying GP [42, 44, 57]. To tackle this issue, our previous work introduced ARJA [57], which

Y. Yuan (✉)
Department of Computer Science and Engineering & Beacon Center, Michigan State University, East Lansing, MI, USA
e-mail: yyuan@cse.msu.edu; yyuan@msu.edu

W. Banzhaf
BEACON Center for the Study of Evolution in Action and Department of Computer Science and Engineering, Michigan State University, East Lansing, MI, USA
e-mail: banzhaf@msu.edu

© Springer Nature Switzerland AG 2020
W. Banzhaf et al. (eds.), *Genetic Programming Theory and Practice XVII*, Genetic and Evolutionary Computation, https://doi.org/10.1007/978-3-030-39958-0_19

uses a novel multi-objective GP approach with better search ability to explore the search space. Although ARJA has achieved much improved performance and also demonstrated its strength in multi-location repair, major challenges [26] still remain for evolutionary software repair.

The first challenge is how to construct a reasonable search space that is more likely to contain correct patches. In this respect, GenProg and ARJA exploit the statement-level *redundancy assumption* [36] (also called *plastic surgery hypothesis* [3]). That is, they only conduct statement-level changes and use existing statements in the buggy program for replacement or insertion. The problem here is that fix statements randomly excerpted from somewhere in the current buggy program may have little pertinence to the likely-buggy statement to be manipulated. Due to this problem, GenProg usually generates patches overfitting the test suite or even fails to fix a bug. To relieve the issue, Kim et al. [20] proposed PAR, which exploits *repair templates* to produce program variants. Each template specifies one type of program transformation and is derived from common fix patterns (e.g., adding a null-pointer checker for an object reference) manually learned from human-written patches. Compared to GenProg, PAR usually works in a more promising search space, since the program transformations performed by PAR are more targeted. Nevertheless, as can be inferred from the results in [57], the redundancy-based approaches can really fix some bugs that cannot be fixed by typical template-based approaches (e.g., PAR and ELIXIR [46]) which implies that combining the redundancy assumption and repair templates to generate fix statements could further improve repair effectiveness.

The second challenge is how to design a search algorithm that can navigate the search space more effectively. The combination of the statement-level redundancy assumption and repair templates will lead to a much larger search space, thereby making this challenge more serious. Recent studies [42, 57] have indicated that compared to using GenProg's patch representation, using a lower-granularity patch representation that decouples the partial information of an edit can significantly improve the search ability of GP in bug repair. However such representations are specially designed for statement-level edits and cannot be directly used for template-based edits (usually occurring at the expression level). Besides the patch representation, the fitness function is another important factor that influences the search ability of GP. In existing evolutionary repair approaches, the fitness function is generally defined based on how many test cases a patched program passes. However this kind of fitness function can only provide a binary signal (i.e, passed or failed) for a test case and cannot measure how close a modified program is to pass a test case. In consequence, there may be a large number of plateaus in the search space [11, 26, 44], thereby trapping GP.

The third challenge is how to alleviate patch overfitting [47]. Evolutionary repair approaches can usually find a number of plausible patches within a computing budget. But most of these patches may be incorrect in general, by just overfitting the given test suite. To pick correct patches more easily, it is necessary to include a post-processing step for these approaches, which can filter out incorrect patches (i.e., *overfit detection*) or rank the plausible patches found (i.e., *patch ranking*).

However, almost all existing evolutionary repair systems, including GenProg, PAR, and ARJA, do not implement such a step.

In this chapter, we describe ARJA-e, a new evolutionary repair system for Java programs that aims to address the above three challenges. To determine a search space that is more likely to contain correct patches, ARJA-e combines two sources of fix ingredients (i.e., the statement-level redundancy assumption and repair templates) with contextual analysis based search space reduction, thereby leveraging their complementary strengths. To encode patches in GP more properly, ARJA-e unifies the edits at different granularities into statement-level edits, and then uses a new lower-granularity patch representation that is characterized by the decoupling of statements for replacement and statements for insertion. Furthermore, ARJA-e uses a finer-grained fitness function that can make full use of semantic information contained in the test suite, which is expected to better guide the search of GP. To alleviate patch overfitting, ARJA-e includes a post-processing tool that can serve the purposes of overfit detection and patch ranking.

19.2   Background and Motivation

19.2.1   Related Work

Our system belongs to the class of evolutionary repair approaches which explore a repair search space using evolutionary algorithms. GenProg [25, 27], PAR [20], GenProg with anti-patterns [48] and ARJA [57] all fall into this category. Their basic ideas have been described in Sect. 19.1. ARJA-e organically combines the characteristic components of all these approaches, making it distinctly different from any of them. Several approaches employ other kinds of search algorithms, instead of EAs, to traverse GenProg's search space (e.g., RSRepair [43] uses random search and AE [50] uses an adaptive search strategy).

Inspired by the idea of using templates [20], some repair approaches (e.g., SPR [31] and ELIXIR [46]) employ a set of richer templates (or code transformations) that are defined manually. Genesis [30] aims to automatically infer such code transformations from successful patches. Cardumen [35] mines repair templates from the program under repair. Similar to these approaches, ARJA-e uses templates extended and enhanced from those in PAR.

Beyond the current buggy program and its associated test suite, some approaches exploit other information to help the repair process. HDRepair [24] uses mined historical bug fixes to guide its random search. ACS [55] uses the information of javadoc comments to rank variables. SearchRepair [19] and ssFix [53] both use existing code from an external code database to find potential repairs.

A number of existing approaches infer semantic specifications from the test cases and then use program synthesis to generate a repair that satisfies the inferred specifications. These are usually categorized as semantics-based approaches. SemFix [41] is a pioneer in this category. Other typical approaches of this kind include

DirectFix [37], QLOSE [8], Angelix [38], Nopol [56], JFix [22] and S3 [23]. Recently, machine learning techniques have been used in software repair. Prophet [32] uses a probabilistic model to rank the candidate patches over the search space of SPR. DeepFix [14] uses deep learning to fix common programming errors.

19.2.2 Motivating Examples

In this subsection, we take real bugs as examples to illustrate the key insights motivating the design of ARJA-e.

Figure 19.1 shows the human-written patch for bug Math85 from the Defects4J [18] dataset. To correctly fix this bug, only a slight modification is required (i.e., change >= to >), as shown in Fig. 19.1. However, redundancy-based approaches (e.g., GenProg [25, 27], RSRepair [43] and AE [50]) usually cannot find a correct patch for this bug since the fix statement used for replacement (i.e., if (fa * fb > 0){...}) or semantically equivalent ones do not happen to appear elsewhere in the buggy program. In contrast, some template-based approaches (e.g., jMutRepair [10, 34] and ELIXIR [46]) are very likely to fix the bug correctly since changing of infix boolean operators is a specified repair action in such approaches. In addition, GenProg can easily overfit the given test suite [44] by deleting the whole buggy if statement: if (fa * fb >= 0){...}), leading to a plausible but incorrect patch.

Figure 19.2 shows the human-written patch for bug Math39 from Defects4J. To correctly repair the bug, an if statement with relatively complex control logic should be inserted before the buggy code, as shown in Fig. 19.2. However, for approaches only based on repair templates, the bug is hard to fix correctly, because this fix generally does not belong to a common fix pattern and is difficult to be encoded with templates. In contrast, approaches that exploit the redundancy

```
1   public  static  double[]  bracket (...)  {  ...
2  −    if  (fa * fb >= 0.0) {
3  +    if  (fa * fb > 0.0) {
4        throw  new  ConvergenceException (...) ;  }    ...  }
```

Fig. 19.1 The human-written patch for bug Math85

```
1   public  void  integrate (...)  throws  ...  {  ...
2  +  if  (forward) {
3  +    if ( stepStart + stepSize >= t) { stepSize = t − stepStart ; }
4  +  } else {
5  +    if ( stepStart + stepSize <= t) { stepSize = t − stepStart ; }  }
6     ...  }
```

Fig. 19.2 The human-written patch for bug Math39

assumption can potentially find a correct patch for the bug, because the following
`if` statement

---

if ((forward && (stepStart + stepSize > t)) ‖ ((!forward) && (stepStart + stepSize <
t))) { stepSize = t − stepStart ; }

---

happens to be in the buggy program elsewhere, which is semantically equivalent to
the one inserted by human developers.

From the above examples, it can be seen that redundancy- and template-based
approaches potentially have complementary strengths in bug fixing. We aim to
combine both statement-level redundancy assumption and repair templates, to
generate potential fix ingredients. Such a combination will lead to a much larger
search space, posing a great challenge to the search algorithm. So we will also
introduce several strategies to properly reduce the search space and enhance the
search algorithm with a new lower-granularity patch representation.

19.3   Overview of ARJA-e

The input of ARJA-e is a buggy program associated with a JUnit test suite. ARJA-e
basically aims to make all these test cases pass by modifying the buggy program.
First, we use the fault localization technique called Ochiai [5] to select n likely-
buggy statements (LBSs) with the highest suspiciousness. For the j-th LBS, we
determine three sets denoted by R_j, I_j and O_j. R_j is the set of statements that
can be used to replace the LBS, I_j is the set of statements that can be used for
insertion before the LBS, and O_j is a subset of three operation types: "delete",
"replace" and "insert". To find simpler patches, we uses a multi-objective GP to
explore the determined search space, with the guidance of a finer-grained fitness
function. Through evolutionary search, ARJA-e can usually find a number of
plausible patches. However, many of these patches may overfit the test suite and
would thereby be not correct. To alleviate the patch overfitting issue, we develop
a post-processing tool which can identify overfitting patches or rank the plausible
patches found by ARJA-e.

In the following sections, we will detail how to shape the search space (i.e.,
determine R_j, I_j and O_j, see Sect. 19.4), how to conduct multi-objective search
(see Sect. 19.5) and how to alleviate patch overfitting (see Sect. 19.6).

19.4   Shaping the Search Space

19.4.1   Exploiting the Statement-Level Redundancy
Assumption

For each LBS selected, we first collect the statements within the package where the
LBS resides, and then ignore those statements that are not in-scope at the destination
of the LBS or violates the complier constraints. For each of the remaining statements

(denoted by s), we further check the program context. Our insight is that if replacing the LBS with s is a promising manipulation, s should generally exhibit a certain *similarity* to the LBS; and if it is potentially useful to insert s before the LBS, s should generally have a certain *relevance* to the context surrounding the LBS. In the following, we describe how to quantify such similarity and relevance.

Suppose V_s and V_{LBS} are the sets of variables (including local variables and fields) used by s and the LBS respectively. We define the similarity between s and the LBS as the Jaccard similarity coefficient between sets V_s and V_{LBS}:

$$sim(s, \text{LBS}) = \frac{|V_{\text{LBS}} \cap V_s|}{|V_{\text{LBS}} \cup V_s|} \tag{19.1}$$

Note that when collecting fields used by a statement, we also consider the fields accessed by invoking the methods in the current class.

In the method where the LBS resides, suppose V_{bef} and V_{aft} are the sets of variables used by k statements before and after the LBS, respectively, where k is set to 5 by default. We define the relevance of s to the context of LBS as follows:

$$rel(s, \text{LBS}) = \frac{1}{2}\left(\frac{|V_s \cap V_{\text{bef}}|}{|V_s|} + \frac{|V_s \cap V_{\text{aft}}|}{|V_s|}\right) \tag{19.2}$$

Eq. (19.2) indeed averages the percentages of the variables in V_s that are covered by V_{bef} and V_{aft}.

If $|V_{\text{LBS}} \cup V_s| = 0$, $sim(s, \text{LBS})$ is set to 1, and if $|V_s| = 0$, $rel(s, \text{LBS})$ is set to 0. So $sim(s, \text{LBS}) \in [0, 1]$ and $rel(s, \text{LBS}) \in [0, 1]$. Only when $sim(s, \text{LBS}) > \beta_{\text{sim}}$ can s be put into R_j (i.e., the set of candidate statements for replacement), and only when $rel(s, \text{LBS}) > \beta_{\text{rel}}$ can s be put into I_j (i.e., the set of candidate statements for insertion), where β_{sim} and β_{rel} are predetermined threshold parameters.

19.4.2 Exploiting Repair Templates

In ARJA-e, we also use 7 repair templates to manipulate the LBS, which are mainly extended from templates used in PAR. These templates are described in Table 19.1.

Template ER replaces an abstract syntax tree (AST) node element in a LBS with another compatible one. Table 19.2 lists the elements that can be replaced and also shows alternative replacers for each kind of elements. This template generalizes the templates "Parameter Replacer" and "Method Replacer" used in PAR. Several replacement rules are inspired by recent template-based approaches (e.g., replacing a primitive type with widened type follows ELIXIR [46] and replacing x with f(x) follows the transformation schema in REFAZER [45]).

Unlike PAR which applies templates on-the-fly (i.e., during the evolutionary process), ARJA-e executes the above seven repair templates in an offline manner. Specifically, we perform all the possible transformations defined by the templates

Table 19.1 The description of repair templates used in this study

No.	Template name	Description
1	Null pointer checker (NPC)	Add an `if` statement before a LBS to check whether any object reference in this LBS is `null`
2	Range checker (RC)	Add an `if` statement before a LBS to check whether any array or list element access in this LBS exceeds the upper or lower bound
3	Cast checker (CC)	Add an `if` statement before a LBS to assure that the variable or expression to be converted in this LBS is an instance of casting type
4	Divide-by-Zero checker (DC)	Add an `if` statement before a LBS to check whether any divisor in this LBS is 0
5	Method parameter adjuster (MPA)	Add, remove or reorder the method parameters in a LBS if this method has overloaded methods
6	Boolean expression adder or remover (BEAR)	For a condition branch (e.g., `if`), add a term to its predicate (with `&&` or `||`), or remove a term from its predicate
7	Element replacer (ER)	Replace an AST node element (e.g., variable or method name) in a LBS with another one with compatible type

Table 19.2 List of replacement rules for different elements

Element	Format	Replacer
Variable	x	(1) The visible fields or local variables with compatible type (2) A compatible method invocation in the form of $f()$ or $f(x)$
Field access	e.g., this .a	The same as above
Qualified name	a.b	The same as above
Method name	f (...)	The name of another visible method with compatible parameter and return types
Primitive type	e.g., int	A widened type, e.g., float to double
Boolean literal	true or false	The opposite boolean value
Number literal	e.g., 1 or 0.5	Another number literal located in the same method
Infix operators	e.g., + or >	A compatible infix operator, e.g., + to −, > to >=
Prefix/postfix operators	e.g., ++	The opposite prefix/postfix operator, e.g., ++ to −−
Assignment operators	e.g., +=	The opposite assignment operator, e.g., += to −=
Conditional expression	a ? b : c	b or c

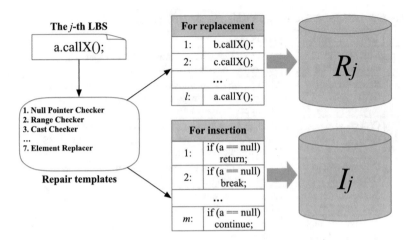

Fig. 19.3 Illustration of the offline execution of templates

for each LBS before searching for patches. Then each LBS can derive a number of new statements, each of which can either replace the LBS or be inserted before it. So various template-based edits (usually at the expression-level) are abstracted into two types of statement-level edits (i.e., replacement and insertion). These statements for replacement and insertion are added into R_j and I_j respectively. For the LBS `a.callX()`, Fig. 19.3 illustrates the way to exploit the templates in ARJA-e. Note that we do not consider similarity and relevance as in Sect. 19.4.1 since the statements generated by templates are highly targeted. Moreover, we only apply a template to a single AST node at a time to avoid combinatorial explosion. For example, we do not simultaneously modify `a` and `callX` in `a.callX()` using the template ER.

19.4.3 Initialization of Operation Types

The deletion operation should be executed carefully because it can easily lead to the following two problems: It can (1) cause a compiler error of the modified code; or (2) generate overfitting patches [44]. To address the first problem, we use the two rules defined in [57], that is, if a LBS is a variable declaration statement or a `return`/`throw` statement which is the last statement of a method not declared `void`, we disable the deletion operation for this LBS. To address the second problem, we use the 5 anti-delete patterns defined in [48]. If a LBS follows any of these patterns, we ignore the deletion operation. For example, according to one of the anti-delete patterns, if a LBS is a control statement (e.g., `if` statement or loops), deletion of the LBS is disallowed.

19.5   Multi-Objective Evolution of Patches

19.5.1   Patch Representation

To encode a patch as a genome in GP, we first number the LBSs and the elements in R_j, I_j and O_j respectively, starting from 1, where $j \in \{1, 2, \ldots, n\}$. All the IDs are fixed throughout the search.

A solution (i.e., a patch) to the program repair problem is encoded as $\mathbf{x} = (\mathbf{b}, \mathbf{u}, \mathbf{p}, \mathbf{q})$, which contains four different parts each being a vector of size n. In the solution $\mathbf{x}$, $b_j \in \{0, 1\}$ indicates whether the j-th LBS is to be edited or not; $u_j \in \{1, 2, \ldots, |O_j|\}$ indicates the u_j-th operation type in O_j is used for the j-th LBS; $p_j \in \{1, 2, \ldots, |R_j|\}$ means that if replace operation is used, the p_j-th statement in R_j will be selected to replace the j-th LBS; and $q_j \in \{1, 2, \ldots, |I_j|\}$ means that if insert operation is used, the q_j-th statement in I_j will be inserted before the j-th LBS. Figure 19.4 illustrates the new lower-granularity patch representation. Suppose the j-th LBS is a.callX(); in this figure, then the edit on the j-th LBS is: replace a.callX(); with b.callX();.

19.5.2   Finer-Grained Fitness Function

To evaluate the fitness of an individual $\mathbf{x}$, we still use a bi-objective function as in the original ARJA [57]. The first objective (i.e., $f_1(\mathbf{x})$) is the patch size, which is exactly the same as that in ARJA. The second objective (i.e., $f_2(\mathbf{x})$) is the weighted failure rate. Different from that in ARJA, we compute $f_2(\mathbf{x})$ through finer-grained analysis of test execution in this study, in order to provide smoother gradient for the genetic search to navigate the search space. Since our repair system targets Java,

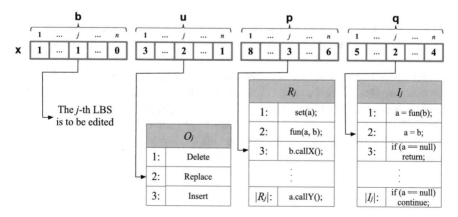

Fig. 19.4 Illustration of the new lower-granularity patch representation

our implementation is based on the JUnit [7] framework. Specifically, we define a metric to measure the degree of violation for each assertion, which we call *assertion distance*. Suppose an assertion (denoted by e) asserting x and y are equal to within a positive δ: assertEquals(x, y, δ), then the assertion distance $d(e)$ is computed as:

$$d(e) = \begin{cases} v(|x - y| - \delta), & |x - y| \geq \delta \\ 0, & |x - y| < \delta \end{cases} \tag{19.3}$$

Here, $v(x)$ is a normalizing function in $[0, 1]$ and we use the one suggested in [1]: $v(x) = x/(x + 1)$.

After executing a program variant $\mathbf{x}$ over a test case t, we can compute a metric $h(\mathbf{x}, t) \in [0, 1]$ to indicate how badly $\mathbf{x}$ fails the test case t by using the collected assertion distances. This metric is defined as follows:

$$h(\mathbf{x}, t) = \frac{\sum_{e \in E(\mathbf{x}, t)} d(e)}{|E(\mathbf{x}, t)|} \tag{19.4}$$

where $E(\mathbf{x}, t)$ is the set of executed assertions by $\mathbf{x}$ over t, and $d(e)$ is the assertion distance for the assertion e. Based on $h(\mathbf{x}, t)$, $f_2(\mathbf{x})$ can be formulated as follows:

$$f_2(\mathbf{x}) = \frac{\sum_{t \in T_{pos}} h(\mathbf{x}, t)}{|T_{pos}|} + w \times \left(\frac{\sum_{t \in T_{neg}} h(\mathbf{x}, t)}{|T_{neg}|} \right) \tag{19.5}$$

where $w \in (0, 1]$ is a parameter that can introduce a bias toward negative test cases.

19.5.3   Genetic Operators

Genetic operators, including crossover and mutation, are used to produce the offspring individuals in GP. Crossover is applied to each part of the patch representation separately, in order to inherit good genetic materials from parents. For all four parts, we employ the half uniform crossover (HUX) operator.

We apply a guided mutation to the information of a single selected LBS. To be specific, we first use roulette wheel selection to choose a LBS, where the j-th LBS is chosen with a probability of $susp_j/\sum_{j=1}^{n} susp_j$; suppose that the j-th LBS is finally selected, then we apply bit flip mutation to b_j and uniform mutation to u_j, p_j and q_j respectively. Figure 19.5 illustrates the crossover and mutation operations, where only a single offspring is shown for brevity.

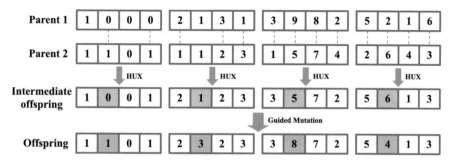

Fig. 19.5 Illustration of the crossover and mutation

19.5.4 Multi-Objective Search

With the patch representation, fitness function and genetic operators designed above, any multi-objective evolutionary algorithm can serve the purpose of searching for patches. In this work, we basically employ NSGA-II [9] as the multi-objective search framework. To initialize the population, we combine the fault localization result and randomness: for the first part (i.e., **b**), b_j is initialized to one with a probability of $susp_j \times \mu$, where $\mu \in (0, 1)$ is a predefined parameter; and the remaining three parts (i.e., **u**, **p**, **q**) are just initialized randomly. After population initialization, the search algorithm iterates over generations until the stopping criterion is satisfied.

19.6 Alleviating Patch Overfitting

19.6.1 Overfit Detection

For overfit detection, we take a buggy program, a set of positive test cases and a plausible patch as input, and determine whether or not this plausible patch is an overfitting patch. Our approach is based on the assumption that the buggy program will perform correctly on the test inputs encoded in the positive test cases.

Figure 19.6 shows the overall process of this approach. First, given a plausible patch and a buggy program, we can localize the methods where the statements will be modified by the patch. Then we instrument the bytecode of these methods in the buggy program. For each method, the instrumentation is conducted at its entry point and all its possible exit points. At the entry point, we inject new bytecode to save the *input* of the method, including all the method parameters and the current object this (i.e., the object whose method is being called), into a file. At each exit point, we inject new bytecode to save the *output* of the method, including the return value, the current object this and the reference-type method parameters, into another file.

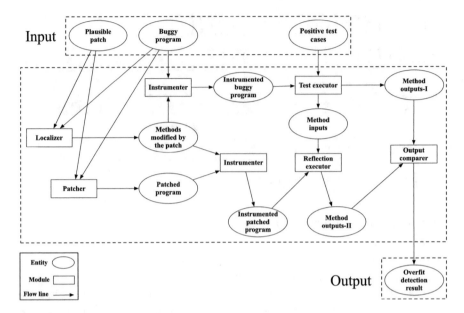

Fig. 19.6 The overview of our overfit detection approach

Note that if a method to be instrumented is a static method, we just ignore the current object. To save the objects, we leverage the Java serialization technique. This technique can convert object state to a byte stream that can be reverted back into a copy of the object.

With the instrumented buggy program, we run the positive test cases so that we can capture a number of input-output pairs for the localized methods. Suppose that there are K such pairs, denoted by a set $PA = \{(In_1, Out_1), (In_2, Out_2), \ldots, (In_K, Out_K)\}$. According to our assumption, all these input-output pairs should reflect the correct program behavior. In order to judge patch correctness, we will feed these inputs $In_1, In_2, \ldots, In_K$ into the corresponding methods in the patched program so as to see whether the correct outputs can be obtained. Specifically, for each input-out pair $(In_i, Out_i) \in PA$ collected previously, we deserialize In_i from the file and use the Java reflection technique to invoke the corresponding method in the instrumented patched program with the deserialized input In_i, so that we can collect the method output Out_i'. Lastly, we compare every Out_i' with the corresponding Out_i, and if there exists any difference, we identify the plausible patch as an overfitting patch that is incorrect.

19.6.2   Patch Ranking

ARJA-e can sometimes output more than one plausible patch (with the same patch size) for a bug. As a post-processing step, we design a heuristic procedure to rank these patches. For this ranking purpose, we first define three metrics for a patch. The first metric, denoted by $Susp$, represents the summation of the suspiciousness for the LBSs modified by the patch. The second metric, denoted by $Dist$, is based on our overfit detection approach. Recall that for the purpose of overfit detection, we only need to know whether there is a difference between Out_i and Out_i', where $i = 1, 2, \ldots, K$. Here we want to quantify such a difference. To do this, we deserialize Out_i and Out_i' and extract all primitive data and string data contained in the two outputs in a recursive way. Similar to the computing of assertion distance, we can easily compute the distance for each corresponding primitive/string data and normalize it to $[0, 1]$. Then $Dist$ is calculated as the average of these normalized distances for all outputs. Before defining the third metric, we determine a preference relation of operation types in our system. We prefer the operation type that is generally less likely to bring in side effects, and the preference relation is: NPC/RC/CC/DC $\prec$ MPA $\prec$ ER $\prec$ BEAR $\prec$ SR/SI $\prec$ SD. Here SR and SI mean statement replacement and insertion based on the redundancy assumption respectively, and SI means statement deletion. The others are all template-based operations that can be referred to in Sect. 19.4.2. We assign a preference score for each operation type: SD is scored 1, SR and SI are scored 2, BEAR is scored 3 and so on. With these scores, the second metric for a patch, denoted by $Pref$, is defined as the sum of scores of operation types contained in the patch. For $Susp$ and $Pref$, larger is better; whereas for $Dist$, smaller is better.

When comparing two patches in the ranking, $Susp$, $Dist$ and $Pref$ are considered in sequence until the two patches can be distinguished. If all the three metrics cannot distinguish the two patches, the patch found earlier is ranked higher.

19.7   Experimental Design

19.7.1   Research Questions

We intend to answer the following research questions:

RQ1: How effective is ARJA-e compared to state-of-the-art repair systems on real bugs?

RQ2: Can ARJA-e fix bugs in a novel way compared to a human developer?

RQ3: How good is our overfit detection approach?

19.7.2   Dataset of Bugs

We perform the empirical evaluation on a database of real bugs, called Defects4J
[18], which has been extensively used for evaluating Java repair systems [6, 33, 46,
53, 55, 57]. We consider four projects in Defects4J, namely Chart, Time, Lang and
Math. Table 19.3 shows the descriptive statistics of the four projects. In total, there
are 224 real bugs: 26 from Chart (C1–C26), 27 from Time (T1–T27), 65 from Lang
(L1–L65) and 106 from Math (M1–M106).

19.7.3   Parameter Setting

Table 19.4 shows the parameter setting for ARJA-e in the experiments. Note that
crossover and mutation operators presented in Sect. 19.5.3 are always executed,
so the probability (i.e., 1) is omitted in this table. Given the stochastic nature of
ARJA-e, we execute five random trials in parallel for each bug. Each trial of ARJA-
e is terminated after it reaches the maximum number of generations (i.e., 50) or its
execution time exceeds 1 h.

Table 19.3 The descriptive statistics of Defects4J dataset

Project	ID	#Bugs	#JUnit tests	Source KLoC	Test KLoC
Chart	C	26	2205	96	50
Time	T	27	4043	28	53
Lang	L	65	2295	22	6
Math	M	106	5246	85	19
Total		224	13,789	231	128

Table 19.4 The parameter setting for ARJA-e

Parameter	Description	Value
N	Population size	40
γ_{min}	Threshold for the suspiciousness	0.1
n_{max}	Maximum number of LBSs considered	60
β_{sim}	Threshold for similarity	0.3
β_{rel}	Threshold for relevance	0.3
w	Refer to Sect. 19.5.2	0.5

19.8   Results and Discussions

19.8.1   Performance Evaluation (RQ1)

To show the superiority of ARJA-e over the state of the art, we compare ARJA-e with 13 existing repair approaches in terms of the number of bugs fixed and correctly fixed. The 13 approaches are jGenProg [33, 34] (an implementation of GenProg for Java), xPAR (a reimplementation of PAR by Le et al. [24]), Nopol [56], HDRepair [24], ACS [55], ssFix [53], JAID [6], ELIXIR [46], ARJA [57], SimFix [17], CAPGEN [52], SOFIX [29] and SKETCHFIX [16], which include almost all published approaches that have ever been tested on Defects4J. Note that here we use a strict criterion for judging whether a bug is correctly fixed by ARJA-e, that is, a bug is regarded as being correctly fixed only when the plausible patch ranked first (using the procedure in Sect. 19.6.2) is correct.

Table 19.5 shows the comparison results. From Table 19.5, we can see that ARJA-e outperforms all other approaches in terms of the number of fixed bugs and correctly fixed bugs. We further compare ARJA-e with ACS, ssFix and SimFix by analyzing the overlaps among their repair results. ACS, ssFix and SimFix are selected because they show prominent performance among the 13 compared approaches and the IDs of (correctly) fixed bugs are available for them [17, 53, 55]. Figure 19.7 shows the intersection of fixed bugs (in Fig. 19.7a) and correctly fixed bugs (in Fig. 19.7b) between ARJA-e, ACS, ssFix and SimFix, using Venn diagrams. From Fig. 19.7a, ARJA-e performs much better than the other three approaches in

Table 19.5 Comparison with existing repair tools in terms of the number of bugs fixed and correctly fixed (Plausible/Correct)

Project	ARJA-e	jGenProg	xPAR	Nopol	HDRepair[a]	ACS	ssFix
Chart	**18/7**	7/0	–/0	6/1	–/2	2/2	7/2
Lang	**28/9**	0/0	–/1	7/3	–/7	4/3	12/5
Math	**51/21**	18/5	–/2	21/1	–/6	16/12	26/7
Time	**9/2**	2/0	–/0	1/0	–/1	1/1	4/0
Total	**106/39**	27/5	–/3	35/5	–/16	23/18	49/14

Project	JAID	ELIXIR	ARJA[b]	SimFix	CAPGEN	SOFIX	SKETCHFIX
Chart	4/2	7/4	9/3	8/4	–/4	–/5	8/6
Lang	8/1	12/8	17/4	13/9	–/5	–/4	4/1
Math	8/1	19/12	29/10	26/14	–/12	–/13	8/7
Time	0/0	3/2	4/1	1/1	–/0	–/1	1/1
Total	20/4	41/26	59/18	48/28	–/21	–/23	21/15

The best results are shown in bold

"–" Means the data is not available since it is not reported by the original authors

[a]HDRepair generated correct patches for 16 bugs, but only 10 of them were ranked first

[b]In ARJA, a bug is regarded as being fixed correctly if one of its plausible patches is identified as correct

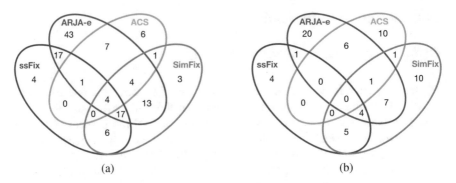

Fig. 19.7 Venn diagram of repaired bugs by ARJA-e, ACS, ssFix and SimFix. (**a**) Test-adequate bug fixing. (**b**) Correct bug fixing

terms of test-adequate bug fixing, and most of the bugs fixed by ACS, ssFix and SimFix can also be fixed by ARJA-e. From Fig. 19.7b, ARJA-e fixes the highest number of bugs correctly (i.e., 39), where 20 bugs cannot be fixed correctly by any of the other three approaches. So ARJA-e indeed complements to the three approaches very well. But it should be noted that the three approaches also show good complementarity to ARJA-e in terms of correct bug fixing. Specifically, ACS, ssFix and SimFix can correctly fix 11, 9 and 16 bugs that cannot be correctly fixed by ARJA-e, respectively. This may be the case because ACS and ssFix are quite different from ARJA-e in technique. ACS aims at performing precise condition synthesis while ssFix uses existing code from a code database. It seems possible to further enhance the performance of ARJA-e by borrowing ideas from ACS and ssFix. For example, we can use a method similar to ACS to generate more accurate conditions for instantiating the template BEAR, or we can reuse the existing code outside the buggy program like ssFix.

In summary, ARJA-e outperforms 13 existing repair approaches by a considerable margin. Specifically, by comparison with the best results, ARJA-e can fix 79.7% more bugs than ARJA (from 59 to 106), and can correctly fix 39.3% more bugs than SimFix (from 28 to 39). Moreover, ARJA-e is an effective approach complementary to the state-of-the-art techniques.

19.8.2   Novelty in Generated Repairs (RQ2)

We found that ARJA-e can fix some bugs in a different way from the human developer. These patches are generally beyond a programmer's expectations, showing the surprising novelty [28]. In the following, we will present case studies to demonstrate this point.

Figure 19.8 shows a correct patch generated by ARJA-e for M94. The following function wants to compute greatest common divisor (GCD) of two integers.

```
1    // MathUtils.java
2    public  static  int  gcd( int  u,  int  v) {
3    −    if  (u * v == 0) {
4    +    if  (u == 0 || v == 0) {   // human−written patch
5    +    if  (sign(u) * v == 0) {  // ARAJ−e patch
6        return  (Math.abs(u) + Math.abs(v));
7        }
8        ...    ...
9    }
```

Fig. 19.8 Human-written patch and correct patch generated by ARJA-e for bug M94

```
1    //   StrBuilder.java
2    public  int  indexOf( String  str ,  int  startIndex ) {  ...
3    −      char[]  thisBuf = buffer ;
4    +      char[]  thisBuf = toCharArray();
5        int  len = thisBuf.length − strLen;
6    −      outer: for ( int  i = startIndex ; i < len; i++) {
7    +      outer: for ( int  i = startIndex ; i <= len; i++) {
8        ... } ... }
```

Fig. 19.9 Correct patch generated by ARJA-e for bug L61

Certainly, if one of the integer is 0, GCD is equal to the sum of the absolute values. The bug is that if u or v is a large integer (e.g., u = 3145728 and v = 294912), then u * v can be equal to 0 by mistake due to overflow. The human will just change u * v == 0 to u == 0 || v ==0. But ARJA-e fixes it in a different way by changing u to sign(u), where sign(u) is the sign function, to avoid the problem leading to the bug.

Figure 19.9 shows the correct patch generated by ARJA-e. A human developer fixes this bug just by replacing line 5 with int len = size - strLen + 1;, where size is the number of characters in the array buffer. Instead, the patch by ARJA-e first replaces buffer in line 3 with toCharArray() which copies all characters in buffer into a new array with length exactly equal to size. Now thisBuff.length is equivalent to size. However, the value of len is still one less than the value it should be, according to the human-written patch. To address this, ARJA-e further changes i < len to i <= len, achieving semantic equivalence.

Figure 19.10 shows the correct patch generated by ARJA-e for bug M56. The human-written patch fixes this bug by firstly deleting lines 3–9 and then replacing line 10 with indices[last] = index - count;. Compared to this human-written patch, the ARJA-e patch just does a slight modification (i.e., replacing idx with MathUtils.sign(idx)). Since idx is positive, its sign MathUtils.sign(idx) is always equal to 1. Hence after line 9, idx is just equal to index - count, where count refers to its initial value at line 3. Consequently, the ARJA-e patch is semantically equivalent to the human-written patch and is therefore correct.

```
 1    // MultidimensionalCounter.java
 2    ...
 3    int idx = 1;
 4    while (count < index) {
 5  −    count += idx;
 6  +    count += MathUtils.sign(idx);
 7       ++idx;
 8    }
 9    −−idx;
10    indices [ last ] = idx;
11    return indices;
```

Fig. 19.10 Correct patch generated by ARJA-e for bug M56

```
 1    // Gamma.java
 2    ...  ...
 3  −    while (Math.abs(an) > epsilon && n < maxIterations) {
 4  +    while (Math.sqrt(an) > epsilon && n < maxIterations) {
 5         n = n + 1.0;
 6         an = an * (x / (a + n));
 7         sum = sum + an;
 8      }
```

Fig. 19.11 Plausible patch generated by ARJA-e for bug M104

Figure 19.11 shows a plausible patch generated by ARJA-e for bug M104. This bug is triggered because the maximum allowed numerical error (MANE) is too large. To fix the bug, the loop should be terminated until `Math.abs(an)` reaches a smaller value. So the human-written patch changes the initial value of of `epsilon` from `10e-9` to `10e-15` in order to ensure a smaller MANE. The ARJA-e patch shown in Fig. 19.11 achieves a similar functionality in a different way, which changes the method invocation `abs` to `sqrt`. Although this patch is not semantically equivalent to the human-written patch, it can make the entire test suite pass and is also indicative of the cause of the bug.

19.8.3   Effectiveness of Overfit Detection (RQ3)

In this subsection, we will evaluate our overfit detection approach described in Sect. 19.6.1. To demonstrate its effectiveness, we compare it with Xiong et al.'s approach (XA) [54], which is currently the state-of-the-art technique for overfit detection and shares certain similarities with our approach. To ensure a fair comparison, we use the version without test case generation for XA. According to [54], this simplified version has already achieved competitive performance compared to the version relying on new test cases.

For the subjects, we consider the first plausible patch found by ARJA-e for each bug (according to RQ1). In addition, we include the patches generated by jGenProg and jKali, which are collected from Martinez et al.'s empirical study [33] on Defects4J. In the end, we collect a dataset of 122 plausible patches by ignoring unsupported patches, where 97 patches are incorrect and 25 patches are correct. The correctness of ARJA-e patches is judged by ourselves, while the correctness of jGenProg and jKali patches is according to Martinez et al.'s analysis [33]. Table 19.6 shows the statistics of this dataset.

Table 19.7 show the comparison results on the dataset per tool. From Table 19.7, for the patches of ARJA-e and jGenProg, our approach can filter out more incorrect patches than XA, while for the patches of jKali, the two approaches can identify the same number of incorrect patches. Moreover, our approach does not filter out any correct patch obtained by jGenProg and jKali, while XA filters out one correct patch (for bug M53) by jGenProg. Note that it was reported in [54] that XA does not exclude any correct patch by jGenProg. This inconsistency may be due to different computing environments. For the patches of ARJA-e, both approaches exclude correct patches by mistake, but our approach only excludes 3 out of 19 correct patches whereas XA excludes 7.

To further understand the performance difference between our approach and XA, Fig. 19.12 shows the intersection of incorrect patches identified by the two approaches. It is interesting to see that our approach complements to XA very well. Specifically, our approach can identify 6 incorrect patches by jGenProg, 4 incorrect patches by jKali and 14 incorrect patches by ARJA-e, respectively, which cannot be identified by XA. In addition, we note that none of the 8 correct patches excluded by XA is also excluded by our approach. Given this strong complementarity, it is very

Table 19.6 Dataset of plausible patches used in RQ3

	ARJA-e		jGenProg		jKali		Total	
Project	Incorrect	Correct	Incorrect	Correct	Incorrect	Correct	Incorrect	Correct
Chart	9	3	6	0	6	0	21	3
Lang	16	4	0	0	0	0	16	4
Math	23	11	13	5	13	1	49	17
Time	7	1	2	0	2	0	11	1
Total	55	19	21	5	21	1	97	25

Table 19.7 Comparison between our approach and Xiong et al.'s approach in overfit detection (The Patches are categorized by repair tools)

Tool	Incorrect	Correct	Incorrect excluded		Correct excluded	
			Our approach	XA	Our approach	XA
ARJA-e	55	19	28(50.91%)	27(49.09%)	3(15.79%)	7(36.84%)
jGenProg	21	5	11(52.38%)	8(38.10%)	0(0.00%)	1(20.00%)
jKali	21	1	9(42.86%)	9(42.86%)	0(0.00%)	0(0.00%)
Total	97	25	48(49.48%)	44(45.36%)	3(12.00%)	8(32.00%)

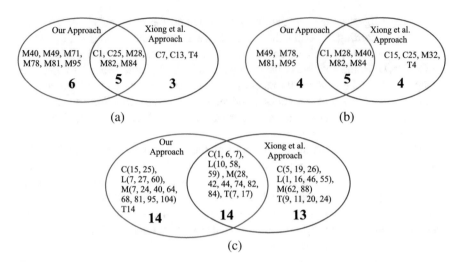

Fig. 19.12 Intersection of incorrect patches identified by our approach and Xiong et al.'s approach. (**a**) jGenProg. (**b**) jKali. (**c**) ARJA-e

promising to further try to improve the accuracy of overfit detection by properly combining the strength of the two approaches.

19.9   Conclusion

In this chapter, we have proposed a new repair system, called ARJA-e, for better evolutionary software repair. By combining two sources of fix ingredients, ARJA-e can conduct complex statement-level transformations, targeted code changes (e.g., adding a null pointer checker), and code changes at a finer-granularity than statement level, which gives ARJA-e great potential to fix various kinds of bugs. To reduce the search space and avoid nonsensical patches, ARJA-e uses a strategy based on a light-weight contextual analysis, which can filter out unpromising replacement and insertion statements, respectively. In order to harness the potential repair power of the search space, ARJA-e first unifies the edits at different granularities into statement-level edits, so as to encode patches in the search space with a lower-granularity patch representation that is characterized by the decoupling of statements for replacement and insertion. With this new patch representation, ARJA-e employs multi-objective GP to navigate the search space. To better guide the search of GP, ARJA-e uses a finer-grained fitness function that can make full use of semantic information provided by existing test cases. Moreover, ARJA-e includes a post-processing tool for alleviating patch overfitting. This tool can serve the purposes of overfit detection and patch ranking.

We have conducted an extensive empirical study on 224 real bugs in Defects4J. The evaluation results show that ARJA-e outperforms 13 existing repair approaches by a considerable margin in terms of both the number of bugs fixed and correctly fixed. Interestingly, we found that ARJA-e can fix some bugs in a creative way, which is usually beyond the exceptions of human programmers. In addition, we have shown that the proposed overfit detection technique shows several advantages over a state-of-the-art approach [54].

In the future, we plan to incorporate additional sources of fix ingredients (e.g., source code repositories [19, 53]) into our repair framework, which may increase the potential for fixing more bugs. Moreover, we would like to investigate new mating and survival selection methods [15, 40, 58] in GP, so as to further improve the evolutionary search algorithm for bug repair.

References

1. Arcuri, A.: It does matter how you normalise the branch distance in search based software testing. In: Proceedings of the Third International Conference on Software Testing, Verification and Validation, pp. 205–214. IEEE (2010)
2. Banzhaf, W., Nordin, P., Keller, R.E., Francone, F.D.: Genetic programming: An introduction, vol. 1. Morgan Kaufmann San Francisco (1998)
3. Barr, E.T., Brun, Y., Devanbu, P., Harman, M., Sarro, F.: The plastic surgery hypothesis. In: Proceedings of the 22nd International Symposium on Foundations of Software Engineering, pp. 306–317. ACM (2014)
4. Brameier, M.F., Banzhaf, W.: Linear genetic programming. Springer Science & Business Media (2007)
5. Campos, J., Riboira, A., Perez, A., Abreu, R.: Gzoltar: An eclipse plug-in for testing and debugging. In: Proceedings of the 27th IEEE/ACM International Conference on Automated Software Engineering, pp. 378–381. ACM (2012)
6. Chen, L., Pei, Y., Furia, C.A.: Contract-based program repair without the contracts. In: Proceedings of the 32nd IEEE/ACM International Conference on Automated Software Engineering, pp. 637–647. IEEE (2017)
7. Contributors, J.: A programmer-oriented testing framework for java (2004). URL https://github.com/junit-team/junit4
8. D'Antoni, L., Samanta, R., Singh, R.: Qlose: Program repair with quantitative objectives. In: Proceedings of International Conference on Computer Aided Verification, pp. 383–401. Springer (2016)
9. Deb, K., Pratap, A., Agarwal, S., Meyarivan, T.: A fast and elitist multiobjective genetic algorithm: NSGA-II. IEEE Transactions on Evolutionary Computation 6(2), 182–197 (2002)
10. Debroy, V., Wong, W.E.: Using mutation to automatically suggest fixes for faulty programs. In: Proceedings of the Third International Conference on Software Testing, Verification and Validation, pp. 65–74. IEEE (2010)
11. Fast, E., Le Goues, C., Forrest, S., Weimer, W.: Designing better fitness functions for automated program repair. In: Proceedings of the 12th Annual Conference on Genetic and Evolutionary Computation, pp. 965–972. ACM (2010)
12. Forrest, S., Nguyen, T., Weimer, W., Le Goues, C.: A genetic programming approach to automated software repair. In: Proceedings of the 11th Annual conference on Genetic and Evolutionary Computation, pp. 947–954. ACM (2009)

13. Gazzola, L., Micucci, D., Mariani, L.: Automatic software repair: A survey. IEEE Transactions on Software Engineering **45**(1), 34–67 (2019)
14. Gupta, R., Pal, S., Kanade, A., Shevade, S.: Deepfix: Fixing common c language errors by deep learning. In: Proceedings of the 31st AAAI Conference on Artificial Intelligence, pp. 1345–1351 (2017)
15. Helmuth, T., Spector, L., Matheson, J.: Solving uncompromising problems with lexicase selection. IEEE Transactions on Evolutionary Computation **19**(5), 630–643 (2015)
16. Hua, J., Zhang, M., Wang, K., Khurshid, S.: Towards practical program repair with on-demand candidate generation. In: Proceedings of the 40th International Conference on Software Engineering, pp. 12–23. ACM (2018)
17. Jiang, J., Xiong, Y., Zhang, H., Gao, Q., Chen, X.: Shaping program repair space with existing patches and similar code. In: Proceedings of the 27th International Symposium on Software Testing and Analysis, pp. 298–309. ACM (2018)
18. Just, R., Jalali, D., Ernst, M.D.: Defects4j: A database of existing faults to enable controlled testing studies for java programs. In: Proceedings of the 2014 International Symposium on Software Testing and Analysis, pp. 437–440. ACM (2014)
19. Ke, Y., Stolee, K.T., Le Goues, C., Brun, Y.: Repairing programs with semantic code search. In: Proceedings of the 30th IEEE/ACM International Conference on Automated Software Engineering, pp. 295–306. IEEE (2015)
20. Kim, D., Nam, J., Song, J., Kim, S.: Automatic patch generation learned from human-written patches. In: Proceedings of the 35th International Conference on Software Engineering, pp. 802–811. IEEE (2013)
21. Koza, J.R.: Genetic programming: on the programming of computers by means of natural selection, vol. 1. MIT press (1992)
22. Le, X.B.D., Chu, D.H., Lo, D., Le Goues, C., Visser, W.: Jfix: Semantics-based repair of java programs via symbolic pathfinder. In: Proceedings of the 26th International Symposium on Software Testing and Analysis, pp. 376–379. ACM (2017)
23. Le, X.B.D., Chu, D.H., Lo, D., Le Goues, C., Visser, W.: S3: syntax-and semantic-guided repair synthesis via programming by examples. In: Proceedings of the 11th Joint Meeting on Foundations of Software Engineering, pp. 593–604. ACM (2017)
24. Le, X.B.D., Lo, D., Le Goues, C.: History driven program repair. In: Proceedings of the 23rd International Conference on Software Analysis, Evolution, and Reengineering, pp. 213–224. IEEE (2016)
25. Le Goues, C., Dewey-Vogt, M., Forrest, S., Weimer, W.: A systematic study of automated program repair: Fixing 55 out of 105 bugs for $8 each. In: Proceedings of the 34th International Conference on Software Engineering, pp. 3–13. IEEE (2012)
26. Le Goues, C., Forrest, S., Weimer, W.: Current challenges in automatic software repair. Software Quality Journal **21**(3), 421–443 (2013)
27. Le Goues, C., Nguyen, T., Forrest, S., Weimer, W.: Genprog: A generic method for automatic software repair. IEEE Transactions on Software Engineering **38**(1), 54–72 (2012)
28. Lehman, J., Clune, J., Misevic, D., Adami, C., Altenberg, L., Beaulieu, J., Bentley, P.J., Bernard, S., Beslon, G., Bryson, D.M., et al.: The surprising creativity of digital evolution: A collection of anecdotes from the evolutionary computation and artificial life research communities. arXiv preprint arXiv:1803.03453 (2018)
29. Liu, X., Zhong, H.: Mining stackoverflow for program repair. In: Proceedings of 25th International Conference on Software Analysis, Evolution and Reengineering, pp. 118–129. IEEE (2018)
30. Long, F., Amidon, P., Rinard, M.: Automatic inference of code transforms for patch generation. In: Proceedings of the 11th Joint Meeting on Foundations of Software Engineering, pp. 727–739. ACM (2017)
31. Long, F., Rinard, M.: Staged program repair with condition synthesis. In: Proceedings of the 10th Joint Meeting on Foundations of Software Engineering, pp. 166–178. ACM (2015)
32. Long, F., Rinard, M.: Automatic patch generation by learning correct code. ACM SIGPLAN Notices **51**(1), 298–312 (2016)

33. Martinez, M., Durieux, T., Sommerard, R., Xuan, J., Monperrus, M.: Automatic repair of real bugs in java: A large-scale experiment on the defects4j dataset. Empirical Software Engineering **22**(4), 1936–1964 (2017)
34. Martinez, M., Monperrus, M.: Astor: A program repair library for java. In: Proceedings of the 25th International Symposium on Software Testing and Analysis, pp. 441–444. ACM (2016)
35. Martinez, M., Monperrus, M.: Ultra-large repair search space with automatically mined templates: the cardumen mode of astor. In: International Symposium on Search Based Software Engineering, pp. 65–86. Springer (2018)
36. Martinez, M., Weimer, W., Monperrus, M.: Do the fix ingredients already exist? an empirical inquiry into the redundancy assumptions of program repair approaches. In: Companion Proceedings of the 36th International Conference on Software Engineering, pp. 492–495. ACM (2014)
37. Mechtaev, S., Yi, J., Roychoudhury, A.: Directfix: Looking for simple program repairs. In: Proceedings of the 37th International Conference on Software Engineering, pp. 448–458. IEEE Press (2015)
38. Mechtaev, S., Yi, J., Roychoudhury, A.: Angelix: Scalable multiline program patch synthesis via symbolic analysis. In: Proceedings of the 38th International Conference on Software Engineering, pp. 691–701. ACM (2016)
39. Monperrus, M.: Automatic software repair: A bibliography. ACM Computing Surveys **51**(1), 17 (2018)
40. Mouret, J.B., Clune, J.: Illuminating search spaces by mapping elites. arXiv preprint arXiv:1504.04909 (2015)
41. Nguyen, H.D.T., Qi, D., Roychoudhury, A., Chandra, S.: Semfix: Program repair via semantic analysis. In: Proceedings of the 35th International Conference on Software Engineering, pp. 772–781. IEEE (2013)
42. Oliveira, V.P.L., de Souza, E.F., Le Goues, C., Camilo-Junior, C.G.: Improved representation and genetic operators for linear genetic programming for automated program repair. Empirical Software Engineering **23**(5), 2980–3006 (2018)
43. Qi, Y., Mao, X., Lei, Y., Dai, Z., Wang, C.: The strength of random search on automated program repair. In: Proceedings of the 36th International Conference on Software Engineering, pp. 254–265. ACM (2014)
44. Qi, Z., Long, F., Achour, S., Rinard, M.: An analysis of patch plausibility and correctness for generate-and-validate patch generation systems. In: Proceedings of the 2015 International Symposium on Software Testing and Analysis, pp. 24–36. ACM (2015)
45. Rolim, R., Soares, G., D'Antoni, L., Polozov, O., Gulwani, S., Gheyi, R., Suzuki, R., Hartmann, B.: Learning syntactic program transformations from examples. In: Proceedings of the 39th International Conference on Software Engineering, pp. 404–415. IEEE Press (2017)
46. Saha, R.K., Lyu, Y., Yoshida, H., Prasad, M.R.: Elixir: Effective object-oriented program repair. In: Proceedings of the 32nd IEEE/ACM International Conference on Automated Software Engineering, pp. 648–659. IEEE (2017)
47. Smith, E.K., Barr, E.T., Le Goues, C., Brun, Y.: Is the cure worse than the disease? overfitting in automated program repair. In: Proceedings of the 10th Joint Meeting on Foundations of Software Engineering, pp. 532–543. ACM (2015)
48. Tan, S.H., Yoshida, H., Prasad, M.R., Roychoudhury, A.: Anti-patterns in search-based program repair. In: Proceedings of the 24th International Symposium on Foundations of Software Engineering, pp. 727–738. ACM (2016)
49. Weimer, W., Forrest, S., Le Goues, C., Nguyen, T.: Automatic program repair with evolutionary computation. Communications of the ACM **53**(5), 109–116 (2010)
50. Weimer, W., Fry, Z.P., Forrest, S.: Leveraging program equivalence for adaptive program repair: Models and first results. In: Proceedings of the 28th International Conference on Automated Software Engineering, pp. 356–366. IEEE (2013)

51. Weimer, W., Nguyen, T., Le Goues, C., Forrest, S.: Automatically finding patches using genetic programming. In: Proceedings of the 31st International Conference on Software Engineering, pp. 364–374. IEEE (2009)
52. Wen, M., Chen, J., Wu, R., Hao, D., Cheung, S.C.: Context-aware patch generation for better automated program repair. In: Proceedings of the 40th International Conference on Software Engineering, pp. 1–11. ACM (2018)
53. Xin, Q., Reiss, S.P.: Leveraging syntax-related code for automated program repair. In: Proceedings of the 32nd IEEE/ACM International Conference on Automated Software Engineering, pp. 660–670. IEEE Press (2017)
54. Xiong, Y., Liu, X., Zeng, M., Zhang, L., Huang, G.: Identifying patch correctness in test-based program repair. In: Proceedings of the 40th International Conference on Software Engineering, pp. 789–799. ACM (2018)
55. Xiong, Y., Wang, J., Yan, R., Zhang, J., Han, S., Huang, G., Zhang, L.: Precise condition synthesis for program repair. In: Proceedings of the 39th International Conference on Software Engineering, pp. 416–426. IEEE Press (2017)
56. Xuan, J., Martinez, M., Demarco, F., Clement, M., Marcote, S.L., Durieux, T., Le Berre, D., Monperrus, M.: Nopol: Automatic repair of conditional statement bugs in java programs. IEEE Transactions on Software Engineering 43(1), 34–55 (2017)
57. Yuan, Y., Banzhaf, W.: ARJA: Automated repair of java programs via multi-objective genetic programming. IEEE Transactions on Software Engineering (2018). https://doi.org/10.1109/TSE.2018.2874648
58. Yuan, Y., Xu, H., Wang, B., Yao, X.: A new dominance relation-based evolutionary algorithm for many-objective optimization. IEEE Transactions on Evolutionary Computation 20(1), 16–37 (2016)

Index

© Springer Nature Switzerland AG 2020
W. Banzhaf et al. (eds.), *Genetic Programming Theory and Practice XVII*, Genetic
and Evolutionary Computation, https://doi.org/10.1007/978-3-030-39958-0

Printed in the United States
by Baker & Taylor Publisher Services